Wetland Biodiversity and Ecosystem Conservation

Wetland Biodiversity and Ecosystem Conservation

Guest Editors

Lele Liu
Yaolin Guo
Youzheng Zhang
Weihua Guo

MDPI

Basel • Beijing • Wuhan • Barcelona • Belgrade • Novi Sad • Cluj • Manchester

Guest Editors
Lele Liu
School of Life Sciences
Shandong University
Jinan
China

Yaolin Guo
School of Renewable
Natural Resources
Louisiana State University
Baton Rouge
USA

Youzheng Zhang
Second Institute
of Oceanography
Ministry of Natural Resources
Hangzhou
China

Weihua Guo
School of Life Sciences
Shandong University
Jinan
China

Editorial Office
MDPI AG
Grosspeteranlage 5
4052 Basel, Switzerland

This is a reprint of the Special Issue, published open access by the journal *Diversity* (ISSN 1424-2818), freely accessible at: https://www.mdpi.com/journal/diversity/special_issues/27QT8CPZF8.

For citation purposes, cite each article independently as indicated on the article page online and as indicated below:

Lastname, A.A.; Lastname, B.B. Article Title. *Journal Name* **Year**, *Volume Number*, Page Range.

ISBN 978-3-7258-8391-2 (Hbk)
ISBN 978-3-7258-8392-9 (PDF)
https://doi.org/10.3390/books978-3-7258-8392-9

Cover image courtesy of Lele Liu

Contents

About the Editors

Lele Liu

Lele Liu is an Associate Research Fellow at the School of Life Sciences, Shandong University, supported by the university's Future Plan for Young Scholars. His research interests center on intraspecific variation and the functional cascading mechanisms of wetland ecosystem engineer plants, particularly the common reed *Phragmites australis*, as well as biodiversity assessment, ecological monitoring, and the soilification of industrial and agricultural solid wastes. He has led projects funded by the National Natural Science Foundation of China (including a General Program and a Young Scientists Fund). Dr. Liu has published extensively on plant eco-evolutionary responses to environmental stressors and serves as a Guest Editor or Young Editorial Board Member for several journals, in addition to being an active reviewer for more than twenty international publications. He is supported by the Cyrus Tang Foundation through the Tang Scholar Program.

Yaolin Guo

Yaolin Guo is a Postdoctoral Research Associate in coastal ecology at the Louisiana State University Agricultural Center. He earned his Ph.D. in Ecology from Fudan University and completed part of his doctoral training at the University of Tübingen. His research aims to reveal the ecological mechanisms underlying coastal wetland responses to global change pressures through field surveys, quantitative modeling, and data synthesis. Building on this mechanistic understanding, he seeks to bridge ecological theory with real-world applications to advance coastal wetland conservation, habitat restoration, sustainable management, and policy development through interdisciplinary collaborations.

Youzheng Zhang

Youzheng Zhang is an Associate Professor at Second Institute of Oceanography (SIO), Ministry of Natural Resource, China. His research focuses on coastal wetland ecology, biological invasions, soil nematode communities, litter decomposition, and biogeographical patterns, particularly involving the invasive *Spartina alterniflora* and native *Phragmites australis*. He has published extensively in leading journals such as *Ecology*, *Functional Ecology*, *Diversity and Distributions*, *Journal of Biogeography* and *Ecosystems Health and Sustainability*. He has reviewed more than 20 papers from about 10 journals and has acted as Editor, Youth Editor, and/or Guest Editor for several journals. Additionally, he also conducts experimental studies investigating the geographic patterns of root litter decomposition in *Spartina alterniflora* and *Phragmites australis* across coastal wetlands.

Weihua Guo

Weihua Guo is a Professor and Doctoral Supervisor at the School of Life Sciences, Shandong University, and serves as the Deputy Director of the Institute of Eco-Environmental Forensics of Shandong University. Her research addresses plant ecological differentiation and biodiversity maintenance mechanisms, conservation and ecological adaptability assessment of biological resources, and key technologies for ecosystem restoration and quality enhancement. She has led major projects including a key program of the Regional Development Joint Fund of the National Natural Science Foundation of China. Professor Guo is a Foreign Academician of the Russian Academy of Natural Sciences, a recipient of the Shandong Provincial Teaching Master Award, and holds numerous leadership roles such as Vice Chair of the China Chapter of the Society of Wetland Scientists and Executive Council Member of the Ecological Society of China. Her work integrates genetic, species, and ecosystem perspectives to advance wetland conservation and restoration.

Preface

This Reprint assembles the core contributions of the Special Issue "Wetland Biodiversity and Ecosystem Conservation", published in the journal *Diversity*. The subject of this collection pertains to the scientific understanding of wetland ecosystems as complex, multi-level systems that sustain biodiversity and provide essential services to both nature and human society. The scope spans genetic and population viability, species-level responses to anthropogenic stressors, ecosystem-level cascading interactions, and innovative restoration and management practices. Its aim is to move beyond single-species- or area-based conservation targets toward an integrated framework that explicitly links genetic diversity, species composition, and ecosystem functioning.

The motivation for compiling this Reprint arises from the accelerating degradation of wetlands worldwide—ecosystems that cover only a small fraction of the land surface yet store a disproportionate share of terrestrial carbon and support a vast array of biodiversity. Despite their ecological and societal importance, wetland conservation efforts have often been fragmented, overlooking the interdependent mechanisms that maintain resilience. The research featured here addresses critical gaps: from the role of phenotypic plasticity in wetland plants and the population viability of endangered species such as the Yangtze finless porpoise, to the type-specific effects of microplastic pollution on native sedges and the contrasting impacts of different management regimes on waterbird communities. These studies collectively demonstrate that effective action requires integrating genetic-, species-, and ecosystem-based perspectives.

This Reprint is addressed to researchers, conservation practitioners, environmental managers, and policymakers who seek evidence-based strategies for wetland protection and restoration. It also serves graduate students and advanced readers interested in global change biology, invasion ecology, and the design of nature-based solutions. By presenting both original research and systematic reviews, this volume aims to translate scientific insights into actionable conservation measures: from population viability modelling and trait-based biodiversity monitoring to the formal recognition of Other Effective area-based Conservation Measures (OECMs). We hope that the findings and ideas gathered here will inspire continued collaboration across disciplines and help bend the curve of wetland biodiversity loss in a rapidly changing world.

Lele Liu, Yaolin Guo, Youzheng Zhang, and Weihua Guo
Guest Editors

diversity

MDPI

Editorial

Wetland Biodiversity and Ecosystem Conservation: Integrating Genetic, Species, and Ecosystem Perspectives for Effective Action

Lele Liu [1,*], Yaolin Guo [2], Youzheng Zhang [3,*] and Weihua Guo [1]

1 Key Laboratory of Ecological Prewarning, Protection and Restoration of Bohai Sea, Ministry of Natural Resource, School of Life Sciences, Shandong University, Qingdao 266237, China; whguo@sdu.edu.cn
2 School of Renewable Natural Resources, Louisiana State University, Baton Rouge, LA 70803, USA; yaolinguo22@gmail.com
3 Second Institute of Oceanography, Ministry of Natural Resources, Hangzhou 310012, China
* Correspondence: liulele@sdu.edu.cn (L.L.); zhangyz@sio.org.cn (Y.Z.)

Wetlands are among the most productive yet most rapidly degrading ecosystems on Earth [1–3]. Covering only 5–8% of the land surface, they store ~30% of terrestrial soil carbon and support 40% of global biodiversity [4,5]. The provision of essential services, from water purification and flood control to carbon sequestration and climate resilience, depends on maintaining intact biodiversity across multiple levels of organisation [6,7]. However, conservation efforts have often focused on single species or simple area targets, neglecting the complex, multi-scale interactions that sustain wetland functions. Effective wetland conservation must simultaneously address genetic and population viability, species-level stressors, ecosystem-level cascades, and innovative restoration approaches [8]. This Special Issue highlights key emerging insights and outlines priority directions for science and action.

At the genetic and population level, a growing body of evidence indicates that habitat diversity and environmental gradients are as important as standing genetic variation in sustaining functional diversity and population resilience [9–12]. For the widespread wetland grass *Phragmites australis*, common garden experiments across eastern China show that phenotypic plasticity, rather than genetic variation alone, drives growth and decomposition [9]. For the Yangtze finless porpoise in Poyang Lake, population viability analysis projects a >90% decline in genetic diversity over the last 100 years and an extinction probability of 0.241 under baseline scenarios, despite recent population increases [10]. Key parameters, including breeding rate, sex ratio at birth, and gene flow, underscore the importance of connectivity and demographic management. Together, these studies demonstrate that environmental conditions and genetic diversity are interdependent: environmental heterogeneity shapes the expression of genetic potential, while genetic diversity provides the raw material for long-term adaptation [13].

At the species level, anthropogenic stressors such as microplastic pollution [14,15] and human land-use intensification [1,16] profoundly alter wetland plant and animal communities. Microplastic types exert type-specific and concentration-dependent effects on the native sedge *Scirpus mariqueter*: PET and PE suppress belowground biomass by more than 40% even at 0.1% concentration, whereas PP can increase seed production [15]. Such differential responses indicate that risk assessments must move beyond total microplastic mass to polymer identity. Human activities further shape species assemblages: intensive aquaculture and floating photovoltaic systems on coal mining subsidence wetlands reduce waterbird diversity and eliminate threatened species, whereas ecologically managed or unutilised subsidence wetlands can serve as valuable alternative habitats [16]. Citizen science platforms such as iNaturalist are increasingly valuable for tracking wetland species distributions and phenology on a large scale, complementing targeted research [17]. These examples collectively show that species-level conservation requires explicit consideration of pollutant chemistry and the intensity of human use.

At the ecosystem level, cascading effects link hydrology, plant nutrition, animal communities, and ecosystem services [6,7,18]. A systematic review of 76 studies finds that biodiversity monitoring in constructed wetlands remains heavily biased towards microbial communities using generic indices (e.g., Shannon–Wiener, Chao1), while plants and animals are rarely linked to treatment performance or ecosystem services [19]. This gap calls for standardised, trait-based and phylogenetically informed frameworks (e.g., Rao's quadratic entropy, Faith's phylogenetic diversity) that directly connect biodiversity to functions such as nutrient removal and carbon sequestration. Recognising and measuring these ecosystem-level interactions is a prerequisite for moving beyond static inventories to dynamic functional assessments.

Turning to restoration and management practices, the reviewed studies offer concrete lessons for improving conservation outcomes. For coal mining subsidence wetlands, long-term waterbird surveys demonstrate that "ecological aquaculture" and unutilised areas provide far greater conservation value than intensive development, suggesting that management regimes are a powerful lever even in heavily modified landscapes [16]. Such areas exemplify Other Effective area-based Conservation Measures (OECMs)—sites outside formal protected areas that deliver long-term in situ biodiversity conservation [20]. For threatened species, comprehensive protection combining total fishing bans, individual rescue, and population interchange reduces extinction probability from 0.241 to 0.0028 in the Yangtze finless porpoise, proving that aggressive action works [10]. The key message is that wetland restoration must shift from a narrow focus on area-based targets to a broader portfolio of active, context-specific, and multi-objective management strategies.

In conclusion, the science of wetland biodiversity conservation is moving decisively towards integration across genetic, species, and ecosystem levels. Building on the insights from this Special Issue, future research should prioritise three frontiers. First, explore the origin and evolution of key functional traits and biotic interactions in wetland organisms, from phenotypic plasticity to trophic cascades [21,22]. Second, predict the responses and adaptation of wetland biodiversity to future global change, including climate warming, microplastic pollution, and land-use intensification [23,24]. Third, develop synergistic mechanisms that couple wetland ecological conservation and restoration with regional socioeconomic development [25,26]. Achieving such synergy requires not only advancing academia, government, business, and the millions of people whose livelihoods depend on wetlands [27], but also harnessing citizen science for large-scale biodiversity monitoring [17] and formally integrating OECMs into national and corporate conservation strategies [20].

The window for action is narrowing, but the tools and insights now available, from population viability modelling to adaptive management, offer a realistic pathway to bend the curve of wetland loss. It is time to move beyond description to implementation by embedding genetic, species, and ecosystem perspectives into national biodiversity strategies, corporate environmental reporting, and on-ground restoration projects. Only then can wetlands fulfil their role as life-support infrastructure for both people and nature.

Author Contributions: All authors managed this editorial, wrote the manuscript and contributed to the revision. All authors have read and agreed to the published version of the manuscript.

Acknowledgments: As Guest Editors of this Special Issue, titled "Wetland Biodiversity and Ecosystem Conservation", we would like to express our deep appreciation to all authors whose valuable works have been published in this issue and have contributed to its success. Our sincere thanks also go to all the academic editors and reviewers whose efforts were essential to the completion of this Special Issue. Lele Liu is supported by the Cyrus Tang Foundation through the Tang Scholar Programme.

Conflicts of Interest: The authors declare no conflicts of interest.

References

1. Fluet-Chouinard, E.; Stocker, B.D.; Zhang, Z.; Malhotra, A.; Melton, J.R.; Poulter, B.; Kaplan, J.O.; Goldewijk, K.K.; Siebert, S.; Minayeva, T.; et al. Extensive Global Wetland Loss over the Past Three Centuries. *Nature* **2023**, *614*, 281–286. [CrossRef]
2. He, Q.; Li, Z.; Daleo, P.; Lefcheck, J.S.; Thomsen, M.S.; Adams, J.B.; Bouma, T.J. Coastal Wetland Resilience through Local, Regional and Global Conservation. *Nat. Rev. Biodivers.* **2025**, *1*, 50–67. [CrossRef]
3. Shen, X.; Jiang, M.; Lu, X.; Thompson, J.R. Protect and Restore Small Wetlands. *Science* **2024**, *384*, 1415. [CrossRef]
4. Wetlands International. *Wetlands at the Heart of Biodiversity and Sustainable Development*; Wetlands International: Ede, The Netherlands, 2025.
5. Stewart, A.J.; Halabisky, M.; Babcock, C.; Butman, D.E.; D'Amore, D.V.; Moskal, L.M. Revealing the Hidden Carbon in Forested Wetland Soils. *Nat. Commun.* **2024**, *15*, 726. [CrossRef]
6. Ren, L.; Jensen, K.; Porada, P.; Mueller, P. Biota-mediated Carbon Cycling—A Synthesis of Biotic-interaction Controls on Blue Carbon. *Ecol. Lett.* **2022**, *25*, 521–540. [CrossRef] [PubMed]
7. Liu, L.; Du, N.; Eller, F.; Ye, S.; Li, X.; Wei, J.; Guo, Y.; Brix, H.; Guo, W. Ecological Mechanisms of Carbon Sequestration in Vegetated Coastal Wetland Ecosystem: Exploring the Roles of Biodiversity and Environmental Changes. *J. Mar. Environ. Eng.* **2025**, *12*, 35–47. [CrossRef]
8. Liu, Y.; Wang, J.; Wu, Z.; Li, S.; Dai, W. Identification of Biodiversity Priority Conservation Areas in China by Integrating Genetic, Species and Ecosystem Diversity. *Biol. Conserv.* **2024**, *300*, 110854. [CrossRef]
9. Wei, W.; Wen, Q.; Zhu, H.; Song, H.; Zhang, X.; Sheng, W.; Xie, L.; Guo, X.; Guo, Y.; Ye, S.; et al. Effects of Phenotypic Plasticity and Genetic Variation on Plant Growth and Litter Decomposition in a Widespread Wetland Grass. *Diversity* **2025**, *17*, 282. [CrossRef]
10. Wu, B.; Wang, W.; Wang, Y.; Zhang, Z. Population Viability Analysis Revealed the Vulnerability of Yangtze Finless Porpoise (*Neophocaena asiaeorientalis*) in Poyang Lake. *Diversity* **2025**, *17*, 410. [CrossRef]
11. Li, X.; Guo, T.; Mu, Q.; Li, X.; Yu, J. Genomic and Environmental Determinants and Their Interplay Underlying Phenotypic Plasticity. *Proc. Natl. Acad. Sci. USA* **2018**, *115*, 6679–6684. [CrossRef]
12. Wei, J.; Guo, T.; Mu, Q.; Alladassi, B.M.E.; Mural, R.V.; Boyles, R.E.; Hoffmann, L.; Hayes, C.M.; Sigmon, B.; Thompson, A.M.; et al. Genetic and Environmental Patterns Underlying Phenotypic Plasticity in Flowering Time and Plant Height in Sorghum. *Plant Cell Environ.* **2025**, *48*, 2727–2738. [CrossRef] [PubMed]
13. Mastretta-Yanes, A.; Suárez, S.; Jordan, R.; Hoban, S.; Da Silva, J.M.; Castillo-Reina, L.; Heuertz, M.; Ishihama, F.; Köppä, V.; Laikre, L.; et al. Guideline Materials and Documentation for the Genetic Diversity Indicators of the Monitoring Framework for the Kunming-Montreal Global Biodiversity Framework. *Biodiv. Inf.* **2024**, *18*, 24–27. [CrossRef]
14. Anekwe, I.M.S.; Mkhize, M.M.; Zhou, H.; Akpasi, S.O.; Mkhize, T.T.G. Ocean Plastic Pollution: Sources, Effects, Challenges and Mitigation Measures. *J. Mar. Environ. Eng.* **2025**, *11*, 303–325. [CrossRef]
15. Jiang, P.; Gao, J.; Li, J.; Wu, M.; Shao, X.; Li, N. Microplastics Alter Growth and Reproduction Strategy of Scirpus Mariqueter by Modifying Soil Nutrient Availability. *Diversity* **2025**, *17*, 472. [CrossRef]
16. Yang, S.; Cao, K.; Wang, Y.; Shen, W.; Lin, T.; Liu, N.; Li, J.; Ji, L.; Chen, H.; Xu, Y.; et al. Impact of Distinct Management Regimes on Wintering Waterbird Communities in China's Coal Mining Subsidence Wetlands. *Diversity* **2026**, *18*, 146. [CrossRef]
17. Cui, S.; Guo, H.; Pugliese, L.; Kragh, G.; Mena, S.; Wu, S. Citizen Science Powers Wetland Restoration. *Environ. Sci. Ecotech.* **2026**, *29*, 100656. [CrossRef]
18. Meng, Z.; Lv, L.; Xu, J.; He, M.; Zhou, M.; Wei, Y. Cascading Effects of Microplastic-Pollutant Co-Exposure in Coastal Wetland Soil-Plant-Microbe Systems: Mitigation Potential through Nutrient Management. *J. Clean. Prod.* **2026**, *538*, 147447. [CrossRef]
19. Uy, M.J.; Robles, M.E.; Oh, Y.; Haque, M.T.; Mueca, C.C.; Kim, L.-H. Biodiversity Monitoring in Constructed Wetlands: A Systematic Review of Assessment Methods and Ecosystem Functions. *Diversity* **2025**, *17*, 367. [CrossRef]
20. Gurney, G.G.; Darling, E.S.; Ahmadia, G.N.; Agostini, V.N.; Ban, N.C.; Blythe, J.; Claudet, J.; Epstein, G.; Estradivari; Himes-Cornell, A.; et al. Biodiversity Needs Every Tool in the Box: Use OECMs. *Nature* **2021**, *595*, 646–649. [CrossRef]
21. Vahsen, M.L.; Blum, M.J.; Megonigal, J.P.; Emrich, S.J.; Holmquist, J.R.; Stiller, B.; Todd-Brown, K.E.O.; McLachlan, J.S. Rapid Plant Trait Evolution Can Alter Coastal Wetland Resilience to Sea Level Rise. *Science* **2023**, *379*, 393–398. [CrossRef]
22. Ngeve, M.N. Genetic Diversity Must Be Explicitly Recognized in Ecological Restoration. *Nat. Clim. Change* **2025**, *15*, 908–909. [CrossRef]
23. Li, C.; Liu, J.; Rillig, M.C.; Bank, M.S.; Fantke, P.; Zhu, D.; Zhu, Y.-G.; Jin, L.N. What Harmful Microbes Are Lurking in the World's 7 Billion Tonnes of Plastic Waste? *Nature* **2024**, *634*, 30–32. [CrossRef] [PubMed]

https://doi.org/10.3390/d18050309

24. Lv, X.; Lin, A.; Cui, X.; Li, Y.; Guo, Z.; Tan, X.; Duan, G. The Interplay Between Climate Warming Driven by Greenhouse Gas Emissions and the Ecotoxicological Effects of Microplastics: Insights from a Meta-Analysis. *Glob. Change Biol.* **2025**, *31*, e70348. [CrossRef]
25. Guo, J.; Yue, H.; Cao, Z.; Cao, S. Incorporating Socioeconomic-Ecological System to Promote Sustainable Restoration: A Case Study in Changting, China. *Ecol. Eng.* **2026**, *222*, 107815. [CrossRef]
26. Prasanya, J.; Kanmani, S.; Senthil Kumar, P. A Review of the Wetland's Restoration Mechanisms and Its Economic and Social Benefits. *Water Pract. Technol.* **2024**, *19*, 4355–4377. [CrossRef]
27. Jiang, W.; Zhang, Z.; Ling, Z.; Deng, Y. Experience and Future Research Trends of Wetland Protection and Restoration in China. *J. Geogr. Sci.* **2024**, *34*, 229–251. [CrossRef]

diversity

MDPI

Article

Patterns in Understorey Vegetation of a Semi-Arid Terminal Wetland over 20 Years in Response to Flood and Drought

Rebekah Grieger [1,*], Jaiden Johnston-Bates [1], Andres Sutton [2] and Samantha J. Capon [3]

1 Australian Rivers Institute, Griffith University, Nathan, QLD 4111, Australia; jaiden.johnston-bates@griffithuni.edu.au
2 Centre for Ecosystem Science (CES), School of Biological Earth and Environmental Science (BEES), University of New South Wales, Sydney, NSW 2052, Australia; andres.sutton@unsw.edu.au
3 Griffith Institute for Human and Environmental Resilience, School of Environment and Science, Griffith University, Nathan, QLD 4111, Australia; s.capon@griffith.edu.au
* Correspondence: r.grieger@griffith.edu.au

Abstract

Floodplains are key components of inland river systems of Australia with floodplain vegetation playing important roles in habitat provision, nutrient cycling, and supporting strong cultural values. These vegetation communities are highly dynamic, particularly in response to flooding. However, decades of water development and highly managed water resources are linked to wetland habitat decline in this region. We explored patterns of vegetation response to flooding over twenty years at the Narran Lakes Ramsar site, a terminal floodplain wetland system in the northern Murray–Darling Basin, Australia. We collated data from previous monitoring efforts and resampled permanent plots for understorey vegetation structure and composition. Three flood events were surveyed over a 20-year period, with each event surveyed on two occasions first, following initial drawdown (minimal standing water) and a second survey under dry or drier conditions (~6 months after the recession of floodwaters). Overall, we observed a high diversity of native plant species (~110 species) in understorey communities across the wetland and high compositional turnover both between flood events and within years (i.e., paired surveys). Notably, vegetation cover, but not species richness, was greatest in the 2023 survey following the largest of the three flood events investigated. Understorey composition was strongly driven by inundation regimes, particularly the duration of recent inundation, and the number of wet and dry years prior. Large flood events are critical for supporting vegetation resilience in these systems, increasingly so under a drier climate and with stretched water resources. Continued long-term monitoring of vegetation through flood cycles at the Narran Lakes will be critical to understanding ecological responses to longer-term changes in climate and hydrology to inform adaptive water management and maintain the values of this Ramsar site.

Keywords: floodplain wetland; long-term monitoring; non-woody vegetation; community dynamics; wetland management

1. Introduction

Inland floodplain wetlands are highly dynamic ecosystems, both spatially and temporally, largely driven by hydrology. Periods of flooding punctuate otherwise dry conditions, which range from regular periods of dry to atypical periods of severe drought [1]. Temporal and spatial variation in inland floodplain wetland vegetation occurs due to variations in flow [2,3]. Changes in the frequency, duration, magnitude, timing, regularity and rate of change of flow events influences species composition [4,5]. Floodplain wetland vegetation is adapted to such variable conditions and many taxa exhibit traits which enable their persistence through dry periods or otherwise unfavourable conditions (e.g., persistent seed banks and physical dormancy [6,7]. Natural flow regimes have been widely altered in large river systems globally for anthropogenic uses such as for agriculture, mining and for drinking water [8,9]. The capacity of individuals, species, and communities to tolerate drought or flooding is critically important where altered hydrological regimes expose plants to highly variable and uncharacteristic moisture conditions [10]. Floodplain wetland vegetation is typically considered resilient to these naturally large perturbations in hydrology. However, the limits to their resilience under altered flow regimes and climatic extremes remain unknown.

The capacity of vegetation communities to recover following natural or anthropogenically driven disturbances is a key component of wetland resilience [11]. A general definition of ecological resilience is the ability of a system to maintain or restore community structure and function when faced with disturbance [12]. Inland floodplain wetland vegetation communities tend to exhibit both wet and dry phases in terms of species representation and productivity [13], but can rapidly reorder from a dry to a wet phase after a flood event [14] which contributes to their overall resilience. Each consecutive phase may not share the same mix of species but rather shows similarities in total plant cover and the expression of functional groups (e.g., annual forbs, perennial shrubs) [5,15]. This dynamism in vegetation composition and productivity suggests that an extreme external disturbance would be required to transition the system into an alternate state. Th ability of inland floodplain wetlands to switch between phases is underpinned by life-history strategies of plant species, including the formation of persistent soil seed banks, phenological flexibility, and traits that enable establishment under the pressure of, or following, disturbance [4,16].

Reordering of vegetation between wet and dry phases and over time can be measured through temporal species turnover, i.e., the species which are gained or lost over time [17,18]. Alternating wet and dry cycles in semi-arid wetlands drive species turnover and replacement, particularly among annual herbaceous species that respond rapidly to water and resource fluctuations [19]. This turnover is dictated by the filtering of species, where environmental conditions exclude those lacking traits suited to the prevailing conditions [20]. For example, flood pulses trigger germination and growth of a range of moisture-dependent taxa [21], while intervening dry periods favour terrestrial, stress-tolerant species capable of persisting under conditions of low moisture availability [22–24]. Stochastic recruitment events following an influx of moisture availability further contribute to temporal variability in community composition [25]. Across temporal scales, cyclical, seasonal and variable periods of hydrology drive short-term shifts in community structure which subsequently accumulate to produce broader long-term trends in the composition and structure of vegetation communities [26]. Long-term vegetation trajectories in terminal wetlands therefore reflect the cumulative impact of repeated wet–dry transitions [27,28] and contribute to the overall resilience of such systems in terms of their capacity to respond to changing conditions and to maintain functionality, e.g., biomass production.

The development of water resources and climate change pose significant disruption to hydrological processes that sustain inland floodplain wetlands [8,29,30]. Upstream, the extraction, regulation and alteration of natural flow regimes have already driven changes to the frequency, duration, and magnitude of overbank flooding across many semi-arid systems [8,29,30]. In the northern Murray–Darling Basin, reduced surface water availability and reduced connection between river and floodplain systems has occurred as a result of damming, extraction, and the reduction in small-to-medium flow events [31]. These

https://doi.org/10.3390/d18050274

changes have contributed to the dieback of tree populations [32], reductions in the extent of non-woody wetlands, and changes to the composition of soil seed banks [23,33]. Climate change can compound impacts of flow alteration by intensifying hydrological extremes, lengthening drought periods, and increasing evapotranspiration [34,35]. Future climate modelling suggests reductions in flow up to 8% with up to a 40% decrease in rainfall runoff in regions of the northern basin by 2030 [18,35,36]. Managing available water resources to achieve environmental and economic outcomes will be vital under increasingly water-scarce conditions.

Understanding patterns of vegetation dynamics and wetland condition over time is important for managing wetlands and water resources into the future. Monitoring of riparian and wetland conditions has been growing in Australia, particularly since 2012 under government-funded programmes such as Flow-MER (Flow-Monitoring Evaluation and Research program) and LTIM (Long-Term Intervention Monitoring) which have been established to assess the outcomes of environmental watering actions for target biota [37,38]. However, much focus has been on systems in the southern Murray–Darling Basin where streamflow is greater, more consistent, and water inputs are highly managed through dam releases and regulation infrastructure [39]. In contrast, river systems within the northern basin have more variable flow regimes, fewer large storages, and less regulation [39]. Several monitoring projects within the northern basin have been conducted over shorter timescales (i.e., 2–3 years up to 10 years) [40]; however, understanding of key ecological processes, such as response to altered flow regimes, restoration, or fire, can occur over much longer timeframes [41]. In floodplain wetlands, vegetation responds to hydrological regimes which can vary in the timing, duration, frequency, and rate of change, such that systems can be derived from the unique combination of flow regimes over history. Limited work exists exploring vegetation responses to repeated flood events through time [37] and this lack of long-term data is identified as a key limiting factor to adequate water management [15,42,43]. Satellite imagery can provide insight to landscape scale dynamics through the use of vegetation and water indices [15,42,43] such as the normalised difference vegetation index (NDVI), enhanced vegetation index (EVI), and normalised difference water index (NDWI), but these are not able to provide information regarding the compositional characteristics of vegetation communities.

Here, we explore patterns in understorey vegetation structure and composition in response to three flood events, interspersed by drought, across a 20-year period in the semi-arid wetland complex of the Narran Lakes. More specifically, we examine patterns in understorey vegetation structure and composition between and within flood event survey periods (i.e., paired surveys following drawdown and following six months of subsequent drying) to determine the role of flow in short- and long-term vegetation community dynamics for this semi-arid wetland. We also explore dynamics of floodplain wetland vegetation communities over this 20-year period through the lens of local and landscape-scale species turnover and the role of flood events in compositional change. Based on the existing knowledge of inland floodplain wetland vegetation dynamics, we expected to see high plant cover and greater species richness, particularly for annual species, in drawdown surveys and a reduction in these metrics following drying. Temporal changes are less well known; however, we expected that flood events resulting in similar hydrological conditions would result in similar vegetation responses.

2. Materials and Methods

2.1. Study Area

The Narran Lakes wetland complex (herein Narran), Dharriwaa in the language of the Yuwaalarayy/Euahlayi First Nations people, is a terminal floodplain system of the Narran

https://doi.org/10.3390/d18050274

River in the northern Murray–Darling Basin (Figure 1; ref. [44]). The wetland complex, covering an area of at least 40,900 ha, comprises several lakes (Clear Lake, Back Lake, North Arm, Narran Lake), and areas of vegetated wetlands. The main Narran River channel flows through the northern wetland section, diverting into the northern lakes before spilling into Narran Lake proper and the southern wetlands and floodplains [44,45]. Only in rare, very large floods does the Narran River connect through the wetland to the Barwon and Bokhara Rivers (south west). More commonly, the northern lakes dry out in approximately 4–8 months through evaporation and underground seepage [46]. A portion of the northern wetlands (8447 ha) within the Narran Lake Nature Reserve is listed as a Ramsar wetland.

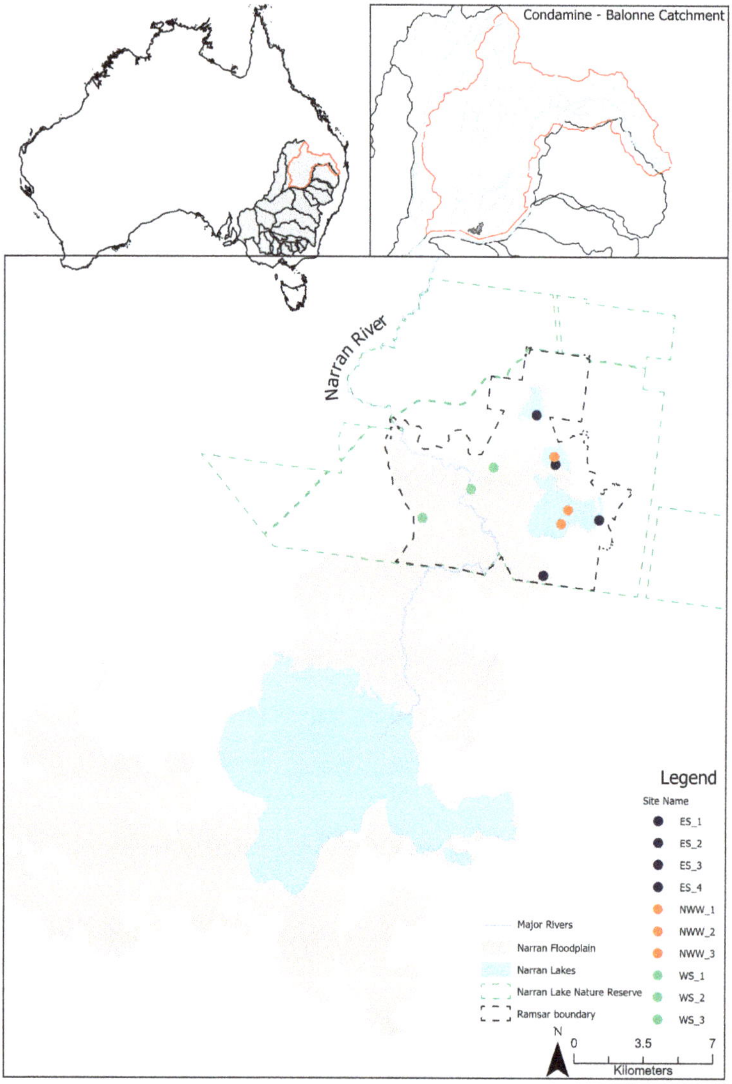

Figure 1. Map of Narran Lakes field survey sites in relation to the Narran Lakes wetland complex (**main map**), the Condamine–Balonne River catchment (**top right**), and the Murray–Darling Basin (**top left**). Site colour relates to the regions of ES: eastern shrubland, NWW: non-woody wetland, WS: western shrubland.

https://doi.org/10.3390/d18050274

The climate is semi-arid with low annual rainfall (440 mm long-term annual average), hot summers (average maximum temperatures between 21 and 36 °C) and mild winters (3.6–18 °C [47]). Rainfall is highly variable; however, widespread flooding of the wetland is typically driven by monsoonal summer rainfall events in the upper Condamine–Balonne catchment [48]. Historically, the northern lakes would fill every 1.5–2.5 years, flood events were long and slow to recede, while river flow was highly variable and frequently experienced cease-to-flow and dry conditions. Upstream water resource development (i.e., dams, diversions, irrigation) has resulted in reduced mean annual flow and reduced flood duration and frequency, especially for smaller and moderate inflow events, which are critical for the wetlands [45]. Recent (2000–2024) mean annual flow for the Narran River is approximately 90,000 ML, although it is highly variable and annual flows of up to 370,000 ML have been recorded and periods of extended nil flows have also been recorded [49].

2.2. *Vegetation Communities*

Three broad vegetation classes dominate the wetland ecosystems of Narran Lakes [50]: floodplain woodlands, lignum shrublands and non-woody wetlands. Floodplain woodlands are typically dominated by *Eucalyptus camaldulensis* (river red gum), *E. coolabah* (coolabah), a sub-canopy of *Acacia stenophylla* (river coobah) and *Eremophila bignoniiflora* (emu bush), and an understorey of mixed grasses, forbs, and shrubs. Lignum shrublands are dominated by varying densities of the large, clump-forming shrub *Duma florulenta* (tangled lignum), and a variable understorey of inundation-responding forbs, sedges, and grasses with a sparse canopy of *A. stenophylla*. Lignum shrublands are key habitats for waterbirds which nest on the large woody structure of lignum during floods. Non-woody wetlands typically occur along lake margins or lakebeds when water recedes and are dominated by mixed forbs, sedges and grasses, typically annual species which respond rapidly to the drawdown conditions. This study focuses on vegetation dynamics within lignum shrublands and non-woody wetlands as the responses of understorey communities are comparable in these areas which lack a substantial tree canopy.

2.3. *Survey Design*

Field data were collected across six survey periods between 2004 and 2024 which represented paired surveys across three flood events: first after initial drawdown of floodwaters and a second survey after six months of drying (Table 1, Figure 2). As the focus of this study we selected ten sites across the wetland which were surveyed in all flood periods of 2004–2005 (flood 1 [45]), 2008–2009 (flood 2), and 2023–2024 (flood 3, see [51]). Sites represented two broad vegetation types, a variety of flood frequencies (i.e., 4–17 years flooded of 20 years), and three geographical regions (non-woody wetland: 3 sites; lignum shrubland: 7 sites; Figure 1). Additional lignum shrubland and non-woody wetland sites monitored more broadly within the wetland complex, although not included in this analysis, also fall within this range of flood frequency. We have excluded sites in tree-dominated vegetation classes due to the effect of canopy cover on understorey responses and the lack of long-term monitoring sites [51].

https://doi.org/10.3390/d18050274

Table 1. Antecedent hydrological conditions for six months preceding field surveys. Rainfall and temperature data for Lightning Ridge post office (station 48243), flow data for Wilby Wilby gauge on the Narran River (gauge 422016), drawdown from Clear Lake and inundation extent determined from LandSat Water Observations (Digital Earth Australia).

Survey	Season	Temperature Minimum–Maximum	Total Rainfall (mm)	Mean Daily Flow (ML/d)	Cumulative Flow (ML)	Days After Cease to Flow	Months After Drying of Clear Lake	Inundation Extent (ha)	Classification
Flood 1									
November 2004	Early summer (flood Feb–July 2004)	9.6 (±5.3)–23.6 (±5.7)	~160	0	0	~200	1 (Oct 04)	2624	Drawdown
May 2005	Autumn	18.8 (±4.3)–32.8 (±4.5)	~260	~20	~430	~380	7 (Oct 04)	99	Dry
Flood 2									
November 2008	Early summer (flood Jan–March 2008)	9.4 (±6.2)–24.1 (±5.7)	~250	0	~5	~190	0	1278	Drawdown
April 2009	Autumn	18.5 (±4.5)–32.8 (±4.8)	~370	~110	~1970	~70	5 (Dec 08)	1470	Dry
Flood 3									
August 2023	Winter (flood 2022–Jan 2023)	12.1 (±6.9)–25.9 (±6.7)	~90	~20	~4290	~160	0	770 (February 2023, maximum 4550 in 2022)	Drawdown
February 2024	Late summer	16.9 (±6.3)–32.2 (±5.6)	~330	~30	~8700	~320 (3 days of flow > 1000 ML as surveying commenced)	5 (Sept 23) filling	1703 (late Feb 2024)	Dry

https://doi.org/10.3390/d18050274

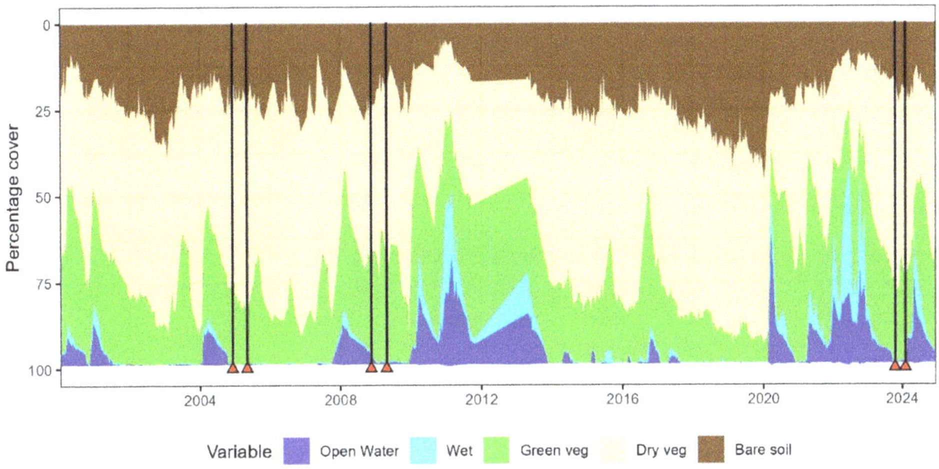

Figure 2. Fractional cover of water, vegetation, and soil for Narran Lakes Nature Reserve showing the timing of vegetation surveys after the drawdown of floodwaters as red triangles. Data generated in Digital Earth Australia sandbox using the Wetland Insights Tool (version 4.0) [52].

2.4. Antecedent Environmental Conditions

Across the 20-year study period, hydro-climatic conditions were highly variable capturing a period of significant drought (2001–2009) and extensive flooding (2020–2024). Conditions preceding the 2023–2024 surveys were significantly wetter and for a longer sustained period than the conditions preceding the earlier surveys which followed small inundation events (Table 1, Figure 2, Supplementary Figure S1). Surveys in November 2004 and 2008, and August 2023, represented the period after drawdown of Clear Lake when flows had ceased for approximately 180 days. However, rainfall prior to the 2008 survey supplemented water levels in Clear Lake while rainfall was low preceding August 2023 (Table 1). Surveys in May 2005, April 2009, and February 2024 represented the period after 6 months of drying, although rainfall prior to May 2005 was lower. In February 2024, no inflows were recorded for close to one year, however recent local and upstream rainfall provided increased flows into Clear Lake during the survey (Table 1). In November 2023, between the surveys for flood 3, a wildfire burned three shrubland sites on the western floodplain, all of which experienced low-to-moderate burn severities according to national mapping [46].

2.5. Field Data Collection

During each vegetation survey we assessed understorey vegetation communities within a 20 × 20 m (0.04 ha) plot at each site. We assessed understorey (i.e., less than 1 m height) vegetation cover in 10 randomly distributed 1 × 1 m quadrats within the larger 0.04 ha site, recording the percentage cover for each species.

Each species observed was identified to the finest taxonomic level possible (genus or species) in the field or using images and samples following the nomenclature of the Australian Plant Census and descriptions of PlantNet [53] and Cunningham et al. [54]. Species were later categorised by plant functional group (i.e., tree, shrub, forb, grass, sedge, fern), life history (i.e., annual, perennial), and endemicity (i.e., native, non-native). Species names were checked against the Australian Plant Census to update older observations with current nomenclature and ensure consistency across the surveys.

https://doi.org/10.3390/d18050274

2.6. Environmental Data

Spatial data for surface water inundation in target areas (i.e., a 30 m radius from each vegetation survey site) was sourced from LandSat 5, 7, 8 and 9 satellites through the Digital Earth Australia sandbox. Images were first filtered to remove timesteps with less than 90% coverage of the target area and cloud cover. Relative area of inundation was determined for each timestep using the Fisher Water Index [55] as the percentage of the area with values greater than −10.63. This threshold was defined using Receiver Operating Characteristic curves [55,56] and reference inundation maps of Narran Lakes floodplain between 2022 and 2024 [57,58]. For each water year (i.e., July to June), we then calculated the start day of inundation events (day of year), and the duration of the inundation event at each site. These metrics were then summarised across the study period (2004–2024) to capture long-term flood frequency and duration as well as short-term effects (Table 2).

Table 2. Summary of environmental variables and flood history metrics calculated for each site during all surveys from field data and the Fisher Water Index applied to LandSat satellite images during the study period 2000–2024.

Name	Abbreviation	Description
Long-term flood frequency (2000–2024)	Flood frequency	Frequency of flood events across the study period, i.e., number of years where a flood event occurred at each site
Long-term annual flood duration	LT_Duration	Annual inundation duration averaged across study period
Years dry	Dry	Number of consecutive years prior to each survey that a site was dry (i.e., no inundation was recorded)
Consecutive inundation	Consecutive flood	Number of consecutive years prior to each survey that a site was inundated
Previous year flooding	FP_N, FP_Y	Binary factor of flooding in the year prior to each survey
Duration inundated	Duration	Number of consecutive days in a water year that inundation occurred at each site
Shrub cover	Shrub	Average shrub cover in each site calculated from the cover in 10 1 × 1 m quadrats
Burn history	Burned	Binary factor representing site status as burned or not burned in November 2023 determined from mapping of fire extent
Season	Season	Categorical variable representing the season during each survey (Table 1)

2.7. Data Analysis

To describe the floristic composition of understorey wetland vegetation of NLNR and differences over time, we summarised observed species by family, genera, plant growth form, life history and endemicity as total species richness and average vegetation cover. We also identified the number of surveys and floods (i.e., paired surveys) each species and genera was observed in, the number of species unique in each survey, and the number of new species to each survey. For each site we then we calculated species richness as the number of different species observed within the 10 quadrats. We also calculated the site's average vegetation cover for all species, growth forms, and native and exotic species.

We assessed temporal species turnover using the approach of Baselga and Orme [59] with the Sorensen family of indices, whereby beta diversity (i.e., the ratio of gamma and alpha diversity) is decomposed to turnover (i.e., the Simpson dissimilarity, species replacement) and nestedness (i.e., the nestedness of Sorensen, a subset of the original

https://doi.org/10.3390/d18050274

species pool). We first presence-/absence-transformed the species observation matrix then calculated all site × survey pairwise comparisons using the function beta.multi from the betapart R package (version 1.6.1) [60]. We then filtered the resulting dataset to only include temporal comparisons by site (i.e., excluding comparisons of site A to site B at any time) and plotted the mean of comparisons across all surveys and floods.

To explore patterns in species composition within and between survey periods, habitat types and in relation to flood history variables, we used redundancy analysis (RDA). We first Hellinger-transformed species cover data and added a dummy species (x = 1) to all rows to account for zero-inflated species composition data. We standardised all numeric explanatory variables to zero-mean and unit variance using decostand (vegan R package version 2.7-1 [61]) with the standardise method and checked for collinearity using variance inflation factor (VIF), discarding variables with VIF > 10 (Supplementary Table S2). We also included the factorial variable Survey Date and Season to capture temporal and seasonal variation and binary variables for flooding in the previous year and burned in the November 2023 fire (Table 2). We used forward stepwise model selection to reduce the number of model terms. Model and variable significance was tested with analysis of variance (anova.cca; vegan), following the protocol of Borcard et al. [62].

The results of RDA were extracted and plotted with ggplot2, where sites were plotted as points, numerical environmental variables were plotted as vectors and factorial environmental variables were plotted as points. Using scaling 2, the length and direction of environmental vectors reflect their importance to the ordination axes and each other (i.e., vectors in opposite directions are negatively correlated [62]).

To further describe variation in species composition we used the multivariate homogeneity of group dispersion method [63]. We calculated beta-dispersion (betadisper; vegan R package) on a Simpson dissimilarity matrix of species composition for all site × survey combinations, then tested for differences is dispersion between sites using analysis of variance.

3. Results

3.1. General Floristics

At least 110 taxa (87 genera) were observed in the understorey across the ten study sites across all surveys, representing 40 families, a large proportion of which were native (~78%) or exhibited a forb growth form (~72%). Native species of the family Chenopodiaceae were highly represented (12 species), followed by 11 native and four exotic species of the family Asteraceae, seven native Poaceae species, six species of Malvaceae (three native, three exotic), and five native Cyperaceae species. Exotic species were present in the families Asteraceae (four species (sp.)), Malvaceae (three sp.), Verbenaceae (three sp.), Boraginaceae (two sp.), and one species each of Carophyllaceae, Convulvulaceae, Fabaceae, Primulaceae, and Rosaceae.

The annual forb (38 native sp., 14 exotic) and perennial forb (23 native sp., four exotic) plant functional groups were the dominant plant forms across all surveys. Annual and perennial monocots (i.e., grasses and sedges) were represented by five and eight species respectively. Shrubs were represented by nine species and two tree species (*Eucalyptus camaldulensis* and *Acacia stenophylla*) were present as seedlings in the groundcover strata. Commonly observed exotic species included *Cuscuta campestris* (six surveys), *Conyza bonariensis* (five surveys), while *Abutilon theophrastii* and *Verbena macrostachya* occurred in four surveys. The prevalence of these exotic species within the studied sites has changed over time, most notably for *C. bonariensis* which was only recorded in one site in Nov–04 but increased to five sites during Aug–23 and Feb–24 (Supplementary Table S1). In contrast,

V. macrostachya was observed in six sites during Nov–04 which reduced to three sites by Feb–24.

3.2. Changes in Diversity in Response to Flood Events/Pulses

Only six species were recorded from these sites in all surveys (Supplementary Table S1) including two perennial forbs (*Calotis scapigera* and *Marsilea drummondii*), the exotic annual forb *Cascuta campestris*, perennial monocot *Sporobolus mitchellii*, and shrub *Duma florulenta*. Thirty species, including five exotic species, were observed across all floods (i.e., at least one of the paired flood surveys), half of which were annual forbs (four exotic), followed by seven perennial forbs (one exotic), three perennial monocots, three shrubs, and two annual monocots. Twenty-four species (three exotic) were observed in two of the three floods, while 56 species (10 exotic) were only observed in one flood, 80% of which were only observed in one survey. The greatest number of exotic species were observed during November 2004 (26 sp., three unique) followed by August 2023 (21 sp., one unique; Table 3).

Table 3. Summary of the taxa observed in understorey vegetation surveys across three floods and six surveys at Narran Lakes. Values in brackets are the native (N) and exotic (E) taxa breakdown of each category.

Survey Period	Native Taxa	Exotic Taxa	Unknown	Unique Observations	New Observations	Total No. of Taxa
Flood 1 (04–05)	53	11	3	17 (N: 13, E: 4)	67	67
Nov–04	50	11	3	14 (N: 10, E: 4)	64	64
May–05	22	2	1	2 (N: 2)	3 (N: 3)	25
Flood 2 (08–09)	52	8	4	16 (N: 13, E: 3)	20	64
Nov–08	47	8	4	11 (N: 9, E: 2)	18 (N: 15, E: 3)	59
Apr–09	29	4	3	1 (N: 1)	2 (N: 2)	36
Flood 3 (23–24)	51	10	2	22 (N: 18, E: 4)	23	63
Aug–23	39	9	1	12 (N: 9, E: 3)	15 (N: 11, E: 4)	49
Feb–24	28	5	1	7 (N: 7)	7 (N: 7)	34

No species were exclusively observed in multiple dry surveys (i.e., May–05, Apr–09, Feb–24), although 10 species were observed in a single dry survey, seven of which were recorded in Feb–24. In contrast, several species were observed exclusively in drawdown surveys (i.e., Nov–04, Nov–08, Aug–23) including nine annual forbs and two annual monocots (*Lachnagrostis filiformis, Isolepis australis*). New species were observed during all surveys, with greater numbers of new and unique species (i.e., only observed in one survey) drawdown surveys (Figure 3A,B). Over time, species richness varied, particularly between the flood pairs with richness greatest in drawdown surveys. Overall, a marginal decline in the number of species observed is apparent over time when considering the paired flood events (Figure 3C,D).

https://doi.org/10.3390/d18050274

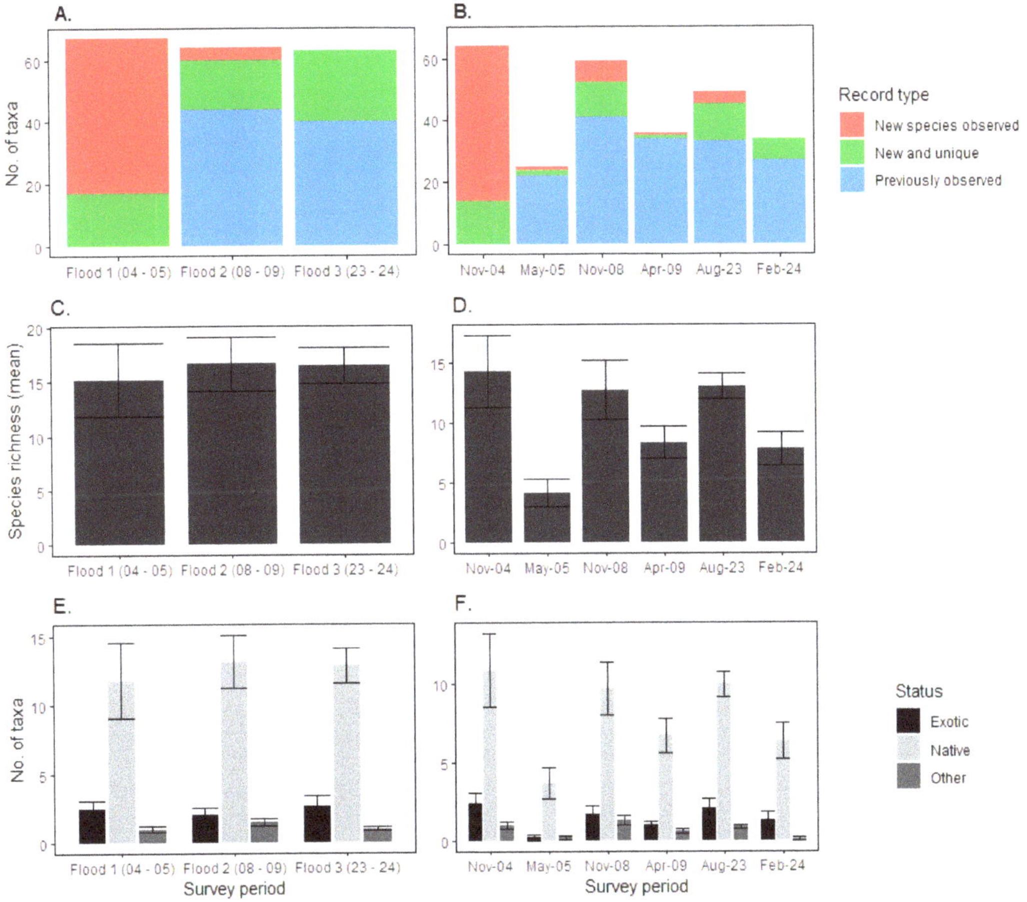

Figure 3. Metrics of understorey wetland taxa over time. (**A**,**B**) Total species richness across all sites by flood (**A**) and by survey (**B**), showing stacked values of new observations, new and unique observations (i.e., new in that survey and not in any other survey), and previously observed. (**C**,**D**) Average species richness per site over time by flood (**C**) and by survey (**D**). (**E**,**F**) Average number of native and exotic taxa per site over time by flood (**E**) and by survey (**F**). Error bars represent standard error from the mean.

3.3. Patterns in Vegetation Structure over Time

Total vegetation cover, functional group cover, and native or exotic taxa cover varied substantially over time (Figure 4). In general, total vegetation cover was greatest during drawdown surveys and dominated by annual forbs. Total cover in dry surveys was the lowest in May–05, intermediate in April–09 and high in Feb–24, higher even than total cover in the earlier drawdown surveys (Figure 4A). Perennial sub-shrub and perennial shrub cover was greatest in Aug–23 and Feb–24 compared to previous surveys. Similarly, perennial monocot cover also showed greater cover in Feb–24, while the cover of annual forbs, perennial forbs, and annual monocots was reduced from Aug–23 (Figure 4A). Cover of native and exotic taxa also varied across surveys, with greater cover of native and exotic taxa during drawdown surveys compared to dry surveys (Figure 4B). Exotic taxa showed greater cover during the Aug–23 survey but was similar to that observed during earlier surveys again by Feb–24.

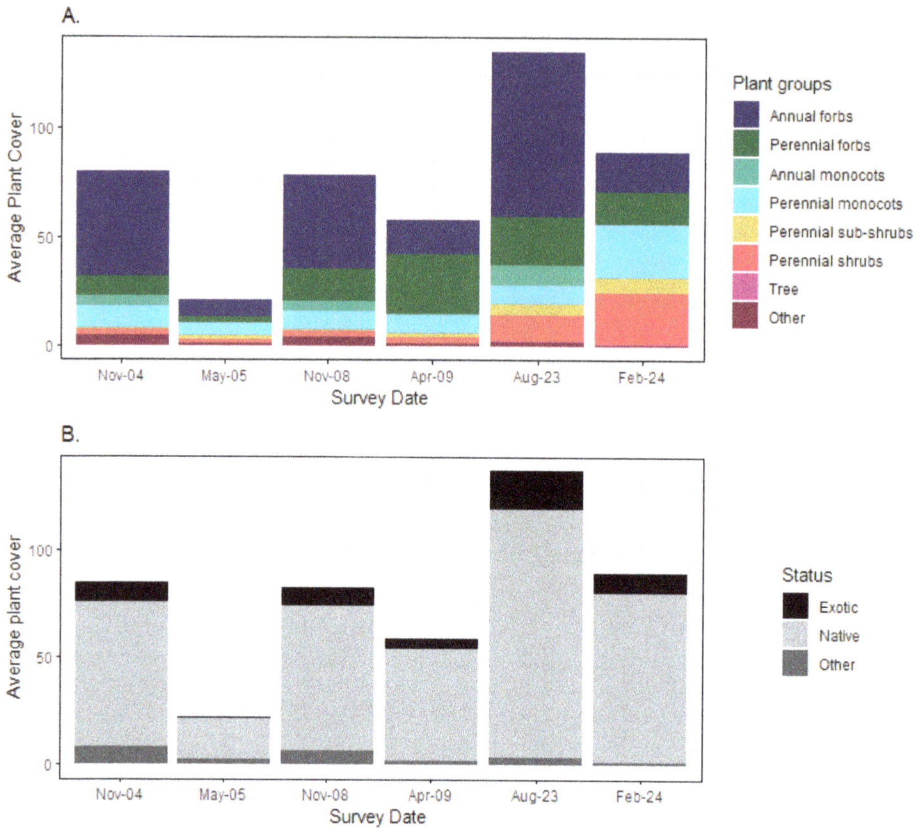

Figure 4. Stacked barplot of average vegetation cover per site. (**A**) Cover of plant functional groups over time. (**B**) Cover of native and exotic taxa over time. Total bar height represents total plant cover average by site. Cover values can be greater than 100 to account for taxa overlapping.

3.4. Change in Composition over Time

Species turnover of understorey taxa within sites, as represented by binary Simpson dissimilarity, generally increased over time, but varied between sites (Figure 5). Western floodplain sites (WS_1, WS_2) exhibited the greatest increase in turnover, particularly with substantial differences in surveys 5 and 6 compared to all others. Turnover at the sites NWW_1, NWW_2, and WS_3 followed a similar pattern and showed a substantial difference particularly in relation to survey 2 (May–05). The sites ES_1 and ES_4 exhibited moderate turnover for all surveys. Northern wetland sites (i.e., ES_2, ES_3 and NWW_3) also increased in turnover over time, peaking in survey 5 (Aug–23, Figure 5). The pattern of increasing turnover with time is also present when comparing flood events (i.e., species turnover between floods for composition combined by survey pairs) where composition during the first and second floods are similar while the composition during flood 3 shows greater difference to both earlier events (Figure 5). The alternate is true for WS_3 where composition during flood 3 is equal to both earlier events while flood 2 is different from flood 1.

https://doi.org/10.3390/d18050274

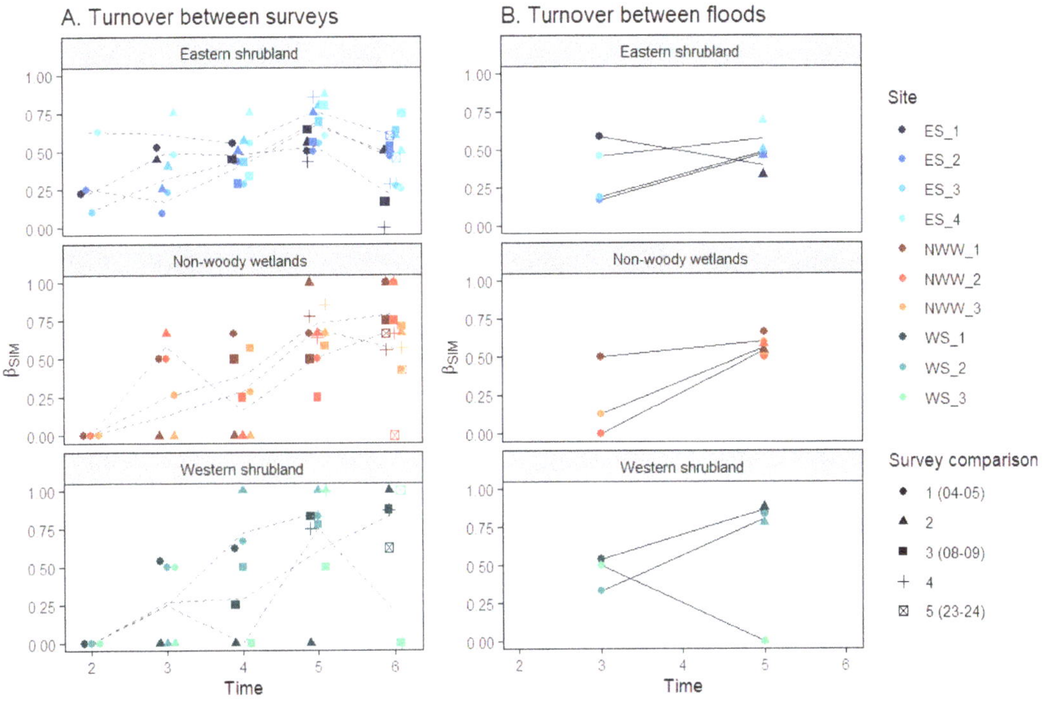

Figure 5. Turnover of presence-/absence-transformed species composition over time by site as represented by Simpson dissimilarity of pairwise comparisons to all previous surveys. (**A**) Turnover between all surveys, (**B**) turnover between floods (i.e., change between flood 1 (04–05) and flood 2 (08–09) shown for time 3). Shapes represent the survey order of the pairwise comparison to all previous surveys (i.e., circle represents the comparison of subsequent surveys to the first survey). Line represents the mean turnover value at each time point.

3.5. Drivers of Compositional Change

Redundancy analysis identified a significant relationship between forward selected historical flooding metrics and understorey species composition which explained approximately 21% (adjusted R^2) of variation in composition ($F = 2.73$, $p = 0.001$, Figure 6, Supplementary Table S2). In general, axis one (RDA1) represented a gradient of short-term flood history associated with the duration of inundation ($F = 2.25$, $p = 0.004$) and greater long-term flood frequency ($F = 8.58$, $p = 0.001$) on the right-hand side (Figure 6A). Axis two (RDA2) partially represented a gradient of short-term flood history and survey dates, extending from surveys with greater consecutive years of inundation ($F = 1.69$, $p = 0.032$) on the positive axis direction to less frequent inundation in the negative (Figure 6A). Survey date was also a significant contributor to the model ($F = 2.07$, $p = 0.001$) highlighting the compositional variation between individual surveys (Figure 6A) and flood events (Figure 6B). Composition within sites also varied over time (Figure 6B), tending to be more homogenous in 23–24 surveys compared to 04–05 surveys and 08–09 surveys which appeared to form two distinct community types of eastern wetland sites (non-woody wetland (NWW) and eastern shrubland (ES)) on the right-hand side, and western shrubland (WS) sites on the left-hand side (Figure 6B) during the earlier surveys. Fire in western shrubland sites also contributed to species composition ($F = 1.9$, $p = 0.013$).

https://doi.org/10.3390/d18050274

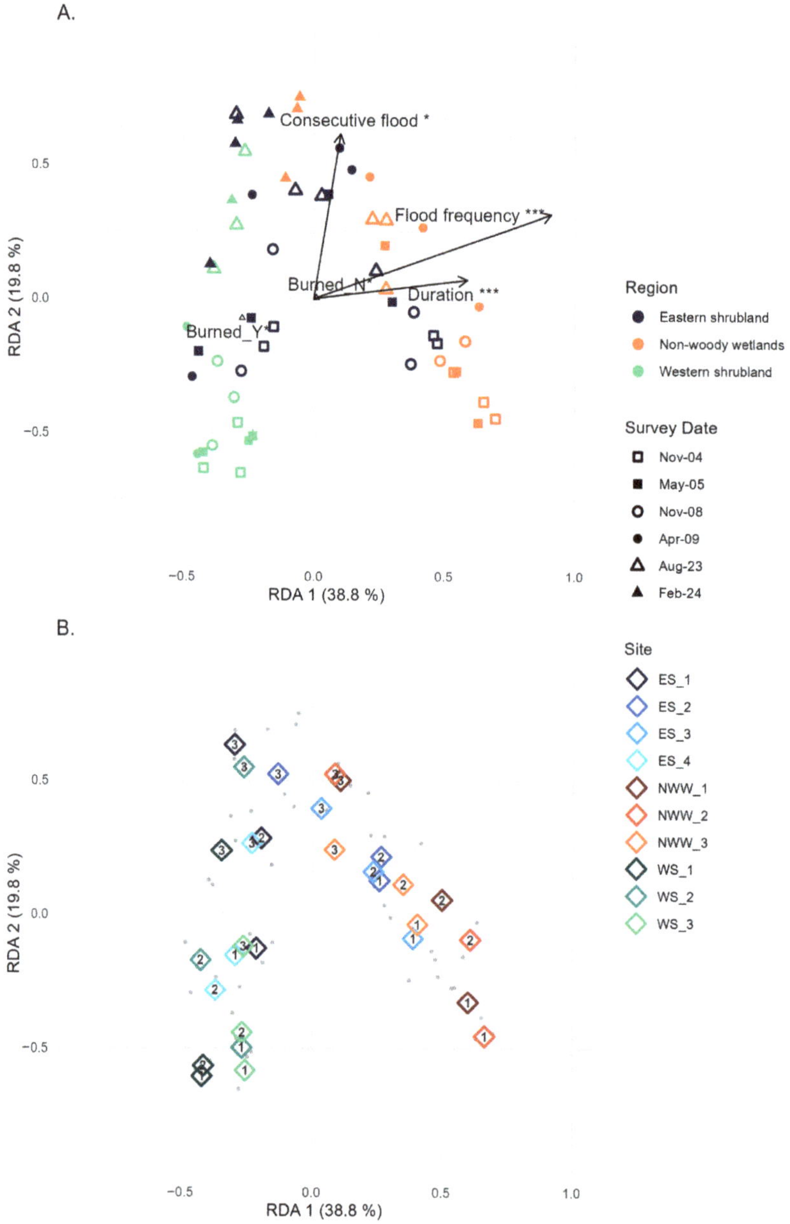

Figure 6. Redundancy analysis biplot for understorey species composition in relation to forward selected environmental variables; axis titles show the proportion of constrained variance explained. (**A**) All sites across surveys showing the forward selected environmental variables. (**B**) Median centroids of sites in paired flood surveys showing trajectory over time as the flood number ordered; all points shown in background. Flood frequency: long-term flood frequency; Duration: duration of inundation in previous year; consecutive flood: consecutive years inundation; Burned_N and _Y: burned in November 2023 no and yes. Asterisks represent significance level of environmental variable (* $p < 0.05$, *** $p < 0.001$). Numbers within points represent the flood order 1: Nov-04–May-05; 2: Nov-08–Apr-09; 3: Aug-23–Feb-24.

https://doi.org/10.3390/d18050274

Dispersion (i.e., the ordination distance between time points) of each site across survey dates (i.e., within site variation) was relatively similar for all sites, suggesting equal and large variation in species composition between surveys regardless of site (Supplementary Table S3).

4. Discussion

Wetland vegetation communities surveyed across the Narran Lakes wetland system have varied substantially in structure and composition across the 20-year study period in response to wetting and drying. Vegetation in lignum shrublands and non-woody wetlands has changed over time both within individual flood periods, as drying occurs, and between successive flood events, at both site and whole-wetland scales. Most notably, species richness and plant cover at a site scale, and the number of new species observed overall, were greater in drawdown surveys than during subsequent dry surveys, highlighting the importance of flood events for replenishing plant biodiversity [64]. High levels of species turnover between survey periods and the discovery of new species during each survey highlights the episodic and dynamic nature of these vegetation communities, with over half of the taxa recorded overall only observed during a single flood event. Annual forbs varied most between surveys, particularly with respect to their cover abundance, reflecting their highly responsive nature to environmental conditions. Understorey species composition was strongly associated with flood history, both short- and long-term, especially the duration of recent flooding, the number of wet years prior, and the frequency of flooding over the study period. Survey date and the recent fire also contributed to species composition, further highlighting the temporal variability of this system. These results together highlight the spatially and temporally dynamic nature of this inland floodplain wetland system and the vital role of flood events (both small and large) in wetland dynamics [14].

Much of the existing work on wetland vegetation in the Murray–Darling Basin has focused on short-term (i.e., up to 5 years) responses to individual flooding events [37]. In contrast, our 20-year dataset captures multiple discrete flood cycles, enabling us to examine not only immediate post-flood responses but also vegetation dynamics over longer timescales, facilitating a comparison of responses across multiple flood events. Across individual events, our findings support expected patterns reported in previous studies (e.g., [6,7,33,40]), such as substantial increases in species richness and plant cover immediately following floodwater drawdown with subsequent declines as a result of drying. Taxa with annual life histories showed the greatest fluctuation in richness and abundance between drawdown and dry surveys. Annuals which maintain a persistent soil seed bank can lay dormant for long dry periods and respond rapidly to flood drawdown, often proliferating under these high-soil-moisture conditions and rapidly completing their life cycles [7,65]. Expression of the bulk of soil seed bank diversity often occurs after floods and typically comprises a highly diverse and abundant assemblage of annual taxa [6]. The dominance of annuals observed in drawdown surveys here suggests that soil seed banks strongly contribute to the diversity of extant assemblages immediately following floodwater drawdown while the perennial taxa become more dominant as the wetland dries out [5,6,23].

A key finding of this study was the exceptionally high species turnover observed across both space and time. New taxa were recorded in every survey, with many species detected only once or restricted to a single flood event (i.e., paired surveys). This raises questions regarding the origin of these newly recorded species and whether they represent recent establishment events or the emergence of individuals from persistent local sources (i.e., soil seed and propagule banks). Two main sources are likely: in situ sources (i.e., soil seed banks, asexual reproduction, dispersal within the wetland complex) and ex situ

https://doi.org/10.3390/d18050274

sources (i.e., dispersal from upstream, transport by animals or wind [66]). Soil seed banks are recognised as important seed stores in dynamic systems, particularly for those which experience hydrological variability like the Narran Lakes and wetlands within the Murray–Darling Basin, where they are often distinct from the extant vegetation but largely comprise annual taxa [6,16,33], which we observed in the drawdown surveys.

Dispersal of taxa within the wetland complex is also likely, with many of the newly observed species noted more widely throughout the studied vegetation types and woodland vegetation types within the wetland, including exotic species [45,51,67]. The large flood event in 2022 likely dispersed propagules throughout the Narran wetland complex, from core, frequently flooded areas to fringing habitats and vice versa, while also providing conditions suitable for germination and establishment of a greater suite of perennials than was achieved in the earlier, smaller floods. Alternatively, ex situ propagule sources such as dispersal from upstream is likely for taxa which are hydrochorous [68,69] or those dispersed by animals which visit the wetland, i.e., waterbirds who nest in the lignum during large flood events [70]. Dispersal and establishment of exotic taxa is commonly noted following flood disturbances [68] and accordingly we recorded greater richness and cover of exotics during drawdown surveys. However, the number of new exotics was similar for all floods and reduced in dry surveys, suggesting the exotics were not able to survive under drying conditions. Similar numbers of exotic plant species have been recorded in other, more northerly MDB Ramsar-listed wetlands [40]; however, their management would require a catchment-wide approach. Further assessment of the soil seed bank assemblage and dispersal patterns is required to understand their roles in species turnover and vegetation dynamics.

Other factors influencing the vegetation responses observed here could include local canopy or shrub layer conditions [71], rainfall, temperature, rate of drying, grazing [65], or other local disturbances (i.e., fire). Temperature and rainfall can alter germination cues and influence the composition of emerging taxa [72], which could have contributed to the different responses between dry surveys where rainfall was higher in the preceding months. Additionally, the cooler winter temperatures experienced during the August 2023 drawdown survey may have resulted in different species being observed, associated with seasonal temperature differences (i.e., a greater number of winter annuals) [72,73] and low rainfall. However, given our study design incorporates flood events, we cannot detangle seasonal effects from hydrology and further study is required to understand vegetation responses following smaller flood events and under different temperature scenarios—although Wen et al. [42] suggests that hydrology would have a greater influence than season on these wetlands given their inherit dynamism and flood-responsive nature. Local canopy conditions (i.e., canopy cover) and the presence of shrub cover can also influence the abundance and diversity of understorey taxa [71,74]. While we did not directly measure these attributes and explored sites without a dominant tree canopy, the record of increased perennial shrub, sub-shrub, and tree taxa within understorey quadrats recorded during the third flood surveys suggests greater cover in other strata which may have influenced the expression of understorey vegetation by providing shade and reducing evaporation, leading to rapid senescence of plant material [74]. Localised disturbances, such as a fire which occurred in November 2023, could also have contributed to the high turnover between the Aug–23 and Feb–24 surveys observed in western shrubland sites by reducing standing plant biomass, removing the germinable soil seed bank and litter seed bank, or triggering the gemination of fire-responding species [75,76]. However, burned sites were compositionally similar to those unburned by Feb–24, suggesting that rainfall supported recovery. Understanding how this event has altered vegetation communities and the role of flooding for restoration requires further investigation; however, Mackay et al. [76], suggest

https://doi.org/10.3390/d18050274

that fire impacts may only be temporary, particularly if followed by inundation, and that inundation history is a stronger driver of composition and structure compared to short-term fire effects.

Floodplain wetland vegetation communities are structured by, and respond to, short- and long-term patterns in flood history. We observed highly variable but native-dominated vegetation communities through space and time which allow the system to persist through abiotic perturbations (i.e., water fluctuations and disturbance events) and respond with a suite of species adapted to specific conditions. Notably, plant cover, but not richness, was greatest during the latest flood event, likely as a result of multiple wet years (Figure 2) supporting the growth of specific species. The smaller flood pulses which occurred prior to vegetation surveys in 2004–05 and 2008–09 maintained heterogenous communities across the eastern and western floodplains. The larger flood event surveyed in 2023–24, both in terms of inundation extent and duration, resulted in large turnover across most sites and a more homogenous community assemblage at the whole-wetland scale. These results highlight the importance of variable flood events for maintaining a mosaic of diverse wetland vegetation communities [77–79], further supporting long-term resilience. Smaller events provide an important resource pulse to the core wetland area which can drive divergence in species composition between the core area and those on the periphery. Larger events can reset wetland systems [24] by facilitating the distribution and germination of taxa across all areas of the wetland, resulting in a more homogenous assemblage at an ecosystem scale. Subsequent drying or smaller flood events which only reach core areas would likely result again in diverging communities across the wetland system [24]. This variability in the magnitude, duration, and frequency of flood events drives the distribution and resilience of vegetation communities in arid floodplain wetlands at local and landscape scales [79]. Heterogeneity within and between vegetation communities at the landscape scale [78] and over time contributes to their resilience by providing a large species and resource pool from which local extirpations can be replenished, particularly during flood events.

Floodplain wetland plant communities are inherently dynamic, switching between wet and dry states [13,14], which contributes to their overall resilience. However, this process is challenged under extreme drying, altered flow regimes, and climate change. If a drier climate develops in the Narran Lakes region, as expected under many climate change scenarios [35,80], significant ecological impacts are likely to follow, including dieback of canopy trees, altered plant community composition through local extirpations or invasions, and the degradation of vegetation condition [80]. Significant drying could also facilitate encroachment by terrestrial vegetation [78,81] or an increase in the cover of woody species as we have observed. If drying is coupled with anthropogenic pressures of land modification and water extraction, there may be a reduced capacity to return such terrestrialised areas to a wetland state without management intervention. Despite experiencing a significant drought event in 2013–2019 (Figure 2), vegetation communities studied here maintained high native richness following the flood with minimal change in the richness and cover of exotics species, further highlighting their ability to recover when water returns. Flow management, through prescribed environmental flows that mimic more natural flow patterns, can support biodiversity and ecological processes [82]. The strategic implementation of environmental flow regimes can be used to simulate the natural variability of both short and long inundation events, both of which we have demonstrated are pivotal to the resilience of floodplain wetland vegetation communities [14,21]. Specifically, such flooding events nourish extant vegetation by meeting moisture requirements, stimulate germination from seed banks and facilitate seed dispersal via hydrochory, and provide connectivity between disjointed habitats [7,16,69]. Managed environmental flows are mechanisms proven to

https://doi.org/10.3390/d18050274

achieve beneficial outcomes for wetland environments in many regions of the Murray–Darling Basin, including the Narran Lakes [67]. Despite challenges in environmental water delivery for this region, such as highly variable flows, small managed water storages, large private storages, and flow-contingent environmental water licences [83], the use of environmental flows in conjunction with natural flow events is highly valuable. We strongly recommend the continued use of environmental watering actions, both through flow protection and managed releases from public and private storages, in this system. These actions, and monitoring of their outcomes, are particularly important in the face of a drying climate; however, we highlight the need for variability in managed flow regimes to maintain vegetation mosaics [84].

Understanding of wetland vegetation patterns and responses to flow events over long periods is lacking, particularly for the northern Murray–Darling Basin where observational evidence is rarely reported over more than 5 years [37]. While there is a growing body of work, much of the reporting tends to identify changes across a single flood event or responses to a drought breaking [85]. Long-term monitoring is particularly important for understanding changes to vegetation structure and dynamics of woody taxa [41,81], which in turn influences the understorey community, although we have not included tree-dominated sites in our analysis. Here, we instead present evidence across multiple flood events and suggest that 'vital sign' monitoring [40] in conjunction with reactionary monitoring provides the most holistic view of vegetation resilience and responses. We also acknowledge that funding schemes and personnel changes can hamper efforts of long-term monitoring; there is much-needed value and certainty in management decisions that can result from this understanding [86]. Although our results align with other studies of understorey vegetation in semi-arid wetlands [6,7,33,40], caution in interpretation is required given the small number of sites analysed and the small proportion of variance in species composition captured by the redundancy model. Additionally, the results presented are the culmination of multiple disturbance events (i.e., floods, droughts, fire) and rather than attempting to draw out the individual effects we instead suggest that this presents a realistic view of wetland vegetation responses in variable systems, which are likely to become more variable under future climate projections [39].

This 20-year study demonstrates that wetland vegetation at the Narran Lakes is inherently dynamic, structured by the magnitude, duration, and sequencing of flood events. Large inundation events acted as system-wide reset mechanisms, reshaping composition and redistributing species across the landscape. Exceptionally high species turnover and continual recruitment from seed banks and dispersal pathways underscore the pulse-driven nature of these communities. Through these many fluctuations the lignum shrublands and non-woody wetlands of the Narran Lakes system have demonstrated a high level of dynamism by maintaining substantial native biodiversity and functional composition, despite challenges of altered flow regimes, drought, and fire. Maintaining variability in flood regimes, rather than simply restoring water, is therefore fundamental to sustaining biodiversity, ecosystem function, and vegetation responses in arid wetlands under a changing climate.

Supplementary Materials: The following supporting information can be downloaded at: https://www.mdpi.com/article/10.3390/d18050274/s1, Figure S1: Rainfall and river flow history for January 2000–February 2024. Table S1: Summary of species observed across the period 2004—2024 at selected long term monitoring sites within the Narran Lake Nature Reserve. Table S2: Environmental variables used to model variation in species composition through redundancy analysis showing variance inflation factor, constrained variance and term significance. Table S3: Average dispersion of sites over time calculated from Simpson dissimilarity of species cover for understorey vegetation.

https://doi.org/10.3390/d18050274

Author Contributions: Conceptualization, R.G. and S.J.C.; Methodology, R.G., A.S. and S.J.C.; Formal analysis, R.G.; Data curation, R.G., J.J.-B., A.S. and S.J.C.; Writing—original draft, R.G. and J.J.-B.; Writing—review & editing, all authors; Funding acquisition, S.J.C. All authors have read and agreed to the published version of the manuscript.

Funding: This work was funded by the Australian Government Department of Climate Change Energy Environment and Water through the Commonwealth Environmental Water Holder. Earlier work contributing data to this paper was funded through the Murray Darling Basin Authority and the Commonwealth Environmental Water Holder.

Data Availability Statement: Data is contained within the article or Supplementary Material. Further inquiries can be directed to the corresponding authors.

Acknowledgments: We acknowledge the traditional owners of the land on which this study was conducted, the Yuwaalarayy/Euahlayi peoples, who have a deep and continuing cultural connection with the land and waters. We thank the many volunteers who contributed to field surveys and we acknowledge New South Wales Parks and Wildlife and the Narran Lakes Joint Management Committee for logistical support and cultural knowledge sharing. We also acknowledge Peta Zivec for review of an earlier version of this manuscript and the anonymous reviewers for their contributions.

Conflicts of Interest: The authors declare no conflict of interest.

References

1. Jiao, T.; Williams, C.A.; Rogan, J.; De Kauwe, M.G.; Medlyn, B.E. Drought Impacts on Australian Vegetation During the Millennium Drought Measured with Multisource Spaceborne Remote Sensing. *J. Geophys. Res. Biogeosciences* **2020**, *125*, e2019JG005145. [CrossRef]
2. Ballinger, A.; Nally, R.M. The Landscape Context of Flooding in the Murray–Darling Basin. In *Advances in Ecological Research*; Academic Press: Cambridge, MA, USA, 2006; Volume 39, pp. 85–105.
3. Sheldon, F.; Boulton, A.J.; Puckridge, J.T. Conservation value of variable connectivity: Aquatic invertebrate assemblages of channel and floodplain habitats of a central Australian arid-zone river, Cooper Creek. *Biol. Conserv.* **2002**, *103*, 13–31. [CrossRef]
4. Casanova, M.T.; Brock, M.A. How do depth, duration and frequency of flooding influence the establishment of wetland plant communities? *Plant Ecol.* **2000**, *147*, 237–250. [CrossRef]
5. Wassens, S.; Ning, N.; Hardwick, L.; Bino, G.; Maguire, J. Long-term changes in freshwater aquatic plant communities following extreme drought. *Hydrobiologia* **2017**, *799*, 233–247. [CrossRef]
6. Capon, S.J.; Reid, M.A. Vegetation resilience to mega-drought along a typical floodplain gradient of the southern Murray-Darling Basin, Australia. *J. Veg. Sci.* **2016**, *27*, 926–937. [CrossRef]
7. James, C.S.; Capon, S.J.; White, M.G.; Rayburg, S.C.; Thoms, M.C. Spatial variability of the soil seed bank in a heterogeneous ephemeral wetland system in semi-arid Australia. *Plant Ecol.* **2007**, *190*, 205–217. [CrossRef]
8. Nilsson, C.; Reidy, C.A.; Dynesius, M.; Revenga, C. Fragmentation and Flow Regulation of the World's Large River Systems. *Science* **2005**, *308*, 405–408. [CrossRef]
9. Grill, G.; Lehner, B.; Thieme, M.; Geenen, B.; Tickner, D.; Antonelli, F.; Babu, S.; Borrelli, P.; Cheng, L.; Crochetiere, H.; et al. Mapping the world's free-flowing rivers. *Nature* **2019**, *569*, 215–221, Erratum in *Nature* **2019**, *572*, E9.. [CrossRef]
10. Niinemets, Ü. Responses of forest trees to single and multiple environmental stresses from seedlings to mature plants: Past stress history, stress interactions, tolerance and acclimation. *For. Ecol. Manag.* **2010**, *260*, 1623–1639. [CrossRef]
11. Middleton, B.A. Soil seed banks and the potential restoration of forested wetlands after farming. *J. Appl. Ecol.* **2003**, *40*, 1025–1034. [CrossRef]
12. Holling, C.S. Resilience and Stability of Ecological Systems. *Annu. Rev. Ecol. Evol. Syst.* **1973**, *4*, 1–23. [CrossRef]
13. Thapa, R.; Thoms, M.C.; Parsons, M.; Reid, M. Adaptive cycles of floodplain vegetation response to flooding and drying. *Earth Surf. Dynam.* **2016**, *4*, 175–191. [CrossRef]
14. Bunn, S.E.; Thoms, M.C.; Hamilton, S.K.; Capon, S.J. Flow variability in dryland rivers: Boom, bust and the bits in between. *River Res. Appl.* **2006**, *22*, 179–186. [CrossRef]
15. Thapa, R.; Thoms, M.C.; Parsons, M. The response of dryland floodplain vegetation productivity to flooding and drying. *J. Arid Environ.* **2016**, *129*, 42–55. [CrossRef]
16. Kehr, J.M.; Merritt, D.M.; Stromberg, J.C. Linkages between primary seed dispersal, hydrochory and flood timing in a semi-arid region river. *J. Veg. Sci.* **2014**, *25*, 287–300. [CrossRef]

https://doi.org/10.3390/d18050274

17. Dornelas, M.; Magurran, A.E.; Buckland, S.T.; Chao, A.; Chazdon, R.L.; Colwell, R.K.; Curtis, T.; Gaston, K.J.; Gotelli, N.J.; Kosnik, M.A.; et al. Quantifying temporal change in biodiversity: Challenges and opportunities. *Proc. Biol. Sci.* **2013**, *280*, 20121931. [CrossRef] [PubMed]
18. Shimadzu, H.; Dornelas, M.; Magurran, A.E. Measuring temporal turnover in ecological communities. *Methods Ecol. Evol.* **2015**, *6*, 1384–1394. [CrossRef]
19. Stott, I.; Franco, M.; Carslake, D.; Townley, S.; Hodgson, D. Boom or bust? A comparative analysis of transient population dynamics in plants. *J. Ecol.* **2010**, *98*, 302–311. [CrossRef]
20. Portela, A.P.; Durance, I.; Vieira, C.; Honrado, J. Environmental filtering and environmental stress shape regional patterns of riparian community assembly and functional diversity. *Freshw. Biol.* **2023**, *68*, 1428–1441. [CrossRef]
21. Brock, M.; Casanova, M. Plant life at the edge of wetlands: Ecological responses to wetting and drying patterns. *Front. Ecol. Res. Build. Links* **1997**, 181–192. [CrossRef]
22. Bried, J.T.; Jog, S.K.; Dzialowski, A.R.; Davis, C.A. Potential Vegetation Criteria for Identifying Reference-Quality Wetlands in the South-Central United States. *Wetlands* **2014**, *34*, 1159–1169. [CrossRef]
23. Nicol, J.M.; Ganf, G.G.; Pelton, G.A. Seed banks of a southern Australian wetland: The influence of water regime on the final floristic composition. *Plant Ecol.* **2003**, *168*, 191–205. [CrossRef]
24. Reynolds, L.V.; Cooper, D.J.; Hobbs, N.T. Drivers of riparian tree invasion on a desert stream. *River Res. Appl.* **2014**, *30*, 60–70. [CrossRef]
25. Gui, H.; Hou, L.; Wang, J.; Dong, X.; Han, S. Flood changed the community composition and increased the importance of stochastic process of vegetation and seed bank in a riparian ecosystem of the Yellow River. *Ecol. Indic.* **2023**, *154*, 110505. [CrossRef]
26. Poff, N.L.; Allan, J.D.; Bain, M.B.; Karr, J.R.; Prestegaard, K.L.; Richter, B.D.; Sparks, R.E.; Stromberg, J.C. The Natural Flow Regime. *BioScience* **1997**, *47*, 769–784. [CrossRef]
27. Capon, S.J. Plant community responses to wetting and drying in a large arid floodplain. *River Res. Appl.* **2003**, *19*, 509–520. [CrossRef]
28. Busch, D.E.; Loftus, W.F.; Bass, O.L. Long-term hydrologic effects on marsh plant community structure in the southern Everglades. *Wetlands* **1998**, *18*, 230–241. [CrossRef]
29. Sheldon, F.; Rocheta, E.; Steinfeld, C.; Colloff, M.J.; Moggridge, B.; Carmody, E.; Hillman, T.; Kingsford, R.T.; Pittock, J. Are environmental water requirements being met in the Murray–Darling Basin, Australia? *Mar. Freshw. Res.* **2024**, *75*, MF23172. [CrossRef]
30. Davies, B.R.; Thoms, M.C.; Walker, K.F.; O'Keeffe, J.H.; Gore, J.A. Dryland Rivers: Their Ecology, Conservation and Management. In *The Rivers Handbook*; Blackwell Scientific: Oxford, UK, 1994; pp. 484–511.
31. Kingsford, R.T. Ecological impacts of dams, water diversions and river management on floodplain wetlands in Australia. *Austral Ecol.* **2000**, *25*, 109–127. [CrossRef]
32. Catelotti, K.; Kingsford, R.T.; Bino, G.; Bacon, P. Inundation requirements for persistence and recovery of river red gums (Eucalyptus camaldulensis) in semi-arid Australia. *Biol. Conserv.* **2015**, *184*, 346–356. [CrossRef]
33. Reid, M.; Capon, S. Role of the soil seed bank in vegetation responses to environmental flows on a drought-affected floodplain. *River Syst.* **2011**, *19*, 249–259. [CrossRef]
34. Döll, P.; Zhang, J. Impact of climate change on freshwater ecosystems: A global-scale analysis of ecologically relevant river flow alterations. *Hydrol. Earth Syst. Sci.* **2010**, *14*, 783–799. [CrossRef]
35. Murray-Darling Basin Authority. *The Future of Climate and Water Availability in the Murray-Darling Basin*; Murray-Darling Basin Authority: Canberra, Australia, 2025.
36. CSIRO. *Water Availability in the Condamine-Balonne: A Report to the Australian Government from the CSIRO Murray-Darling Basin Sustainable Yeilds Project*; CSIRO: Canberra, Australia, 2008.
37. Colloff, M.J.; Caley, P.; Saintilan, N.; Pollino, C.A.; Crossman, N.D. Long-term ecological trends of flow-dependent ecosystems in a major regulated river basin. *Mar. Freshw. Res.* **2015**, *66*, 957–969. [CrossRef]
38. Gawne, B.; Hale, J.; Stewardson, M.J.; Webb, J.A.; Ryder, D.S.; Brooks, S.S.; Campbell, C.J.; Capon, S.J.; Everingham, P.; Grace, M.R.; et al. Monitoring of environmental flow outcomes in a large river basin: The Commonwealth Environmental Water Holder's long-term intervention in the Murray–Darling Basin, Australia. *River Res. Appl.* **2020**, *36*, 630–644. [CrossRef]
39. Chiew, F.; Zheng, H.; Post, D.; Robertson, D.; Rojas, R. Hydroclimate trends and future projections in the Murray-Darling Basin. In *Technical Report for the Murray-Darling Basin Authority*; CSIRO: Canberra, Australia, 2022.
40. Mason, T.J.; Honeysett, J.; Thomas, R.F.; Popovic, G.C.; Hosking, T.; Shelly, D.J.; Bowen, S. Monitoring vital signs: Wetland vegetation responses to hydrological resources in the Macquarie Marshes NSW, Australia. *Austral Ecol.* **2022**, *47*, 1296–1314. [CrossRef]
41. Lindenmayer, D.B.; Lavery, T.; Scheele, B.C. Why We Need to Invest in Large-Scale, Long-Term Monitoring Programs in Landscape Ecology and Conservation Biology. *Curr. Landsc. Ecol. Rep.* **2022**, *7*, 137–146. [CrossRef]

https://doi.org/10.3390/d18050274

42. Wen, L.; Mason, T.J.; Ryan, S.; Ling, J.E.; Saintilan, N.; Rodriguez, J. Monitoring long-term vegetation condition dynamics in persistent semi-arid wetland communities using time series of Landsat data. *Sci. Total Environ.* **2023**, *905*, 167212. [CrossRef]
43. Sims, N.C.; Colloff, M.J. Remote sensing of vegetation responses to flooding of a semi-arid floodplain: Implications for monitoring ecological effects of environmental flows. *Ecol. Indic.* **2012**, *18*, 387–391. [CrossRef]
44. Butcher, R.; Hale, J.; Capon, S.J.; Thoms, M.C. Ecological character description for Narran Lake Nature Reserve Ramsar Site Ecological Character Description. In *Report to the Department of Sustainability, Environment, Water, Population and Communities*; Commonwealth of Australia: Canberra, Australia, 2011.
45. Thoms, M.; Capon, S.; James, C.; Padgham, M.; Rayburg, S. *The Narran Ecosystem Project: The Response of a Terminal Wetland System to Variable Wetting and Drying. Final Report to the Murray–Darling Basin Commission*; Murray-Darling Basin Commission: Canberra, Australia, 2007.
46. Thomas, R.; Karunaratne, S.; Heath, J.; Kuo, W. *Assessment of Site-Specific Flow Indicator Inundated Areas in the Vegetation of Narran Lakes*; NSW Office of Environment and Heritage: Parramatta, Australia, 2016.
47. BOM. Climate Data Online, Daily Rainfall for Lightning Ridge (48243). Available online: https://www.bom.gov.au/climate/data/?ref=ftr (accessed on 28 March 2024).
48. Rayburg, S.; Thoms, M. A coupled hydraulic–hydrologic modelling approach to deriving a water balance model for a complex floodplain wetland system. *Hydrol. Res.* **2009**, *40*, 364–379. [CrossRef]
49. BOM. Water Data Online, Narran River at Wilby Wilby (Belvedere) 244016. Available online: https://www.bom.gov.au/waterdata/ (accessed on 6 June 2024).
50. McGann, T.D.; Kingswood, R.; Bell, D. Vegetation of Narran Lake Nature Reserve, North Western Plains, New South Wales. *Cunninghamia* **2001**, *7*, 43–63.
51. Grieger, R.; Johnston-Bates, J.; Capon, S. *Long-Term Wetland Vegetation Dynamics of Dharriwaa-Narran Lakes Nature Reserve: In Report for the Commonwealth Environmental Water Holder*; Department of Climate Change, Energy, the Environment and Water: Brisbane, Australia, 2024.
52. Dunn, B.; Ai, E.; Alger, M.J.; Fanson, B.; Fickas, K.C.; Krause, C.E.; Lymburner, L.; Nanson, R.; Papas, P.; Ronan, M.; et al. Wetlands Insight Tool: Characterising the Surface Water and Vegetation Cover Dynamics of Individual Wetlands Using Multidecadal Landsat Satellite Data. *Wetlands* **2023**, *43*, 37. [CrossRef]
53. Royal Botanic Gardens and Domain Trust. PlantNet (The NSW Plant Information Network System). Available online: https://plantnet.rbgsyd.nsw.gov.au (accessed on 6 June 2024).
54. Cunningham, G.M.; Mulham, W.E.; Milthorpe, P.L.; Leigh, J.H. *Plants of Western New South Wales*; CSIRO Publishing: Sydney, Australia, 2011.
55. Fisher, A.; Flood, N.; Danaher, T. Comparing Landsat water index methods for automated water classification in eastern Australia. *Remote Sens. Environ.* **2016**, *175*, 167–182. [CrossRef]
56. Fawcett, T. An introduction to ROC analysis. *Pattern Recognit. Lett.* **2006**, *27*, 861–874. [CrossRef]
57. Heath, J.T.; Grimmett, L.; Gopalakrishnan, T.; Thomas, R.F.; Lenehan, J. Integrating Sentinel 2 Imagery with High-Resolution Elevation Data for Automated Inundation Monitoring in Vegetated Floodplain Wetlands. *Remote Sens.* **2024**, *16*, 2434. [CrossRef]
58. State Government of NSW; NSW DCCEEW. *Inundation Maps for NSW Inland Floodplain Wetlands*; New South Wales Department of Climate Change Energy the Environment and Water: Sydney, Australia, 2020.
59. Baselga, A.; Orme, C.D.L. betapart: An R package for the study of beta diversity. *Methods Ecol. Evol.* **2012**, *3*, 808–812. [CrossRef]
60. Baselga, A. Partitioning the turnover and nestedness components of beta diversity. *Glob. Ecol. Biogeogr.* **2010**, *19*, 134–143. [CrossRef]
61. Oksanen, J.; Simpson, G.L.; Blanchet, F.G.; Kindt, R.; Legendre, P.; Minchin, P.R.; O'Hara, R.B.; Solymos, P.; Stevens, M.H.H.; Szoecs, E.; et al. *vegan: Community Ecology Package*. CRAN R Package, 2025. Available online: https://cran.r-project.org/web/packages/vegan/index.html (accessed on 29 April 2026). [CrossRef]
62. Borcard, D.; Gillet, F.; Legendre, P. Spatial Analysis of Ecological Data. In *Numerical Ecology with R*; Borcard, D., Gillet, F., Legendre, P., Eds.; Springer International Publishing: Cham, Switzerland, 2018; pp. 299–367.
63. Anderson, M.J.; Ellingsen, K.E.; McArdle, B.H. Multivariate dispersion as a measure of beta diversity. *Ecol. Lett.* **2006**, *9*, 683–693. [CrossRef] [PubMed]
64. Leigh, C.; Sheldon, F.; Kingsford, R.T.; Arthington, A.H. Sequential floods drive 'booms' and wetland persistence in dryland rivers: A synthesis. *Mar. Freshw. Res.* **2010**, *61*, 896–908. [CrossRef]
65. Reid, M.A.; Ogden, R.; Thoms, M.C. The influence of flood frequency, geomorphic setting and grazing on plant communities and plant biomass on a large dryland floodplain. *J. Arid Environ.* **2011**, *75*, 815–826. [CrossRef]
66. Zivec, P.; Johnston-Bates, J. Seed rain as a propagule source for restoration of semi-arid floodplain old fields. *Appl. Veg. Sci.* **2024**, *27*, e70001. [CrossRef]

https://doi.org/10.3390/d18050274

67. Vincent, B.; Bowen, S.; Southwell, M. Vegetation Community and Tree Stand Condition Assessment of the Flood-Dependent vegetation of Narran Lakes Nature Reserve, 2020–2021. In *A Report to the Commonwealth Environmental Water Office*; Department of Climate Change, Energy, the Environment and Water: Canberra, Australia, 2022.
68. Catford, J.A.; Jansson, R. Drowned, buried and carried away: Effects of plant traits on the distribution of native and alien species in riparian ecosystems. *New Phytol.* **2014**, *204*, 19–36. [CrossRef]
69. Hyslop, J.; Trowsdale, S. A review of hydrochory (seed dispersal by water) with implications for riparian rehabilitation. *J. Hydrol.* **2012**, *51*, 137–152.
70. Brandis, K.; Bino, G.; Spencer, J.; Ramp, D.; Kingsford, R. Decline in colonial waterbird breeding highlights loss of Ramsar wetland function. *Biol. Conserv.* **2018**, *225*, 22–30. [CrossRef]
71. Capon, S.J.; Balcombe, S.R.; McBroom, J. Environmental watering for vegetation diversity outcomes must account for local canopy conditions. *Ecohydrology* **2017**, *10*, e1859. [CrossRef]
72. Ooi, M.K.J.; Auld, T.D.; Denham, A.J. Climate change and bet-hedging: Interactions between increased soil temperatures and seed bank persistence. *Glob. Change Biol.* **2009**, *15*, 2375–2386. [CrossRef]
73. Capon, S.J.; Brock, M.A. Flooding, soil seed bank dynamics and vegetation resilience of a hydrologically variable desert floodplain. *Freshw. Biol.* **2006**, *51*, 206–223. [CrossRef]
74. James, C.S.; Capon, S.J.; Quinn, G.P. Nurse plant effects of a dominant shrub (*Duma florulenta*) on understorey vegetation in a large, semi-arid wetland in relation to flood frequency and drying. *J. Veg. Sci.* **2015**, *26*, 985–994. [CrossRef]
75. Bixby, R.J.; Cooper, S.D.; Gresswell, R.E.; Brown, L.E.; Dahm, C.N.; Dwire, K.A. Fire effects on aquatic ecosystems: An assessment of the current state of the science. *Freshw. Sci.* **2015**, *34*, 1340–1350. [CrossRef]
76. Mackay, K.D.; Vincent, B.; Southwell, M.; Growns, I.; Mika, S. Differential impacts of fire and inundation on a wetland plant community after wildfire. *Mar. Freshw. Res.* **2024**, *75*, MF24040. [CrossRef]
77. Lawson, J.R.; Fryirs, K.A.; Lenz, T.; Leishman, M.R. Heterogeneous flows foster heterogeneous assemblages: Relationships between functional diversity and hydrological heterogeneity in riparian plant communities. *Freshw. Biol.* **2015**, *60*, 2208–2225. [CrossRef]
78. Shilpakar, R.L.; Thoms, M.C.; Reid, M.A. The resilience of a floodplain vegetation landscape. *Landsc. Ecol.* **2021**, *36*, 139–157. [CrossRef]
79. Raulings, E.J.; Morris, K.A.Y.; Roache, M.C.; Boon, P.I. The importance of water regimes operating at small spatial scales for the diversity and structure of wetland vegetation. *Freshw. Biol.* **2010**, *55*, 701–715. [CrossRef]
80. Johnston-Bates, J.; Grieger, R.; Capon, S.J. Predicting trajectories of dryland wetland vegetation transformation under climate change: A case study of the northern Murray–Darling Basin, Australia. *Mar. Freshw. Res.* **2024**, *75*, MF24016. [CrossRef]
81. Saintilan, N.; Asbridge, E.; Lucas, R.; Rogers, K.; Wen, L.; Powell, M.; Colloff, M.J.; Rodriguez, J.F.; Saco, P.M.; Sandi, S.; et al. Australian forested wetlands under climate change: Collapse or proliferation? *Mar. Freshw. Res.* **2021**, *73*, 1255–1262. [CrossRef]
82. Acreman, M.; Arthington, A.H.; Colloff, M.J.; Couch, C.; Crossman, N.D.; Dyer, F.; Overton, I.; Pollino, C.A.; Stewardson, M.J.; Young, W. Environmental flows for natural, hybrid, and novel riverine ecosystems in a changing world. *Front. Ecol. Environ.* **2014**, *12*, 466–473. [CrossRef]
83. DPIE. *Intersecting Streams Long Term Water Plan Parts A and B*; Department of Planning, Industry and Environment: Parramatta, Australia, 2020.
84. Campbell, C.J.; Freestone, F.L.; Duncan, R.P.; Higgisson, W.; Healy, S.J. The more the merrier: Using environmental flows to improve floodplain vegetation condition. *Mar. Freshw. Res.* **2021**, *72*, 1185–1195. [CrossRef]
85. Kingsford, R.T.; Mac Nally, R.; King, A.; Walker, K.F.; Bino, G.; Thompson, R.; Wassens, S.; Humphries, P. A commentary on 'Long-term ecological trends of flow-dependent ecosystems in a major regulated river basin', by Matthew J. Colloff, Peter Caley, Neil Saintilan, Carmel A. Pollino and Neville D. Crossman. *Mar. Freshw. Res.* **2015**, *66*, 970–980. [CrossRef]
86. Dourado, G.F.; Rallings, A.M.; Viers, J.H. Overcoming persistent challenges in putting environmental flow policy into practice: A systematic review and bibliometric analysis. *Environ. Res. Lett.* **2023**, *18*, 043002. [CrossRef]

diversity

Article

Do Ecological Patterns Persist in Highly Impacted Urban Wetlands? A Spatiotemporal Analysis of Aquatic Macrophytes and Limnological Variability in a Peruvian Coastal Wetland

Flavia Valeria Rivera-Cáceda [1], José Antonio Arenas-Ibarra [1,2,*] and Sofía Isabel Urrutia-Ramírez [2]

[1] Faculty of Environmental Science, Marine Biology Department, Universidad Científica del Sur, Lima 15842, Peru; 180000527@cientifica.edu.pe
[2] Faculty of Biological Sciences, Universidad Ricardo Palma, Lima 15039, Peru; sofia.urrutia@urp.edu.pe
* Correspondence: jarenas@cientifica.edu.pe

Abstract

Urban coastal wetlands along the Peruvian Pacific coast are increasingly affected by urban expansion, pollution, and hydrological alterations, compromising their ecological integrity. In this context, the spatiotemporal variation of the aquatic macrophyte community and its relationship with limnological conditions and drivers of change were evaluated in the Santa Rosa wetland (Chancay, Lima). The objective is to evaluate the spatiotemporal variation of the aquatic macrophyte community in the Santa Rosa wetland and analyze its relationship with physicochemical limnological variables and drivers of change. Sampling was conducted during two contrasting hydrological seasons in 2022: T1 (low-water season) and T2 (high-water season), at six sampling points (P1–P6). Physicochemical variables (water depth, temperature, pH, conductivity, total dissolved solids—TDS, total suspended solids—TSS, dissolved oxygen—DO, turbidity, nitrate—NO_3^-, ammonium—NH_4^+, phosphate—PO_4^{3-}, and dissolved organic matter—DOM) were measured, and the relative abundance of aquatic macrophytes was evaluated. Drivers of change were identified through direct observation and a structured matrix, with phosphate a PCoA performed to summarize spatiotemporal trends. Data were analyzed using Principal Component Analysis (PCA), Co-inertia analysis, and Multi-Response Permutation Procedures (MRPP). Significant spatiotemporal variation was observed in physicochemical parameters ($p < 0.05$), with moderate covariation between the two matrices (RV = 0.47). A total of ten aquatic macrophyte species were recorded, with higher abundance of *Pontederia crassipes* and *Pistia stratiotes* in T1, and *Hydrocotyle ranunculoides* and *Bacopa monnieri* in T2. The most relevant drivers of change were solid waste, livestock grazing, organic contamination, and urban expansion. Spatial heterogeneity was observed in the drivers of change affecting the Santa Rosa wetland, forming a mosaic of areas with different impact profiles. Despite multiple anthropogenic pressures, the Santa Rosa wetland maintains a limnological structure and a functionally coupled macrophyte community, suggesting that essential ecological processes are maintained within the temporal scope of this study. The observed covariation between physicochemical conditions and vegetation confirms the persistence of essential ecological processes, even within an altered urban context. This study demonstrates that integrating biotic components, limnological variables, and drivers of change is fundamental to understanding and monitoring the ecological dynamics of urban wetlands along the Peruvian coast.

 https://doi.org/10.3390/d18040214

Keywords: aquatic plants; drivers of change; physical-chemical limnology; spatiotemporal variation; Santa Rosa wetland

1. Introduction

Wetlands are highly productive ecosystems that provide essential ecosystem services, such as hydrological regulation, pollutant purification, and habitat provision for a wide variety of species, particularly adapted to their hydrological dynamics [1]. They also act as carbon sinks and store CO_2 in their biomass and sediments, thereby helping to mitigate climate change [2]. Their ability to retain sediments and nutrients prevents eutrophication, leading them to be referred to as the "kidneys of the landscape" [3]. These ecosystems cover between 8.9% and 9.5% of the Earth's surface and are home to remarkable biological diversity, including endemic and endangered species [4]. However, since 1700, nearly 3.4 million km^2 of wetlands have been lost, mainly due to land use change; in the last 150 years, more than 50% have been degraded by urban and agricultural expansion, while pollution from fertilizers and pesticides has deteriorated water quality and affected biodiversity, including aquatic macrophyte habitats [5–8].

The ecological functioning of wetlands is intrinsically linked to the seasonality of the hydrological cycle, physicochemical limnology of the water, and structure of aquatic macrophyte communities [9–12]. Hydrological regimes regulate seasonal fluctuations in water depth, generating variations in physicochemical limnology, redox conditions, and biogeochemical cycles that control nutrient availability [11,13,14]. Water hydrochemistry and seasonality modulate pH, conductivity, salinity, and temperature, thereby affecting the biotic composition and the capacity of wetlands to purify water, maintain biodiversity, and provide ecosystem services [15,16].

The aquatic macrophyte community includes freshwater photosynthetic organisms that are visible to the naked eye; following the life-form classification of Chambers and Maberly [17], freshwater plants can be categorized into several functional groups based on their habit, position on the water column, and access to resources: emergent species, amphibious and floating-leaved species, pleustophytes, and submerged species. The role of aquatic macrophytes in ion absorption and transport between sediment and the water column is key to the nutrient cycle and carbon dynamics, influencing dissolved oxygen concentration, nutrient availability, and water quality [9,18,19]. Likewise, their primary production supports multiple trophic levels, and their physical structures provide habitats, shelters, and substrates for various organisms [19,20]. This multifunctionality contributes to ecosystem stability and resilience to disturbances [21,22].

The composition, distribution, and abundance of aquatic macrophytes respond to interactions among abiotic, biotic, and anthropogenic factors. Among the main hydrological factors, the intensity, duration, and frequency of high and low water levels determine water volume and habitat availability [12,19,21]. Similarly, variability in nutrients, pH, turbidity, and conductivity, together with biotic processes such as competition and herbivory, regulates the richness and dominance of macrophyte species [18,20,23,24]. Eutrophication, fragmentation, and anthropogenic impacts alter these processes to varying degrees [22,25–27].

The wetlands of the Peruvian Pacific are unique ecosystems in the arid subtropical desert, shaped by the interaction between natural processes, such as marine intrusions, El Niño events (ENSO), and exfiltration from coastal aquifers, and human activities, including urban and agricultural runoff, resulting in a complex and dynamic hydrological regime [28–31]. El Niño events are particularly significant geomorphological drivers in this

https://doi.org/10.3390/d18040214

context, as the marked increase in river discharge and suspended sediment load associated with warm ENSO phases causes overbank flooding that reshapes coastal geomorphology, alters wetland boundaries, and may create new water bodies by inundating coastal depressions [32]. The seasonal hydrology of the Peruvian Pacific coastal wetlands follows a delayed recharge–exfiltration cycle. Rainfall in the Andean headwaters during the austral summer (December–March) recharges the unconfined coastal aquifer via river infiltration and percolation from irrigation canals. Several months later, this groundwater reaches the coast and sustains high water levels between July and September [28–31,33,34]. This lag produces two hydrological phases: a low-water phase (T1), with direct surface inputs and episodic discharge pulses in some sectors, and a high-water phase (T2), when phreatic exfiltration raises the water table and increases water depth across the wetland. The Santa Rosa wetland, the subject of this study, is one of the most important on the central Peruvian coast from a scientific perspective and is characterized by the highest reported floristic diversity among the coastal wetlands of the Department of Lima, with well-structured communities representative of different functional types of macrophytes [35–37]. Although studies have examined the diversity of the Peruvian Pacific wetlands, primarily flora and avifauna [38–40], limnological knowledge, particularly regarding the relationships between water physicochemical parameters and biota, remains poorly resolved. Previous studies addressing the physicochemical characteristics of water in this system have been spatially restricted and temporally discontinuous, precluding robust assessments of seasonal and interannual variability [38–40], and studies that integrate relationships with aquatic biota are mostly limited to benthic invertebrates [41–43]. While some relationships between physicochemical limnology and aquatic macrophytes have been explored [44,45], the role of physicochemical parameters as determinants of the structure of this community and its spatiotemporal dynamics in urban wetlands represents a critical information gap, which is particularly relevant given the increasing degradation of these environments by anthropogenic activities that compromise their diversity and ecosystem services [35,46–49].

In this context, the present study aimed to evaluate the spatiotemporal dynamics of the aquatic macrophyte community in the Santa Rosa wetland during the low-water season (March) and high-water season (September) of 2022, determine its relationship with water physicochemical parameters, and characterize the most important drivers of change. We hypothesize that the contrasting hydrological seasons—low-water season (T1) characterized by reduced water depth and increased ionic concentration, high-water season (T2) driven by aquifer exfiltration with greater depth and ionic dilution—drive significant variation in water physicochemical parameters and in the composition and percent cover of aquatic macrophyte assemblages, with life forms turnover reflecting adaptive responses to these seasonal hydrological filters.

2. Materials and Methods

2.1. Study Area

The Santa Rosa wetland is located in the district of Chancay, Huaral, Lima region, Peru (Figure 1). Bordered by El Cascajo (north) and Salinas (east) hills, the agricultural lands of Peralvillo and Salinas Altas (south), and the Pacific Ocean (west), it includes a main lake of approximately 17.5 ha and secondary wetlands [50,51]. The climate is very dry and semi-warm, with mean annual temperatures ranging from 18 to 19 °C, absolute maxima of 25–28 °C, and minima of 12–15 °C [45,52]. Annual precipitation is low, averaging 18 mm and ranging between 0 and 36 mm; from April to December, light drizzle, persistent cloud cover, and high relative humidity (81–97%) predominate, particularly from June onward. From January to March, solar irradiance increases and south-southwesterly winds prevail under the influence of the Humboldt Current, with mean velocities of 8–12 km/h [45,52].

https://doi.org/10.3390/d18040214

Its hydrological regime depends on an irrigation canal fed by the Chancay River and on groundwater exfiltration, which is recharged by agricultural irrigation [31,50]. The maximum water depth occurs between August and October, then decreases towards austral summer, when the Chancay River floods slightly raise it [42]. Additionally, the wetland receives episodic discharges from an irrigation canal derived from the Chancay River during the austral summer, contributing to short-term hydrological variability [45].

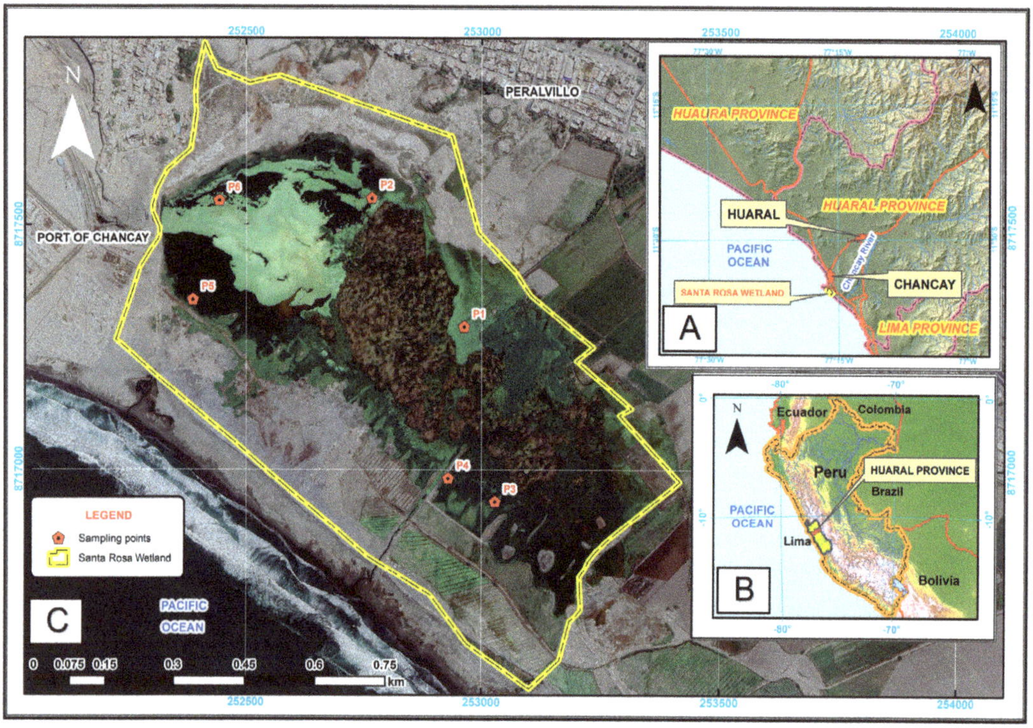

Figure 1. Location of the study area. (**A**,**B**) Geographic location in the province of Huaral, district of Chancay, Lima, Peru. (**C**) Santa Rosa wetland environmental conservation area with sampling points (P1–P6).

The wetland features diverse habitats and plant formations: the main body of water; reed beds dominated by *T. domingensis* and *S. americanus*; beach areas; unvegetated areas; mixed meadows associated with agricultural and urban areas; cleared areas affected by human activities; and shrubby areas [53]. Since 2020, it has been classified as an Environmental Conservation Area under municipal management (Ordinance No. 013-2020-MPH-CM).

2.2. Sampling Design

Six sampling points (P1–P6) were established (Figure 1). Sampling was conducted during two contrasting periods of the hydrological cycle: March 2022 (low-water season, T1) and September 2022 (high-water season, T2).

P1 is located in an area influenced by the irrigation canal derived from the Chancay River, which receives agricultural and urban effluents, with a predominance of *Pontederia crassipes* (Mart.). P2 was located at the eastern end of the main lagoon and is covered with *Bacopa monnieri* (L.) Wettst and *P. stratiotes*. P3 is a seasonally saturated area with heterogeneous and seasonal vegetation of *T. domingensis*, *P. crassipes*, and *S. americanus*. P4 corresponds to a deep drainage channel where *P. crassipes* is conspicuous and the sediments

https://doi.org/10.3390/d18040214

are thick. P5 and P6, at the western and central ends of the lagoon, respectively, are areas with seasonally variable depths and accumulation of organic matter, dominated by *P. stratiotes*, *B. monnieri*, and *H. ranunculoides*.

2.3. *In Situ Analysis of Physicochemical Parameters*

Depth was measured using a graduated ruler with an accuracy of 1 mm. Physicochemical and limnological parameters were recorded in the water surface layer (0–20 cm), following the National Protocol for Water Resource Monitoring [54]. The temperature (°C) (Standard Method 2550) [55], pH (Standard Method 4500-H+) [55], electrical conductivity (μS/cm) [55] and total dissolved solids (TDS) (Standard Method 2510) [55], and dissolved oxygen (DO) (Standard Method 4500-O) [55] were measured using a YSI Pro multiparameter meter (YSI Inc., Yellow Springs, OH, USA). Turbidity was determined using a Lutron LT-TU2016 turbidimeter (Lutron Electronic Enterprise, Taipei, Taiwan) (Standard Method 2130 B) [55], while total suspended solids (TSS) were measured with a Hach DR900 multiparameter portable colorimeter following the manufacturer's photometric method (Hach, Loveland, CO, USA) (Method 8006) [55]. All measurements were taken between 10:00 a.m. and 2:00 p.m. to minimize diurnal variability in physicochemical and limnological parameters. Determinations were made in accordance with the Standard Methods for the Examination of Water and Wastewater [55]. For nutrient analysis, 500 mL of surface water was collected at each sampling point, selecting representative and homogeneous areas and avoiding substrate disturbance. The samples were stored at 4 °C for transport and subsequent laboratory processing. Phosphate concentration was determined using the ascorbic acid method with a Hanna HI3850 photocolorimeter (HANNA instruments, Smithfield, RI, USA) (Standard Method 4500-P E) [55], ammonium using the Nessler method with a Hanna HI733 photocolorimeter (HANNA instruments, Smithfield, RI, USA) (Standard Method 4500-NH3) [55], and nitrate using the cadmium reduction method with a Hach DR900 photocolorimeter (Hach, Loveland, CO, USA) (Standard Method 4500-NO3-E) [55]. Dissolved organic matter (DOM) was determined as volatile dissolved solids by gravimetry, using 0.45 μm glass fiber filters, vacuum filtration, and subsequent incineration at 550 °C in a muffle furnace for 2 h [56].

2.4. *Aquatic Macrophytes Abundance Determination and Sample Collection*

2.4.1. Determination of the Relative Abundance of Aquatic Macrophytes

To estimate the relative abundance of aquatic macrophytes, five 0.5 m^2 quadrats were placed along an approximately 20 m diagonal belt transect aligned with the depth gradient at each sampling point. The transect extended from the deepest area toward the shallower margins on both sides [57]. This setup ensured representation of the macrophyte life-form zones at each site, including pleustophyte-dominated deeper areas, amphibious transitional zones, and emergent shallow areas. The percentage cover of each species was visually and directly estimated within each quadrat by the same observer throughout the study, using a quantitative direct cover estimation method to reduce inter-observer variability [58].

2.4.2. Species Identification of Aquatic Macrophytes

Aquatic macrophytes were collected at each sampling point and preserved in polyethylene bags with formalin. Identification was based on specialized literature [59] and taxonomic checklists [36,37]. The specimens were deposited in the Herbarium of the Natural History Museum of Ricardo Palma University.

2.5. *Identification and Classification of Drivers of Change (DOC)*

Drivers of change (DOC) were assessed following a standardized field protocol proposed by Sarkar et al. [60] and adapted by Aponte et al. [35].

https://doi.org/10.3390/d18040214

In the case of the Santa Rosa wetland, specific modifications were incorporated, and the recommendations from the recently developed methodological framework for coastal wetlands in Peru [47] were also considered in the interpretation of the data. DOCs were classified as direct (immediate physical and chemical impacts) and indirect (sociopolitical, economic, and cultural factors that modulate direct drivers). Direct DOCs included: uncontrolled urban growth (UUG); agricultural (ADG) and livestock (LDG) degradation; introduction of exotic species (IES); chemical (CHC), organic (ORC), and microbiological (MBC) contamination; domestic (DSW) and construction (CSW) solid waste; landfill and drainage (LFD); presence of farms (POF); effluents (EFL); and other disturbances (undermining or extraction) (DIS). The only indirect DOC considered was poor management and inadequate governance policies (GPI). Identification was carried out through direct observation at the six sampling points with photographic recording and GPS georeferencing, based on parallel research and a literature review. Discrepancies between observers were resolved through field discussion and consensus.

2.6. Data Analysis

Principal coordinate analysis (PCoA) was performed using the vegan package in R [61] to evaluate the spatiotemporal variation of the DOCs. Matrices were constructed from the physicochemical parameters and relative abundances of macrophytes by season (T1 and T2) and sampling point (P1–P6), and were summarized in tables and graphs. The relationship between the environmental and biological matrices was evaluated using Spearman's correlation (ρ), considering the value of the statistic and its significance ($p < 0.05$). Physicochemical data were normalized, and the biotic matrix was transformed using the Hellinger procedure, excluding species with a single occurrence [62].

Principal component analysis (PCA) was used to summarize variability and determine spatiotemporal patterns. The significant axes of the PCA were analyzed using multiple response permutation analysis (MRPP) in Vegan package, considering seasons and sampling points as factors of variability [63,64]. The significant axes of an ordination analysis concentrate on the variability of a data matrix and identify the variables that explain it, that is, those most correlated with the axes, allowing statistical inference about this relationship [62]. The relationship between the matrices was analyzed by co-inertia analysis using the ade4 package [65,66]. Co-inertia assesses the covariation of the significant axes of PCA, quantifying shared inertia through the RV coefficient as a measure of the association between both matrices [62]. All analyses were performed using R and RStudio software [67,68].

3. Results

3.1. Spatiotemporal Variation

3.1.1. Macrophytes

Ten species of aquatic macrophytes were recorded: *H. ranunculoides*, *P. crassipes*, *P. stratiotes*, *Azolla filiculoides* Lam., *B. monnieri*, *Ludwigia peruviana* (L.) H. Hara, *S. americanus*, *L. gibba*, *Nasturtium officinale* W. T Aiton, and *Paspalum vaginatum* Sw. Following the life-form classification of Chambers and Maberly [17], these species represent three functional groups: pleustophytes (*P. crassipes*, *P. stratiotes*, *A. filiculoides*, *L. gibba*), amphibious and floating-leaved species (*H. ranunculoides*, *B. monnieri*, *N. officinale*), and emergent species (*L. peruviana*, *S. americanus*, *P. vaginatum*). Macrophyte coverage generally ranged from 35% to 100%, with significant variation between T1 and T2 at sampling points P1–P6. At P1, the coverage increased from 88% in T1 to 100% in T2. At P2, there was an increase from 85% in T1 to 100% in T2. In contrast, at P3, coverage decreased slightly from 90% in T1 to 85% in T2. At P4, high coverage was observed in both periods

https://doi.org/10.3390/d18040214

(95% in T1 and 100% in T2). At P5, coverage showed notable variation, from 60% in T1 to 35% in T2. Finally, at P6, coverage reached 98% in T1 and decreased to 75% in T2 (Figure 2).

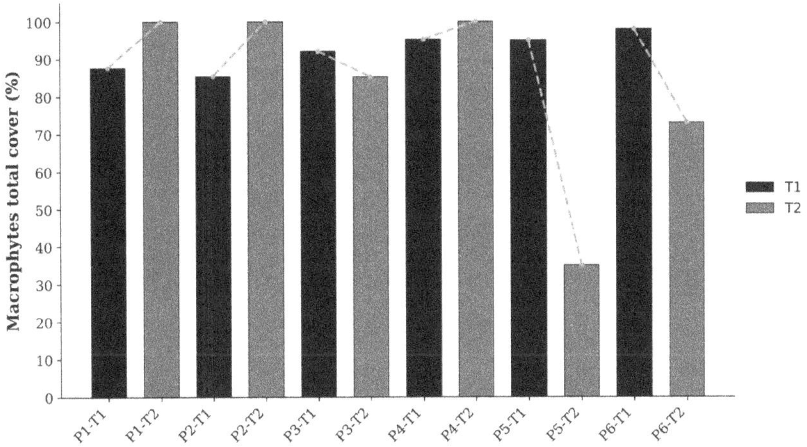

Figure 2. Macrophyte total cover (%) at sampling points (P1–P6) in the Santa Rosa wetland during two seasons: T1 (low-water season) and T2 (high-water season).

Regarding the composition and representativeness of aquatic macrophytes at each sampling point and season, shifts in dominance among life-form groups were observed (Figure 3). At P1, dominance changed from *P. crassipes* (65% in T1) to *H. ranunculoides* (55% in T2) (Figure 3A). At P2, *H. ranunculoides* was dominant in T1 (50%); however, in T2, it decreased to 6% and was replaced by *P. stratiotes* (80%) (Figure 3B). At P3, *B. monnieri* dominated in T1 (50%), whereas in T2, it was absent and dominance shifted to *H. ranunculoides* (40%) together with *P. stratiotes* (30%) (Figure 3C). At P4, the dominance of *P. crassipes* (60%) and *L. peruviana* (35%) in T1 decreased in T2, whereas *A. filiculoides* reached 50% (Figure 3D). At P5, the dominance among *B. monnieri* (40%), *H. ranunculoides* (30%), and *P. stratiotes* (25%) in T1 was reduced in T2 to the exclusive presence of *B. monnieri* (35%) (Figure 3E). Finally, at P6, the shared dominance between *P. stratiotes* (60%) and *H. ranunculoides* (35%) in T1 changed in T2 to the sole presence of *P. stratiotes* (75%) (Figure 3F).

The distribution of macrophyte life forms indicated seasonal changes in species composition (Figure 4). At P1, the assemblage exhibited a transition from pleustophyte dominance in T1 to a more balanced contribution of pleustophytes and amphibious/floating-leaved forms in T2 (Figure 4A). At P2, amphibious/floating-leaved forms dominated in T1, whereas pleustophytes dominated in T2 (Figure 4B). At P3, T1 was characterized by a mixed assemblage, with amphibious/floating-leaved and emergent forms predominating. In contrast, in T2, the macrophytes community transitioned towards a mixed pleustophyte-amphibious/floating-leaved structure, with only a minor contribution of emergent forms (Figure 4C). At P4, pleustophytes and emergent forms co-occurred in T1, but in T2 the assemblage became strongly dominated by pleustophytes, with a smaller contribution of amphibious/floating-leaved forms (Figure 4D). At P5, amphibious/floating-leaved forms dominated in both seasons and were the only life form recorded in T2, where emergent and pleustophyte forms were no longer registered, consistent with the marked reduction in total macrophyte cover at this site (Figure 4E). At P6, pleustophytes dominated in both seasons, increasing in relative abundance in T2, when amphibious/floating-leaved forms were no longer recorded (Figure 4F). In general,

https://doi.org/10.3390/d18040214

emergent forms were limited to P3 and P4 and exhibited greater predominance in T1 than in T2.

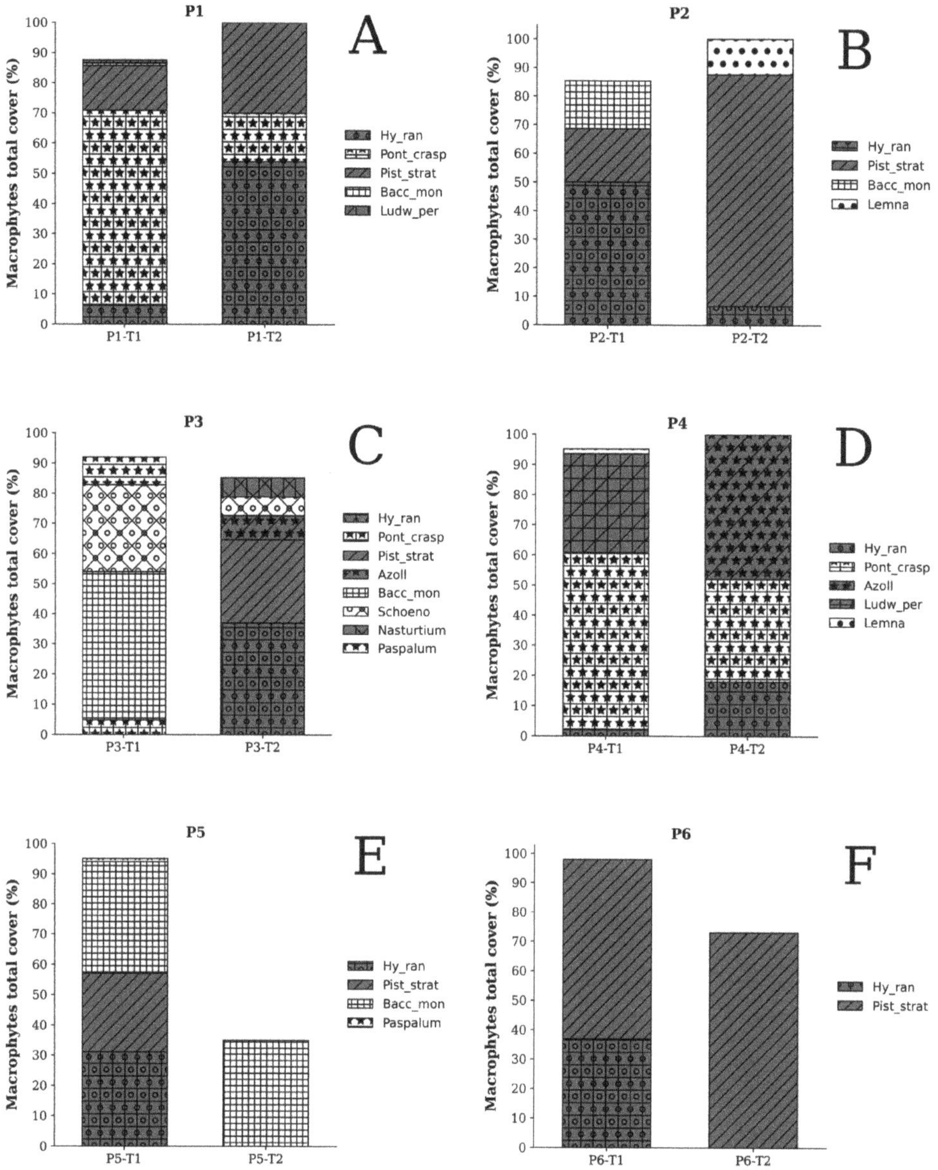

Figure 3. Percentage composition of aquatic macrophytes in Santa Rosa wetland during two seasons (T1: low-water season, T2: high-water season) at sampling points: (**A**) P1, (**B**) P2, (**C**) P3, (**D**) P4, (**E**) P5, and (**F**) P6. Hy_ran: *Hydrocotyle ranunculoides*; Pont_crasp: *Pontederia crassipes*; Pist_strat: *Pistia stratiotes*; Bacc_mon: *Bacopa monnieri*; Ludw_per: *Ludwigia peruviana*; Lemna: *Lemna gibba*; Schoeno: *Schoenoplectus americanus*; Nasturtium: *Nasturtium officinale*; Paspalum: *Paspalum vaginatum*; Azoll: *Azolla filiculoides*.

https://doi.org/10.3390/d18040214

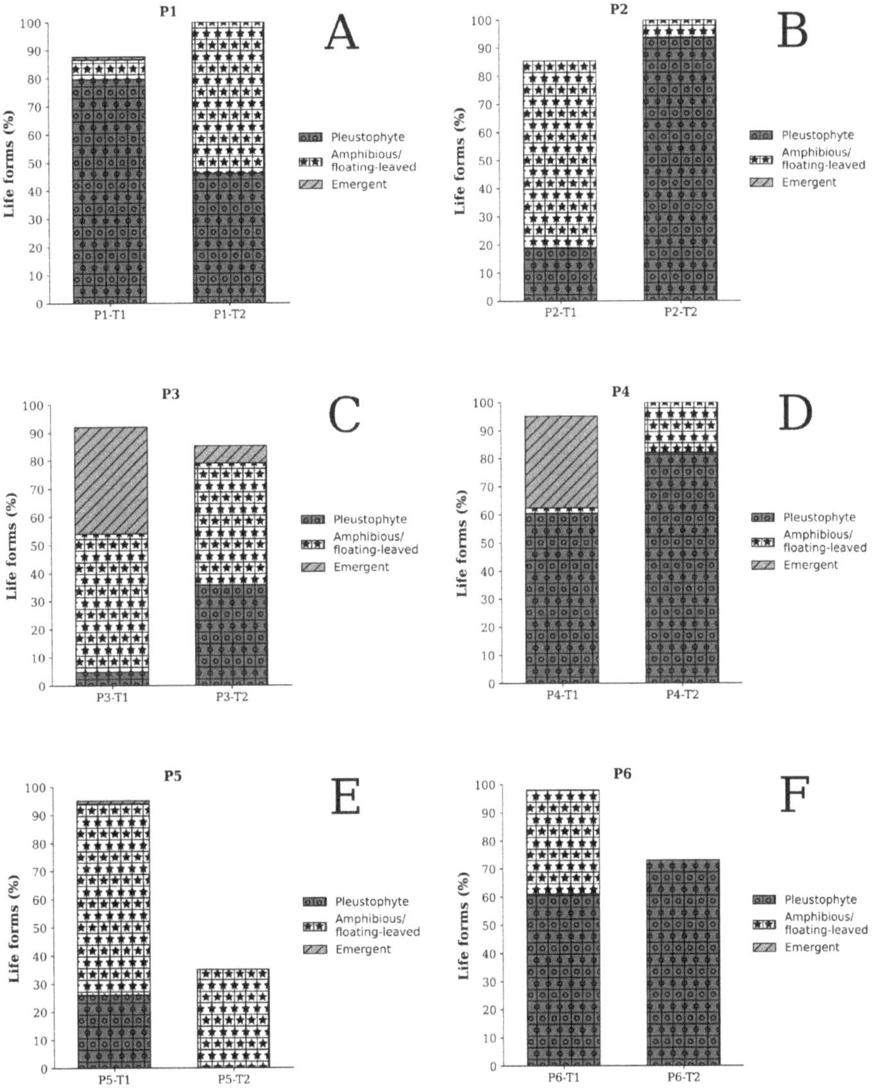

Figure 4. Percentage composition of aquatic macrophytes life forms in Santa Rosa wetland during two seasons (T1: low-water season, T2: high-water season) at sampling points: (**A**) P1, (**B**) P2, (**C**) P3, (**D**) P4, (**E**) P5, and (**F**) P6.

3.1.2. Physicochemical Parameters

Spatiotemporal variation was observed in physicochemical parameters (Supplementary Materials). Depth increased in T2 compared with T1, except at P4, where it decreased. The temperature decreased in T2 (16.7–22.1 °C) compared with T1 (22–25.9 °C). DO increased significantly in T2 (3.2–8.2 mg/L), reaching the maximum value at P5 (8.5 mg/L) and the minimum at P2 (3.2 mg/L). The pH remained alkaline in both seasons, with a slight decrease in T2. Conductivity and TDS decreased in T2, reflecting a seasonal ionic dilution. Similarly, TSS and turbidity decreased, with the highest values recorded at P1, P6, and P7 during T1, and the lowest at P4 in T2. Regarding nutrients, nitrates showed marked spatial variation in both seasons, while ammonium and phosphates tended to increase in

https://doi.org/10.3390/d18040214

T2, with maximum at P1 and P3 for ammonium, and at P3 for phosphates. DOM showed a heterogeneous pattern among sampling points, with increases at P3 and P5 and decreases at P1, P2, and P4 in T2 (Supplementary Materials).

3.2. Spearman's Correlation Matrix and Drivers of Change

3.2.1. Spearman's Correlation Matrix

A positive correlation was observed between electrical conductivity and TDS (ρ = 0.88), as well as between TSS and turbidity (ρ = 0.86), TSS and DOM (ρ = 0.73), and turbidity and DOM (ρ = 0.68). O2 was negatively correlated with temperature (ρ = −0.84). P. crassipes and L. peruviana were positively correlated (ρ = 0.72). Regarding the physicochemical–macrophyte relationship, P. crassipes was negatively correlated with conductivity (ρ = −0.67), TDS (ρ = −0.87), and phosphates (ρ = −0.78) (Figure 5).

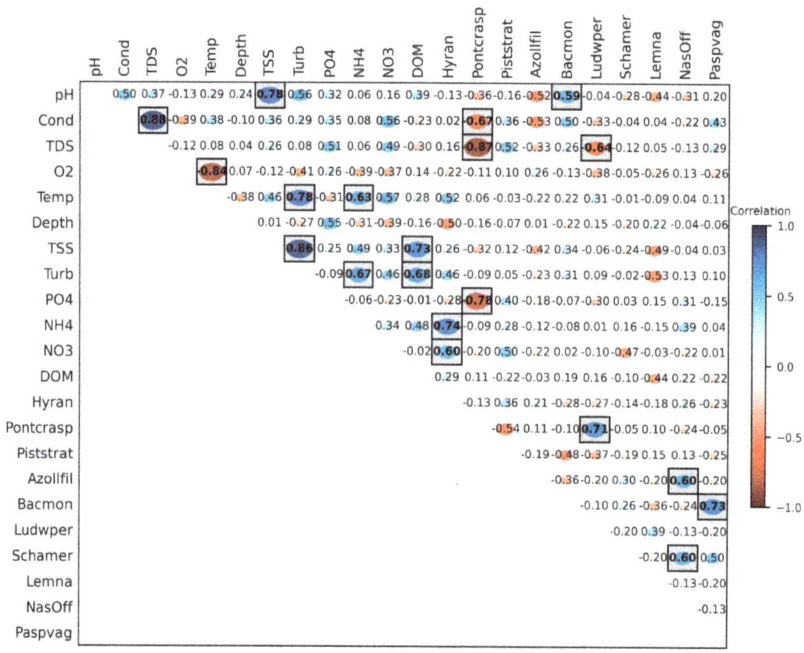

Significance levels are indicated, but correlations should be interpreted cautiously given the sample size (n=12).

Figure 5. Spearman's correlation matrix of association between biotic and abiotic variables. The color scale and ellipse size represent the direction and strength of the correlation coefficient (ρ), respectively. Those values marked marked in bold indicate the level of significance of the correlation ($p < 0.05$). Cond: conductivity; TDS: total dissolved solids; O2: dissolved oxygen; Temp: temperature; TSS: total suspended solids; Turb: turbidity; DOM: dissolved organic matter; Hyran: Hydrocotyle ranunculoides; Pontcrasp: Pontederia crassipes; Piststrat: Pistia stratiotes; Bacmon: Bacopa monnieri; Ludwper: Ludwigia peruviana; Lemna: Lemna gibba; Schamer: Schoenoplectus americanus; NasOff: Nasturtium officinale; Paspvag: Paspalum vaginatum; Azollfil: Azolla filiculoides.

3.2.2. Drivers of Change

The analysis of drivers of change (DOC) in the Santa Rosa Wetland revealed a spatially heterogeneous pattern of impact (Table 1). P4 had the highest number of DOCs (12), followed by P1 and P5 (9 each), P6 (7), and P2 and P3 (6 each). Livestock degradation (LDG), characterized by cattle, sheep, goats, and pig farming, was present at all sampling points. Similarly, the introduction of exotic species (IES), such as dogs and free-grazing livestock (cattle, sheep, and goats), occurred at all locations. Organic contamination (ORC),

https://doi.org/10.3390/d18040214

characterized by food and agricultural waste, livestock manure, urban effluents, as well as microbiological contamination (MBC), indicated by the presence of thermotolerant coliforms, was also recorded at all sampling points. Governance failures (GPI), including poor inter-institutional participation and coordination, which led to deficiencies in the effective protection of wetlands, were considered common across all sampling points. Chemical contamination (CHC), characterized by the use of pesticides and fertilizers, was recorded at P1, P4, P5, and P6. Domestic solid waste (DSW) was found at P4, P5, P6, and P1 during T1, and construction solid waste (CSW) and landfill and drainage (LFD) were recorded at P1, P4, and P5. Other DOCs showed localized distribution: uncontrolled urban growth (UUG) was recorded only at P4; agricultural degradation associated with land use change for agriculture (ADG) at P3 and P4; pig and cattle farms (POF) in P2; effluents (EFL) at P1, and other disturbances (DIS), such as undermining or extraction, only in P4 in T2.

Table 1. Drivers of change identified at the sampling points. UUG: Uncontrolled urban growth; ADG: Agricultural degradation; LDG: Livestock degradation; IES: Introduction of exotic species; CHC: Chemical contamination; ORC: Organic contamination; MBC: Microbiological contamination; DSW: Domestic solid waste; CSW: Construction solid waste; LFD: Landfill and drainage; POF: Presence of farms: EFL: Effluents; GPI: Governance failures associated with poor management and inadequate policies; and DIS: Other disturbances (undermining or extraction). Note: 1 = presence of driver; 0 = absence.

DOC	UUG	ADG	LDG	IES	CHC	ORC	DSW	CSW	LFD	POF	MBC	GPI	EFL	DIS
T2P1	0	0	1	1	1	1	0	1	1	0	1	1	0	0
T2P2	0	0	1	1	0	1	0	0	0	1	1	1	0	0
T2P3	0	1	1	1	0	1	0	0	0	0	1	1	0	0
T2P4	1	1	1	1	1	1	1	1	1	0	1	1	0	1
T2P5	0	0	1	1	1	1	1	1	1	0	1	1	0	0
T2P6	0	0	1	1	1	1	1	0	0	0	1	1	0	0
T1P1	0	0	1	1	1	1	1	1	1	0	1	1	1	0
T1P2	0	0	1	1	0	1	0	0	0	1	1	1	0	0
T1P3	0	1	1	1	0	1	0	0	0	0	1	1	0	0
T1P4	1	1	1	1	1	1	1	1	1	0	1	1	0	0
T1P5	0	0	1	1	1	1	1	1	1	0	1	1	0	0
T1P6	0	0	1	1	1	1	1	0	0	0	1	1	0	0

3.2.3. Principal Coordinates Analysis (PCoA)

The ordination based on the drivers of change (DOC) matrix revealed spatial heterogeneity in the pressures affecting the Santa Rosa Wetland (Figure 6). The first three axes accounted for 80.88% of the variability (PCoA1 = 53.41%, PCoA2 = 17.87%, and PCoA3 = 9.59%). Sampling points formed discrete, well-defined groups, in which the primary source of variation was spatial location rather than the temporal dimension, suggesting that pressure profiles persisted at each point throughout the analyzed period. Axis PCoA1 separated P2 and P3 from the remaining sampling points, while axis PCoA2 distinguished P3 from P4. Four subgroups emerged based on DOC incidence: one comprising P1, P5, and P6, which share a similar DOC profile, and three points (P2, P3, and P4) that clustered separately, each exhibiting a distinctive set of drivers of change. The EFL driver was characteristic of P1-T1, whereas CHC, DSW, and LFD were characteristic of P5 and P1-T2. Regarding POF, it was associated with P2 across both T1 and T2. Lastly, UUG and DIS were associated with P4 (T1 and T2). Collectively, these results indicate that driver variability in the wetland is predominantly spatial, resulting in a mosaic of zones with differentiated impact profiles.

https://doi.org/10.3390/d18040214

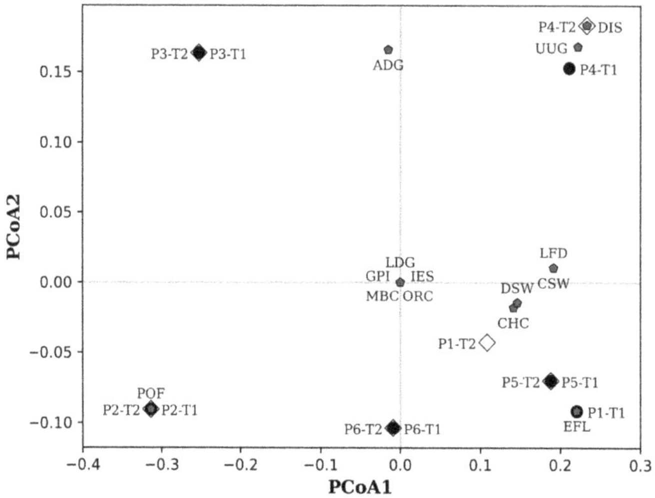

Figure 6. Principal Coordinates Analysis (PCoA) ordination of sampling points based on the drivers of change (DOC). Circles and diamonds represent sampling points across T1, low-water season and T2, high-water season respectively, pentagons represent drivers of change. UUG: Uncontrolled urban growth; ADG: Agricultural degradation; LDG: Livestock degradation; IES: Introduction of exotic species; CHC: Chemical contamination; ORC: Organic contamination; MBC: Microbiological contamination; DSW: Domestic solid waste; CSW: Construction solid waste; LFD: Landfill and drainage; POF: Presence of farms: EFL: Effluents; GPI: Governance failures associated with poor management and inadequate policies; and DIS: Other disturbances (undermining or extraction).

3.3. Relationship Between Biotic and Abiotic Matrices

3.3.1. Principal Component Analysis (PCA)

The first three principal components (PC1, PC2, and PC3) of the principal component analysis (PCA) of the abiotic matrix explained 74.55% of the total variability (PC1: 34.47%; PC2: 21.06%; PC3: 19.02%). PC1 retained the variability in turbidity, temperature, and total suspended solids ($\rho = -0.941$, -0.864, and -0.850, respectively), indicating a temporal gradient. PC2 concentrated the variability of conductivity and TDS ($\rho = 0.794$ and 0.792, respectively), showing a gradient of ionic concentrations between sampling points and seasons. PC3 retained the variability of pH and phosphates ($\rho = -0.750$ and -0.872, respectively), associated with a gradient of acidity and nutrient availability (Figure 7). The PCA of the biotic matrix explained 76.87% of the variability in the first three components (PC1: 31.20%; PC2: 27.15%; and PC3: 18.51%). PC1 retained the variability of *P. stratiotes*, *P. crassipes*, and *L. peruviana* ($\rho = -0.830$, -0.788, and 0.728, respectively), This reflects a primary spatial gradient separating the *Pistia-Hydrocotyle* assemblage—which characterized stations such as P5, P6, and P1-T2, associated with higher nutrient concentrations (as supported by the physicochemical PCA)—from the *P. crassipes-L. peruviana* assemblage characterizing station P4 and P1-T1, linked to deeper and more oxygenated conditions. In PC2, *B. monnieri* and *S. americanus* exhibited concentrated variability ($\rho = -0.891$ and -0.625, respectively), indicating a secondary spatiotemporal pattern with more mixed ecological aspects, involving amphibious (*B. monnieri*) and emergent (*S. americanus*) species that characterized specific points (e.g., P5-T2 and P3-T1) in shallower palustrine environments. PC3 was associated with the occurrence of *A. filiculoides* ($\rho = 0.875$), indicating favorable conditions for this species (Figure 8). Statistically significant differences were found in the MRPPs between seasons (PC1) and sampling points (PC3) of the abiotic matrix ($p < 0.04$ and $p < 0.037$, respectively) and between sampling points for PC2 of the biotic matrix ($p < 0.047$).

https://doi.org/10.3390/d18040214

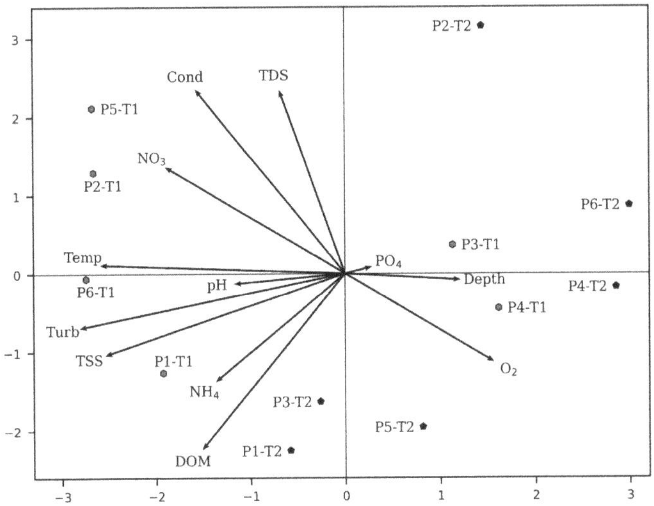

Figure 7. Principal Component Analysis (PCA) for physicochemical variables. Black arrows represent physicochemical variables, hexagons and pentagons represent sampling points across T1, low-water season and T2, high-water season respectively. Cond: conductivity; TDS: total dissolved solids; O_2: dissolved oxygen; Temp: temperature; TSS: total suspended solids; Turb: turbidity; DOM: dissolved organic matter; NO_3: nitrate; PO_4: phosphate; NH_4: ammonium.

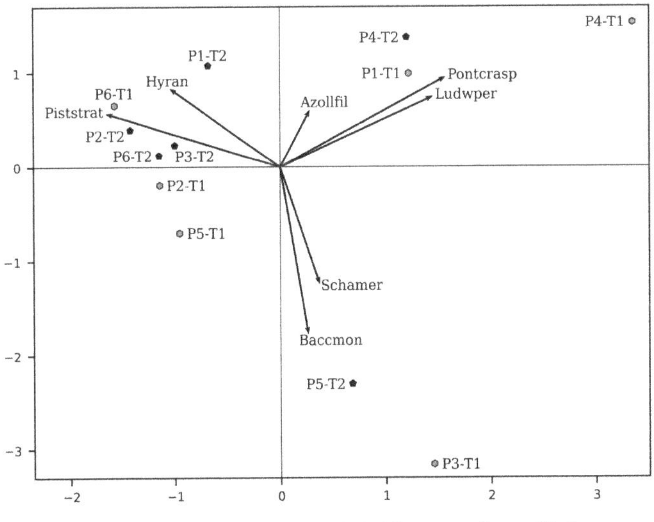

Figure 8. Principal Component Analysis (PCA) for macrophytes. Black arrows represent macrophyte species, hexagons and pentagons represent sampling points across T1, low-water season and T2, high-water season respectively. Hyran: *Hydrocotyle ranunculoides*; Pontcrasp: *Pontederia crassipes*; Piststrat: *Pistia stratiotes*; Azollfil: *Azolla filiculoides*; Baccmon: *Bacopa monnieri*; Ludwper: *Ludwigia peruviana*; Schamer: *Schoenoplectus americanus*.

3.3.2. Co-Inertia

The co-inertia analysis obtained a total inertia of 8.693, retaining 88.97% of the co-inertia (E1: 50.48%, E2: 25.84%, and E3: 12.65%). The correlation (covariation) between both matrices was high for axes 1 and 2, and moderate for axis 3 ($\rho E1 = 0.86$, $\rho E2 = 0.82$, $\rho E3 = 0.67$). An RV coefficient of 0.468 was obtained, representing moderate covariation

https://doi.org/10.3390/d18040214

between the matrices. A main gradient (E1) was observed, defined by TDS, Cond, TSS, Turb, PO_4^{3-}, and NO_3^- (positive correlations) and Depth (negative correlation). E2 showed a gradient of pH and PO_4^{3-} (positive correlation) and NH_4^+ and NO_3^- (negative correlation). The axis structure reflects the pattern observed in the PCA, with the dilution of the main ions increasing with depth. *P. stratiotes* showed covariation with TDS, Cond, NO_3^-, PO_4^{3-}, and pH. *B. monnieri* was associated with pH and PO_4^{3-}, whereas *H. ranunculoides* was more related to shallow environments with high concentrations of NH_4^+, TSS, and Turb. *A. filiculoides* was associated with deeper and less mineralized environments, whereas *P. crassipes* was more related to less mineralized environments (Figure 9). Good agreement was observed between the biotic and abiotic matrices in P3-T2, P2-T2, P6-T1, and P5-T2 (short arrows) (Figure 10), while in P6-T2, P1-T1, P5-T1, P3-T1, and P4-T2, disagreement was observed between both matrices (long arrows). Spatiotemporally, P2-T1, P5-T1, and P6-T1 were associated with the Cond, TDS, pH, and PO_4^{3-} gradient and the presence of *P. stratiotes* and *B. monnieri*; P4-T1 and P4-T2 were associated with low mineralization and the presence of *P. crassipes* and *A. filiculoides*, whereas P1-T2 and P3-T2 were associated with *H. ranunculoides* and high concentrations of NH_4 and turbidity, while P5-T2 aligned with the pH gradient and *B. monnieri*. In general, a pattern of ionic dilution was observed with increasing depth from T1 to T2, along with a temporal gradient in physicochemical limnology and aquatic macrophytes (Figures 9 and 10). Although moderate covariation was observed between the matrices, it was not statistically significant ($p > 0.05$) owing to variability within each sampling point.

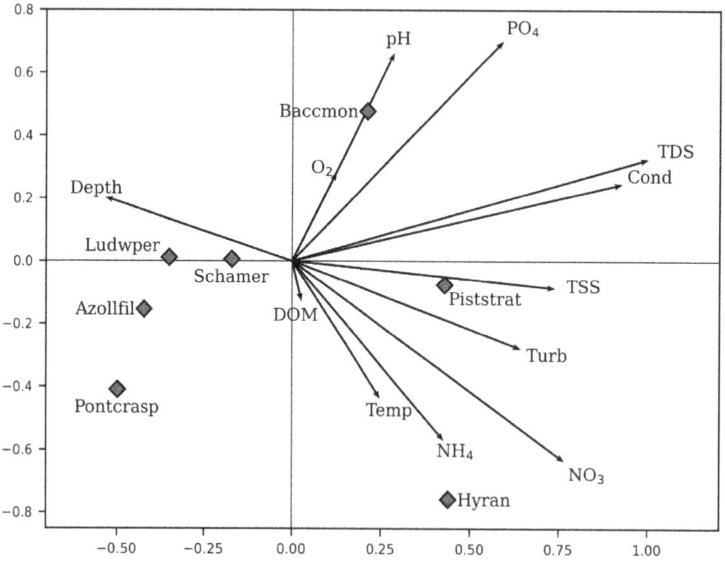

Figure 9. Co-inertia biplot (axes 1–2): covariance-optimized structural matching between limnological variables and macrophyte assemblages. Black arrows represent physicochemical variables, diamonds represent macrophyte species. Arrow length indicates the strength of the contribution to the co-inertia axes. Cond: conductivity; TDS: total dissolved solids; O2: dissolved oxygen; Temp: temperature; TSS: total suspended solids; Turb: turbidity; DOM: dissolved organic matter; NO_3: nitrate; PO_4: phosphate; NH_4: ammonium; Hyran: *Hydrocotyle ranunculoides*; Pontcrasp: *Pontederia crassipes*; Piststrat: *Pistia stratiotes*; Azollfil: *Azolla filiculoides*; Baccmon: *Bacopa monnieri*; Ludwper: *Ludwigia peruviana*; Schamer: *Schoenoplectus americanus*.

https://doi.org/10.3390/d18040214

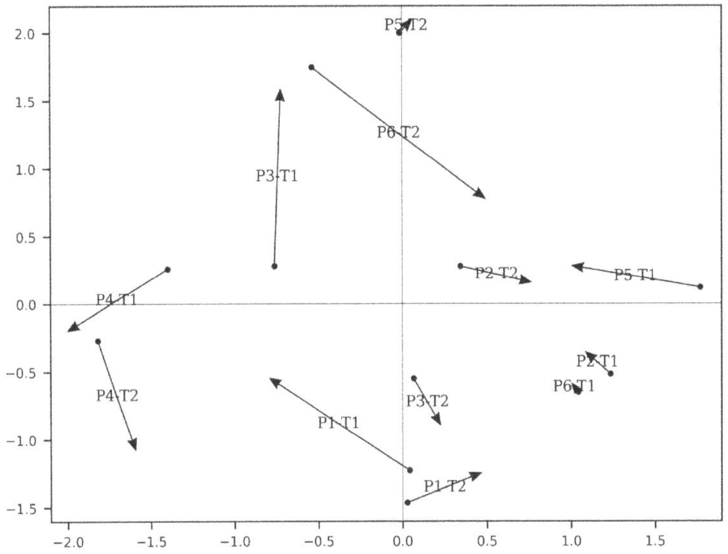

Figure 10. Temporal trajectories showing changes between sampling periods. Arrows connect T1 (low-water season) and T2 (high-water season) positions for each sampling point. Short arrows indicate strong agreement between the biotic and abiotic matrices; long arrows indicate disagreement.

4. Discussion

The ecological dynamics of wetlands are characterized by hierarchical interactions among the hydrological regime, physicochemical limnology, and the structure of the aquatic macrophyte community, with linkages that vary spatially and temporally [9,13,18]. This relationship is not unidirectional; aquatic macrophytes and their effects on physicochemical heterogeneity can feed back into the system, altering or buffering the hydrological regime [17,19,20]. The alternation of these processes defines a wetland as an ecological system [12], and it is within this dynamic that anthropogenic impacts become evident [22,23,69].

The Santa Rosa wetland is severely affected by various drivers of change that compromise its integrity, as are most coastal wetlands on the Lima, which are predominantly urban [35]. However, the fundamental ecological pattern of a wetland, the spatiotemporal variability of physicochemical limnology, and the macrophyte community, modulated by the different phases of hydrological seasonality, remain recognizable, with some reservations, as demonstrated in this study, which is the first of its kind in Peruvian coastal wetlands.

The results show that the Santa Rosa wetland is subject to a combination of direct and indirect drivers of change that act synergistically, generating cumulative pressures on its ecological integrity. The simultaneous presence of factors such as waste accumulation, livestock farming, and the introduction of exotic species has been widely documented as a major cause of tropical wetland degradation [60,70]. These direct drivers act on habitat structure and water quality, whereas indirect factors, such as poor governance and unplanned urban expansion, perpetuate these impacts by weakening institutional capacities to mitigate them [35,71]. Evidence suggests that when these drivers coexist spatially, ecological effects intensify, hindering natural recovery processes [72]. The binary presence/absence coding of drivers of change, adopted following the standardized protocols for tropical and Peruvian coastal wetlands [35,47,60], assigns equal statistical weight to drivers of potentially different magnitude. While the PCoA ordination demonstrates that

https://doi.org/10.3390/d18040214

this approach effectively captures the spatial distribution and co-occurrence of drivers across the wetlandscape, it does not quantify their differential ecological impact. Future studies should incorporate semi-quantitative intensity scales [47] to enable more nuanced assessments of cumulative effects.

In the Santa Rosa wetland, physicochemical limnology showed marked seasonal variations, with higher concentrations of total suspended solids (TSS), turbidity, temperature, conductivity, and total dissolved solids (TDS) during the low-water season (T1), attributable to the influence of the irrigation canal in T1 and the ionic dilution effect associated with greater water depth in T2. This pattern is characteristic of shallow wetlands with seasonal connectivity [13,73]. However, the temporal component of the hydrological regime exerts a more decisive control over the system's limnological dynamics [9,21], which, in the present study, is reflected in increased depth and seasonal ionic dilution. While the atmospheric regime of the central Peruvian coast—persistent cloud cover and reduced evapotranspiration during austral winter versus high solar radiation and increased evaporation during austral summer [34]—modulates the seasonal water balance, it operates as a background climatic condition rather than as a discrete driver of wetland seasonality. In the Santa Rosa wetland, the primary control over limnological variability is the hydrological cycle of aquifer recharge and exfiltration driven by upstream Andean precipitation [42]. Future studies incorporating direct atmospheric measurements would help quantify the relative contribution of evapotranspiration to seasonal water level fluctuations. It is important to note that the recorded nutrient concentrations and physicochemical parameters, which may seem high compared to pristine wetlands, constitute the current reference condition for a system embedded in an agricultural-urban matrix. Rather than anomalies, these values reflect the current hydrological regime and changes in the physicochemical variables that govern the wetland's seasonal variability. The ionic content of this system is modulated by both local processes (natural and anthropogenic) and the specific hydrological characteristics of the Santa Rosa wetland basin. In general, the ionic concentrations (conductivity and TDS) of the Santa Rosa wetland are lower than those of other coastal wetlands, such as the Pantanos de Villa, Humedales de Ventanilla, and wetlands in La Libertad region [38,41,43,44]. The conductivity values recorded in the Santa Rosa wetland (992–3280 μS/cm in T1 and T2) were below the limiting thresholds for dominant floating species, such as *P. stratiotes* and *P. crassipes* [74,75]. Regarding dissolved oxygen (DO), the recording of critical hypoxia thresholds in T1 (<2 mg/L) and recovery to physiologically acceptable levels for macrophytes and biota in general in T2 shows the alternation of states and their amplitudes between seasons [74]. This pattern is particularly evident at certain sampling points, characteristic of structurally complex wetlands with lacustrine and palustrine sectors [18,20]. Persistent hypoxia at specific sampling points results from the synergistic accumulation of oxygen-consuming processes linked to the hydrological transition: residual organic load from agricultural effluents, the turnover from amphibious to pleustophyte growth forms with consequent decomposition of amphibious biomass [76], and the increased surface coverage of interlocking floating macrophytes that restricts atmospheric gas exchange and limits light penetration [77,78]. These coupled oxygen-depleting mechanisms are characteristic of palustrine wetlands subject to fluctuating hydrological regimes [18,20].

Between T1 (low-water season) and T2 (high-water season), seasonal replacement of *P. crassipes* by *H. ranunculoides*, of *P. stratiotes* by *B. monnieri*, and of *B. monnieri* by *P. stratiotes* and *H. ranunculoides* was observed, reflecting a relative increase in marsh species in response to water retreat. However, this pattern varies among sectors of the wetland, where geomorphological heterogeneity and water connectivity modulate the magnitude of the coverage change between sampling point. This pattern coincides with observations in

https://doi.org/10.3390/d18040214

other neotropical and urban aquatic systems, where macrophyte communities maintain a stable functional composition in the face of broad environmental gradients, and the homogenization of physical and chemical conditions limits floristic turnover and favors the dominance of a few species [79]. These results are consistent with studies showing that the seasonal dynamics of aquatic macrophytes in shallow wetlands respond strongly to hydrological regimes, especially the flood and retreat cycle, which favors amphibious and emergent species during periods of decreasing water depth [21,80]. The alternation in dominance between pleustophytes and amphibious/emergent species reflects adaptive mechanisms in response to water variability and habitat availability, which have been documented in neotropical and intertropical systems with marked seasonality [24,73].

The relationship between physicochemical limnology and the distribution of aquatic macrophytes in the Santa Rosa wetland was particularly evident in specific areas. The ionic concentration gradient (pH, conductivity, TDS) was associated with the dominance of *P. stratiotes* and *B. monnieri*, whereas *H. ranunculoides* was limited by increasing depth. The association of *H. ranunculoides* with shallow, turbid, ammonium-rich environments reflects its status as a floating marsh plant with floating or creeping stems in saturated soils [81]. In contrast, *P. crassipes* showed greater affinity for waters with lower pH, TDS, and conductivity. Likewise, a moderate covariation structure was evident between the biotic and abiotic matrices, reflecting specific patterns of change in particular sectors and macrophyte species across seasons and sampling points. Although these patterns were not consistent across the entire matrix, the relationship was not statistically significant. These findings are consistent with previous studies that have documented the sensitivity of aquatic macrophytes to environmental gradients, especially in shallow systems, where fluctuations in pH, nutrients, and conductivity strongly influence species distribution [69]. The specific association of *P. crassipes* with less mineralized waters and *P. stratiotes* with more mineralized environments reflects contrasting ecophysiological strategies in response to environmental conditions [20,22]. In the lagoon sector of the Santa Rosa wetland, Muñoz et al. [82] described the dynamics of *P. stratiotes*, the dominant floating species in that environment, whose growth is influenced by local concentrations of ammonium, phosphate, and phosphorus, highlighting the relevance of physicochemical limnology in the structuring of aquatic macrophytes in wetlands. Likewise, the influence of phosphorus on species assembly coincides with reports indicating that increased nutrient concentrations can induce abrupt changes in the dominance of pleustophyte or amphibious/emergent growth forms [17,25,83,84]. In the Santa Rosa wetland, agricultural activity has been suggested to promote an increase in the coverage of *P. crassipes*, *P. stratiotes*, and *Lemna gibba* [51,85], with high values of chemical oxygen demand (COD), biological oxygen demand (BOD), total nitrogen (TN), and total phosphorus (TP) being recorded [45,86]. The concentrations of nitrate, ammonium, and phosphate in the Santa Rosa Wetland exhibited a spatial and temporal nutrient gradient, associated with variations in macrophyte composition and cover. Although there are no previous records of the degree of trophic status and its progressive increase, the levels recorded in 2022 suggest a system tending toward eutrophication and hypertrophy, with spatial and temporal variability. This condition is not an anomaly, but rather a hydrochemical pattern characteristic of coastal wetlands immersed in agricultural-urban matrices [39,87,88]. The structure of the plant community responds to nutrient availability by assimilating excess nutrients and maintaining ecosystem activity despite anthropogenic pressure [26,89]. Although the results enabled the identification of associations between nutrient gradients and macrophyte distribution, it was not possible to establish causal relationships with respect to the system's eutrophication.

The absence of a statistically significant global RV coefficient ($p > 0.05$) does not indicate a lack of ecological coupling between the biotic and abiotic matrices, but rather reflects

https://doi.org/10.3390/d18040214

the inherent spatial heterogeneity of the Santa Rosa wetland. The site-specific drivers of change and the structure of the co-inertia trajectories indicate that Santa Rosa operates as a wetlandscape [90,91] composed of lacustrine and palustrine functional units [92], each sustaining distinct macrophyte growth forms that alternate in space and time [17] at differentiated rates of change. In this context, biotic–abiotic coupling is not distributed homogeneously across the wetland, but rather concentrated in specific patches with particular functional identity and distinct environmental filters, including water volume and origin, organic and nutrient loading, agricultural activity and drainage dynamics, and mineralization gradients. Additional modulators include preferential herbivory on *P. crassipes* and *H. ranunculoides* [53,93] and sediment inputs from agricultural effluents, all of which contribute to within-site variability that a single global statistic cannot resolve. It should also be noted that the RV coefficient is inherently sensitive to sample size and the number of variables considered [94]. The observed pattern of spatially structured, locally strong associations thus constitutes a key finding in itself, underscoring the need for patch-based monitoring and management approaches in urban wetlands.

It is important to note that 2022 corresponds to the second year of an unusual three-year La Niña event (2020–2023). During La Niña phases, precipitation in the Chancay basin may be reduced, potentially limiting aquifer recharge and altering the wetland's hydrological regime [95]. The spatiotemporal turnover documented in this study therefore reflects a single hydrological cycle under these specific ENSO conditions, and the described patterns should be interpreted in this interannual context. Furthermore, climate change projections for the region suggest potential alterations in precipitation patterns and aquifer recharge dynamics that could directly affect the hydrological driver underlying the observed turnover.

Effective wetland conservation requires an understanding of their ecological functioning and the factors that shape their structure and dynamics, including not only the biotic communities they support but also the subsystems that compose them. Nevertheless, reductionist approaches persist in limnological studies, which tend to focus only on major open-water bodies such as lagoons or shallow lakes, overlooking functionally key environments, such as marshlands, floodplains, seasonal plains, and transitional microhabitats [9,10,69]. Previous limnological studies of Peruvian coastal wetlands have focused on the main lagoons and a few shallow environments, overlooking marsh environments [42,43]. Indeed, coastal wetlands such as Santa Rosa constitute "wetlandscapes," that is, mosaic landscapes composed of multiple functional units with specific physicochemical conditions and differential ecological responses [60,91]. The Santa Rosa wetland exhibits marked internal limnological heterogeneity, with marsh and lake sectors showing contrasting physicochemical dynamics. These sectors can be interpreted as differentiated functional patches, the coexistence of which contributes to the structural and functional stability of the wetland, without implying a formal classification of landscape units.

The results show that different sectors of the Santa Rosa Wetland respond differentially to water quantity and quality, which in turn modulates the structure of the macrophyte community. However, this pattern is only moderately explained by the intrinsic heterogeneity of the wetland and by the effects of drivers of change. This finding contributes to understanding the limnology–biota relationship, an aspect scarcely addressed in the literature on Peruvian wetlands, which has mostly favored descriptive or ecological community approaches without examining functional relationships with limnological dynamics [96,97]. This gap limits the ability to diagnose threats, such as drying, hydrological fragmentation, and climate change, which requires an understanding the coupling between abiotic and biotic matrices [27].

https://doi.org/10.3390/d18040214

This study contributes to the existing body of knowledge by demonstrating that the limnological structure of wetlands cannot be reduced to the main lagoon. Rather, it must be understood as an integrated system, where areas currently undergoing degradation continue to perform key ecological functions. Despite the multiple drivers of change, waste accumulation, livestock farming, introduction of exotic species, and urbanization, the Santa Rosa wetland continues to sustain essential processes such as nutrient retention, habitat provision, and hydrological regulation, demonstrating the persistence of functional integration among temporal dynamics, physicochemical limnology, and aquatic macrophyte communities [60,70]. The findings suggest that the maintenance of critical functions does not depend exclusively on the best preserved areas but also on the interaction between impacted and less impacted sectors, which has direct implications for the design of conservation, restoration, and climate change adaptation strategies [3,12].

Given the increasing degradation of coastal wetlands along the Peruvian Pacific, understanding the relationship among the hydrological cycle, limnology, and aquatic macrophytes, as addressed in this study, is key to redefining conservation priorities, identifying functional thresholds, and guiding interventions toward maintaining fundamental ecological processes, even in highly transformed landscapes. It is also suggested that the most variable sites, or those with a less clear explanatory relationship between biotic and abiotic processes, may be subject to unique local drivers of change. This opens up a particularly important line of research aimed at establishing causal links between the drivers of change and their effect on community-environment coupling. It is recommended that future studies address this question with extended temporal coverage.

5. Conclusions

The Santa Rosa wetland maintains recognizable ecological patterns despite sustained anthropogenic pressures. The seasonal hydrological cycle–contrasting high-water (T2) and low-water (T1) phases driven by aquifer recharge and exfiltration dynamics–remains the primary structuring force of physicochemical limnology. The aquatic macrophyte community responds coherently to seasonal changes in water depth, ionic concentration, and nutrient availability, with a functional turnover between pleustophytes in lacustrine sectors during high water and amphibious/emergent species in palustrine sectors during low water, evidencing functional coupling between biota and environmental conditions. Drivers of change are predominantly spatial in their distribution, generating a mosaic of differentiated impact profiles that modulate this coupling at local scales. By integrating limnological dynamics, macrophyte community structure, and drivers of change—an approach scarcely applied in Peruvian coastal wetlands—these findings provide a scientific foundation for identifying functional thresholds, guiding conservation strategies in increasingly threatened urban wetland systems along the Peruvian Pacific coast.

Supplementary Materials: The following supporting information can be downloaded at: https://www.mdpi.com/article/10.3390/d18040214/s1.

Author Contributions: Conceptualization, J.A.A.-I. and F.V.R.-C.; methodology, J.A.A.-I. and F.V.R.-C.; software, J.A.A.-I.; validation, J.A.A.-I. and S.I.U.-R.; formal analysis, J.A.A.-I.; investigation, F.V.R.-C., S.I.U.-R. and J.A.A.-I.; resources, J.A.A.-I.; data curation, J.A.A.-I. and F.V.R.-C.; writing—original draft preparation, F.V.R.-C. and J.A.A.-I.; writing—review and editing, F.V.R.-C., J.A.A.-I. and S.I.U.-R.; visualization, J.A.A.-I. and F.V.R.-C.; supervision, J.A.A.-I.; project administration, J.A.A.-I.; funding acquisition, J.A.A.-I. All authors have read and agreed to the published version of the manuscript.

Funding: This study was conducted as part of the project: Environmental degradation vs. ecosystem response: A case study in the Santa Rosa wetland from a limnological perspective (004-2020-PRO99), winner of the 2020 Cabieses Teaching Seed Fund Grant from the Universidad Científica del Sur, and is part of Flavia Valeria Rivera Cáceda's Licentiate Thesis in Marine Biology under the guidance of Dr. José Antonio Arenas Ibarra.

Institutional Review Board Statement: Not applicable.

Data Availability Statement: The original contributions presented in this study are included in the article and Supplementary Materials. Further inquiries can be directed to the corresponding authors.

Acknowledgments: Materials and equipment acquired by the project and others loaned for use by Terra Aqua Peru S.A.C. were used for field sampling and laboratory analysis. Chemical determinations were performed at Wet Experimental Laboratory No. 104 of the Universidad Científica del Sur. José Antonio Arenas Ibarra used protected hours for research at the Universidad Científica del Sur, the Institute of Natural Resources and Ecology, and the Ecology Laboratory at the Universidad Ricardo Palma. The facilities of the Ecology Laboratory of the Faculty of Biological Sciences and their infrastructure and research resources were used for the final draft. Access to academic databases at the Universidad Científica del Sur and Universidad Ricardo Palma were used. The authors gratefully acknowledge the National Forest and Wildlife Service (SERFOR) for facilitating the necessary research permits, and Williams Jurado Zevallos of the Santa Rosa Chancay Wetland Surveillance Committee, for providing the community consent to conduct research in the wetland. Special thanks are due to James López Visag, Martín Muñoz, Kaory Tinoco, Bruno Reátegui, and Edwin Pinto for their assistance during fieldwork and sample processing in the laboratory, as well as to Dámaso Ramírez for his guidance on field sampling techniques and his assistance in the taxonomic identification of aquatic macrophytes. We also would like to thank the examination committee of the licentiate thesis of Flavia Valeria Rivera Cáceda, for their suggestions on the thesis from which this article originates. Thanks are likewise extended to the anonymous reviewers of this article, whose observations and comments helped to improve this manuscript.

Conflicts of Interest: The authors declare no conflicts of interest.

Abbreviations

The following abbreviations are used in this manuscript:

Temp	Temperature
TDS	Total dissolved solids
TSS	Total suspended solids
DO	Dissolved oxygen
O2	Dissolved oxygen
Cond	Conductivity
Turb	Turbidity
DOM	Dissolved organic matter
NO_3	Nitrate
PO_4	Phosphate
NH_4	Ammonium
DOC	Drivers of change
PCA	Principal Component Analysis
PCoA	Principal Coordinate Analysis
MRPP	Multiple Response Permutation Procedure
UUG	Uncontrolled urban growth
ADG	Agricultural degradation
LDG	Livestock degradation
IES	Introduction of exotic species
CHC	Chemical contamination
ORC	Organic contamination
DSW	Domestic solid waste
CSW	Construction solid waste

https://doi.org/10.3390/d18040214

LFD	Landfill and drainage
POF	Presence of farms
MBC	Microbiological contamination
GPI	Governance failures associated with poor management and inadequate policies
EFL	Effluents
DIS	Other disturbances (undermining or extraction)
Hy_ran	*Hydrocotyle ranunculoides*
Pont_crasp	*Pontederia crassipes*
Pist_strat	*Pistia stratiotes*
Bacc_mon	*Bacopa monnieri*
Ludw_per	*Ludwigia peruviana*
Lemna	*Lemna gibba*
Schoeno	*Schoenoplectus americanus*
Nasturtium	*Nasturtium officinale*
Paspalum	*Paspalum vaginatum*
Azoll	*Azolla filiculoides*
Hyran	*Hydrocotyle ranunculoides*
Pontcrasp	*Pontederia crassipes*
Piststrat	*Pistia stratiotes*
Azollfil	*Azolla filiculoides*
Baccmon	*Bacopa monnieri*
Ludwper	*Ludwigia peruviana*
Schamer	*Schoenoplectus americanus*
NasOff	*Nasturtium officinale*
Paspvag	*Paspalum vaginatum*

References

1. Grizzetti, B.; Lanzanova, D.; Liquete, C.; Reynaud, A.; Cardoso, A.C. Assessing Water Ecosystem Services for Water Resource Management. *Environ. Sci. Policy* **2016**, *61*, 194–203. [CrossRef]
2. Moomaw, W.R.; Chmura, G.L.; Davies, G.T.; Finlayson, C.M.; Middleton, B.A.; Natali, S.M.; Perry, J.E.; Roulet, N.; Sutton-Grier, A.E. Wetlands in a Changing Climate: Science, Policy and Management. *Wetlands* **2018**, *38*, 183–205. [CrossRef]
3. Mitsch, W.J.; Gosselink, J.G. Wetlands: Human Use and Science. In *Wetlands*, 5th ed.; Wiley: Hoboken, NJ, USA, 2015; pp. 3–26.
4. Davidson, N.C.; Finlayson, C.M. Updating Global Coastal Wetland Areas Presented in Davidson and Finlayson (2018). *Mar. Freshw. Res.* **2019**, *70*, 1195. [CrossRef]
5. Davidson, N.C. How Much Wetland Has the World Lost? Long-Term and Recent Trends in Global Wetland Area. *Mar. Freshw. Res.* **2014**, *65*, 934–941. [CrossRef]
6. Fluet-Chouinard, E.; Stocker, B.D.; Zhang, Z.; Malhotra, A.; Melton, J.R.; Poulter, B.; Kaplan, J.O.; Goldewijk, K.K.; Siebert, S.; Minayeva, T.; et al. Extensive Global Wetland Loss over the Past Three Centuries. *Nature* **2023**, *614*, 281–286. [CrossRef]
7. IPBES. *Global Assessment Report on Biodiversity and Ecosystem Services of the Intergovernmental Science-Policy Platform on Biodiversity and Ecosystem Services*; Zenodo: Bonn, Germany, 2019.
8. Xu, T.; Weng, B.; Yan, D.; Wang, K.; Li, X.; Bi, W.; Li, M.; Cheng, X.; Liu, Y. Wetlands of International Importance: Status, Threats, and Future Protection. *Int. J. Environ. Res. Public Health* **2019**, *16*, 1818. [CrossRef]
9. An, S.; Yin, S.; Verhoeven, J.T.A.; Jeelani, N. Chapter 29—Wetlands. In *Wetzel's Limnology*, 4th ed.; Jones, I.D., Smol, J.P., Eds.; Academic Press: San Diego, CA, USA, 2024; pp. 999–1013.
10. Mitsch, W.J.; Gosselink, J.G. Wetland Biogeochemistry. In *Wetlands*, 5th ed.; Wiley: Hoboken, NJ, USA, 2015; pp. 179–214.
11. Rasmussen, T.C.; Deemy, J.B.; Long, S.L. Wetland Hydrology. In *The Wetland Book: I: Structure and Function, Management, and Methods*; Finlayson, C.M., Everard, M., Irvine, K., McInnes, R.J., Middleton, B.A., van Dam, A.A., Davidson, N.C., Eds.; Springer: Dordrecht, The Netherlands, 2018; pp. 201–216.
12. Tiner, R.W. *Wetland Indicators: A Guide to Wetland Identification, Delineation, Classification, and Mapping*, 2nd ed.; Taylor & Francis: Boca Raton, FL, USA, 2017.
13. Mitsch, W.J.; Gosselink, J.G. Wetland Hydrology. In *Wetlands*, 5th ed.; Wiley: Hoboken, NJ, USA, 2015; pp. 111–160.
14. Reddy, K.R.; DeLaune, R.D. *Biogeochemistry of Wetlands: Science and Applications*; CRC Press: Boca Raton, FL, USA, 2008.

15. Davidson, N.C.; Middleton, B.A.; McInnes, R.J.; Everard, M.; Irvine, K.; Van Dam, A.A.; Finlayson, C.M. Introduction to the Wetland Book 1: Wetland Structure and Function, Management, and Methods. In *The Wetland Book: I: Structure and Function, Management, and Methods*; Springer: Dordrecht, The Netherlands, 2018; pp. 3–14.
16. Ramsar Convention Secretariat. Ramsar Convention on Wetlands. In *Global Wetland Outlook: State of the World's Wetlands and Their Services to People*; Ramsar Convention Secretariat: Gland, Switzerland, 2018.
17. Chambers, P.A.; Maberly, S.C. Chapter 24—Freshwater Plants. In *Wetzel's Limnology*, 4th ed.; Jones, I.D., Smol, J.P., Eds.; Academic Press: San Diego, CA, USA, 2024; pp. 759–816.
18. Bornette, G.; Puijalon, S. Response of Aquatic Plants to Abiotic Factors: A Review. *Aquat. Sci.* **2011**, *73*, 1–14. [CrossRef]
19. Mitsch, W.J.; Gosselink, J.G. Wetland Vegetation and Succession. In *Wetlands*, 5th ed.; Wiley: Hoboken, NJ, USA, 2015; pp. 215–255.
20. Siddiqui, A.J.; Jahan, S.; Adnan, M.; Ashraf, S.A.; Singh, R. Macrophytes and Their Role in Wetland Ecosystems. In *Aquatic Macrophytes: Ecology, Functions and Services*; Kumar, S., Bauddh, K., Singh, R., Kumar, N., Kumar, R., Eds.; Springer Nature: Singapore, 2023; pp. 119–138.
21. Maltchik, L.; Rolon, A.S.; Schott, P. Effects of Hydrological Variation on the Aquatic Plant Community in a Floodplain Palustrine Wetland of Southern Brazil. *Limnology* **2007**, *8*, 23–28. [CrossRef]
22. Freedman, B.; Lacoul, P. Environmental Influences on Aquatic Plants in Freshwater Ecosystems. *Environ. Rev.* **2006**, *14*, 89–136. [CrossRef]
23. Dar, N.A.; Pandit, A.K.; Ganai, B.A. Factors Affecting the Distribution Patterns of Aquatic Macrophytes. *Limnol. Rev.* **2014**, *14*, 75–81. [CrossRef]
24. Rameshkumar, S.; Radhakrishnan, K.; Aanand, S.; Rajaram, R. Influence of Physicochemical Water Quality on Aquatic Macrophyte Diversity in Seasonal Wetlands. *Appl. Water Sci.* **2019**, *9*, 12. [CrossRef]
25. Dubey, D.; Kumar, S.; Dutta, V. Impact of Nutrient Enrichment on Habitat Heterogeneity and Species Richness of Aquatic Macrophytes: Evidence from Freshwater Tropical Lakes of Central Ganga Plain, India. *Int. J. Environ. Sci. Technol.* **2022**, *19*, 5529–5546. [CrossRef]
26. Florescu, L.I.; Catana, R.D.; Mihai, R.A.; Dumitrache, A.C.; Moldoveanu, M.M. Macrophyte Community Distribution in Relation to Anthropization Influences and Phytoplankton Development in an Urban Lake Chain. *Water* **2024**, *16*, 3467. [CrossRef]
27. Madzivanzira, T.C.; Coetzee, J.A.; Dalu, T. Factors Structuring Aquatic Macrophytes. In *Aquatic Macrophytes: Ecology, Functions and Services*; Kumar, S., Bauddh, K., Singh, R., Kumar, N., Kumar, R., Eds.; Springer Nature: Singapore, 2023; pp. 21–52.
28. Cerna-Arrue, A.; Aponte, H. El origen de los humedales peruanos y su relación con los procesos geológico-ambientales de Sudamérica: Una revisión. *Caldasia* **2024**, *46*, 314–325. [CrossRef]
29. Díaz, A.; Sáenz-Zúñiga, L.; Zutta, B. Formación de un humedal en la costa norte del Peru: Estabilidad biofísica y diversidad biológica. *Rev. Peru Biol.* **2021**, *28*, e21132. [CrossRef]
30. Dirección de Geología Ambiental y Riesgo Geológico from INGEMMET. *Estudio Hidrogeológico del Refugio de Vida Silvestre Los Pantanos de Villa, Chorrillos—Lima*; Instituto Geológico, Minero y Metalúrgico—INGEMMET: Lima, Peru, 2019.
31. Verones, F.; Bartl, K.; Pfister, S.; Jiménez Vílchez, R.; Hellweg, S. Modeling the Local Biodiversity Impacts of Agricultural Water Use: Case Study of a Wetland in the Coastal Arid Area of Peru. *Environ. Sci. Technol.* **2012**, *46*, 4966–4974. [CrossRef]
32. Herrera, L.E.; Xu, H. Formación y evolución de la Laguna La Niña 2017, utilizando imágenes satelitales Modis-Aqua. *Bol. Inst. Mar Peru* **2019**, *34*, 105–114.
33. Pulido-Capurro, V.M. Estado actual de la conservación de los hábitats de los Pantanos de Villa, Lima, Peru. *Arnaldoa* **2018**, *25*, 679–702. [CrossRef]
34. Young, K.R. El Ecosistema. In *Los Pantanos de Villa: Biología y Conservación*; Universidad Nacional Mayor de San Marcos, Museo de Historia Natural: Lima, Peru, 1998; pp. 3–12.
35. Aponte, H.; Gonzales, S.; Gomez, A. Impulsores de cambio en los humedales de América Latina: El caso de los humedales costeros de Lima. *South Sustain.* **2020**, *1*, e023. [CrossRef]
36. Aponte, H.; Cano, A. Estudio florístico comparativo de seis humedales de la costa de lima (Peru): Actualización y nuevos retos para su conservación. *Rev. Latinoam. Conserv.* **2013**, *3*, 15–27.
37. Gonzales, S.; Aponte, H.; Cano, A. Update of the Vascular Flora of the Santa Rosa Wetland—Chancay (Lima, Peru). *Arnaldoa* **2019**, *26*, 867–882.
38. Aguirre-Soto, L.; Nieto-Juárez, J.I. Evaluación de La Calidad Del Agua En El Humedal Costero Pantanos de Villa, Chorrillos, Peru. *Tecnia* **2024**, *34*, 11–25. [CrossRef]
39. Benites-Alfaro, E.; Mendivil, V.T.; Farfán, E.E.; Rodriguez, C. Trophic Status Index of the Waters of the Pantanos de Villa Protected Natural Reserve, Chorrillos, Peru. *J. Crit. Rev.* **2020**, *7*, 447–453.
40. Huaman Vilca, S.; Lucen, M.; Paredes Vite, M.; Changanaqui Alfaro, D. Evaluación de la calidad del agua de la Laguna Marvilla en los Pantanos de Villa (Lima—Peru). *South Sustain.* **2020**, *1*, e019. [CrossRef]

https://doi.org/10.3390/d18040214

41. Canales Cuadros, H.; Cabrera Carranza, C.; Arana Maestre, J. Macroinvertebrados acuáticos y calidad de agua en el Área de Conservación Regional Humedales de Ventanilla, Callao. *Rev. Inst. Investig. Fac. Minas Metal Cienc. Geogr.* **2022**, *25*, 295–301. [CrossRef]
42. Castillo-Velásquez, R.-M.; Huamantinco-Araujo, A.-A. Variación espacial de la comunidad de macroinvertebrados acuáticos en la zona litoral del humedal costero Santa Rosa, Lima, Peru. *Rev. Biol. Trop.* **2020**, *68*, 50–68. [CrossRef]
43. Córdova-Tello, C.A.; Huamantinco-Araujo, A.A. Efecto de las variables fisicoquímicas en la comunidad de coleópteros acuáticos en un humedal costero, Lima, Peru. *Acta Biol. Colomb.* **2023**, *28*, 204–219. [CrossRef]
44. Bopp, G.; Peláez, F. Evaluación de La Flora Vascular de Los Humedales Costeros de La Libertad, Peru. *Manglar* **2019**, *16*, 151–156. [CrossRef]
45. Jindo, K.; Morikawa, M. Innovative Feasibility Study for the Reclamation of the Cascajo Wetlands in Peru Utilizing Sustainable Technologies. *Water* **2020**, *12*, 1097. [CrossRef]
46. Aldana-Domínguez, J.; Rojas Quezada, C.; Munizaga, J.; Moschella, P.; Stamm, C.; Martínez, C. Urban Wetland Losses and Land-Use Conservation Challenges in Three Latin American Cities. In *Urban Wetlands in Latin America: Protection, Conservation, Innovation, Restoration, and Community for Sustainable and Water Sensitive Cities*; Rojas Quezada, C., Ed.; Springer Nature: Cham, Switzerland, 2024; pp. 3–18.
47. Aponte, H.; Coello-Sarmiento, M.-P.; Montes-Iturrizaga, D. New Insights About the Drivers of Change in the Coastal Wetlands of Peru: Results of a Rapid Field Survey. *Water* **2025**, *17*, 1473. [CrossRef]
48. Romero-Mariscal, G.; Garcia-Chevesich, P.A.; Morales-Paredes, L.; Arenazas-Rodriguez, A.; Ticona-Quea, J.; Vanzin, G.; Sharp, J.O. Peruvian Wetlands: National Survey, Diagnosis, and Further Steps toward Their Protection. *Sustainability* **2023**, *15*, 8255. [CrossRef]
49. Urbina, D.; Rivera-Cáceda, F.; Aponte, H. ¿Se Están Reduciendo Los Humedales de La Costa Del Pacífico Suramericano? El Caso de Los Humedales de Lima. *Rev. Acad. Colomb. Cienc. Ex. Fis. Nat.* **2022**, *46*, 985–998. [CrossRef]
50. Ramirez, D.W.; Aponte, H.; Cano, A. Flora vascular y vegetación del humedal de Santa Rosa (Chancay, Lima). *Rev. Peru. Biol.* **2011**, *17*, 105–110. [CrossRef]
51. Zamora-Aranda, A.; Aponte, H. Evaluation of Two Ecosystem Services Provided by a Pistia Stratiotes Population on the Pacific Coast of South America. *Biology* **2024**, *13*, 573. [CrossRef]
52. Castillo Polo, L.A. *Humedales Costeros de La Región Lima*, 1st ed.; Gobierno Regional de Lima: Lima, Peru, 2010.
53. Ochoa-Balbacea, D.O.; Gonzales, S.; Apeño, A.; Aponte, H. Hábitats prioritarios en un humedal costero del Pacífico: Propuesta de un índice basado en servicios ecosistémicos. *Rev. Acad. Colomb. Cienc. Exactas Físicas Nat.* **2022**, *46*, 182–191.
54. ANA Protocolo Nacional Para El Monitoreo de La Calidad de Los Recursos Hídricos Superficiales. 2016. Available online: https://sinia.minam.gob.pe/documentos/protocolo-nacional-monitoreo-calidad-recursos-hidricos-superficiales (accessed on 3 February 2022).
55. Baird, R.; Rice, E.W.; Eaton, A.D.; Bridgewater, L.; Federation, W.E. *Standard Methods for the Examination of Water and Wastewater*; American Public Health Association: Washington, DC, USA, 2017.
56. Hauer, F.R.; Lamberti, G.A. (Eds.) *Methods in Stream Ecology*, 2nd ed.; Academic Press/Elsevier: San Diego, CA, USA, 2006.
57. Titus, J.E. Submersed Macrophyte Vegetation and Distribution Within Lakes: Line Transect Sampling. *Lake Reserv. Manag.* **1993**, *7*, 155–164. [CrossRef]
58. Dengler, J.; Dembicz, I. Should We Estimate Plant Cover in Percent or on Ordinal Scales? *Veg. Classif. Surv.* **2023**, *4*, 131–138. [CrossRef]
59. Cirujano, S.; Meco Molina, A.; García Murillo, P. Flora Acuática Española. Hidrófitos Vasculares. Available online: https://bibdigital.rjb.csic.es/records/item/16128-flora-acuatica-espanola-hidrofitos-vasculares?offset=1 (accessed on 28 March 2025).
60. Sarkar, P.; Salami, M.; Githiora, Y.; Vieira, R.; Navarro, A.; Clavijo, D.; Padgurschi, M. A Conceptual Model to Understand the Drivers of Change in Tropical Wetlands: A Comparative Assessment in India and Brazil. *Biota Neotrop.* **2020**, *20*, e20190913. [CrossRef]
61. Oksanen, J.; Simpson, G.L.; Blanchet, F.G.; Kindt, R.; Legendre, P.; Minchin, P.R.; O'Hara, R.B.; Solymos, P.; Stevens, M.H.H.; Szoecs, E.; et al. Vegan: Community Ecology Package. Available online: https://cran.r-project.org/package=vegan (accessed on 12 December 2025).
62. Borcard, D.; Gillet, F.; Legendre, P. *Numerical Ecology with R*; Springer International Publishing: Cham, Switzerland, 2018.
63. Legendre, P.; Legendre, L. *Numerical Ecology*, 3rd English ed.; Developments in Environmental Modelling; Elsevier: Amsterdam, The Netherlands, 2012.
64. Mielke, P.W.; Berry, K.J. Description of MRPP. In *Permutation Methods: A Distance Function Approach*; Mielke, P.W., Berry, K.J., Eds.; Springer: New York, NY, USA, 2001; pp. 9–65.
65. Doledec, S.; Chessel, D. Co-Inertia Analysis: An Alternative Method for Studying Species-Environment Relationships. *Freshw. Biol.* **1994**, *31*, 277–294. [CrossRef]

https://doi.org/10.3390/d18040214

66. Dray, S.; Dufour, A.-B.; Thioulouse, J.; Chessel, D.; Jombart, T.; Pavoine, S.; Lobry, J.R.; Ollier, S.; Borcard, D.; Legendre, P.; et al. Ade4: Analysis of Ecological Data: Exploratory and Euclidean Methods in Environmental Sciences R package version 1.7-23. Available online: https://cran.r-project.org/package=ade4 (accessed on 12 December 2025).
67. R Core Team. *R: A Language and Environment for Statistical Computing*; R Core Team: Vienna, Austria, 2020.
68. RStudio Team. *RStudio: Integrated Development for R*; RStudio: Vienna, Austria, 2020.
69. Agidie, A.; Wondie, A.; Beneberu, G. Wetland Disturbance Level and Macrophyte Composition in Natural Wetlands and Paddy Fields in the Eastern Part of Lake Tana, Ethiopia. *Heliyon* **2024**, *10*, e40701. [CrossRef]
70. Galatowitsch, S.M. Natural and Anthropogenic Drivers of Wetland Change. In *The Wetland Book*; Finlayson, C.M., Milton, G.R., Prentice, R.C., Davidson, N.C., Eds.; Springer: Dordrecht, The Netherlands, 2018; pp. 359–367.
71. Gomez Vilca, A.J.; Aponte, H.; Gonzales, S. ¿Cómo Proteger Los Humedales Costeros Peruanos? Una Respuesta a Partir de Un Modelo Conceptual de Sus Impulsores de Cambio. *Bol. Investig. Mar. Costeras* **2023**, *52*, 125–142. [CrossRef]
72. Pacioglu, O.; Tușa, I.M.; Sidoroff, M.E.; Ițcuș, C. The Best Management Practices in Agriculture for Protection of Inland Water Ecosystems. In *Encyclopedia of Inland Waters*, 2nd ed.; Mehner, T., Tockner, K., Eds.; Elsevier: Oxford, UK, 2022; pp. 58–67.
73. Meerhoff, M.; Beklioğlu, M. Chapter 26—Shallow Lakes and Ponds. In *Wetzel's Limnology*, 4th ed.; Jones, I.D., Smol, J.P., Eds.; Academic Press: San Diego, CA, USA, 2024; pp. 859–892.
74. Hadad, H.R.; Mufarrege, M.M.; Di Luca, G.A.; Maine, M.A. Salinity and pH Effects on Floating and Emergent Macrophytes in a Constructed Wetland. *Water Sci. Technol.* **2018**, *2017*, 270–275. [CrossRef]
75. Haller, W.T.; Sutton, D.L.; Barlowe, W.C. Effects of Salinity on Growth of Several Aquatic Macrophytes. *Ecology* **1974**, *55*, 891–894. [CrossRef]
76. Bianchini, I., Jr.; Cunha-Santino, M.; Panhota, R. Oxygen Uptake from Aquatic Macrophyte Decomposition from Piraju Reservoir (Piraju, SP, Brazil). *Braz. J. Biol.* **2011**, *71*, 27–35. [CrossRef]
77. Poi de Neiff, A.; Neiff, J.J.; Casco, S.L. Leaf litter decomposition in three wetland types of the Paraná River floodplain. *Wetlands* **2006**, *26*, 558–566. [CrossRef]
78. Bunch, A.J.; Allen, M.S.; Gwinn, D.C. Spatial and Temporal Hypoxia Dynamics in Dense Emergent Macrophytes in a Florida Lake. *Wetlands* **2010**, *30*, 429–435. [CrossRef]
79. Pereto, S.C.A.S.; Padial, A.A. Macrophyte Functional Composition Is Stable across a Strong Environmental Gradient of a Neotropical Floodplain. *Acta Bot. Bras.* **2021**, *35*, 62–69. [CrossRef]
80. Rolon, A.; Homem, H.; Maltchik, L. Aquatic Macrophytes in Natural and Managed Wetlands of Rio Grande Do Sul State, Southern Brazil. *Acta Limnol. Bras.* **2010**, *22*, 133–146. [CrossRef]
81. Amaral, M.; Bittrich, V.; Faria, A.; Anderson, L.; Aona, L. *Guia de Campo Para Plantas Aquáticas e Palustres Do Estado de Sao Paulo; Série Manuais Práticos em Biologia*; Holos, Editora: Sao Paulo, Brazil, 2008.
82. Muñoz, M.; Tinoco, K.; Arenas, J.; Aponte, H. Dinámica de La Planta Acuática Pistia Stratiotes y Su Incorporación a Un Sistema de Producción de Compost. *Technol. Water Sci.* **2026**, *17*, 77–123. [CrossRef]
83. Rejmankova, E. The Role of Macrophytes in Wetland Ecosystems. *J. Ecol. Field Biol.* **2011**, *34*, 333–345. [CrossRef]
84. Saluja, R.; Garg, J.K. Macrophyte Species Composition and Structure along Littoral Region in Relation to Limnological Variables of a Tropical Wetland Ecosystem. *Chem. Ecol.* **2017**, *33*, 499–515. [CrossRef]
85. Aponte, H.; Jiménez, R.; Alcántara, B. Challenges for Management and Conservation of Santa Rosa Wetland (Lima-Peru). *Cientifica* **2012**, *9*, 257–264.
86. Loayza-Basan, K.; Castillejos Paniagua, K.; Mestas Valero, R.; Quiliche Duran, J.P.J. Estudio de la Teledetección y Caracterización Fisicoquímica del Humedal "El Cascajo", Santa Rosa, Chancay-Lima, Peru. *Infinitum* **2017**, *7*, 52–58. [CrossRef]
87. Leiva, J.V.Y.; Ibarra, J.A. Evaluación de la calidad del agua en el canal Surco de la Zona de Reglamentación Especial del Humedal Los Pantanos de Villa (Lima, Peru). *South Sustain.* **2024**, *5*, e090. [CrossRef]
88. Mendoza-Caballero, W.; Quiroz-Santos, R.L. Evaluación de los estados tróficos de la laguna principal del área de conservación regional Albufera de Medio Mundo, Huaura—Lima. *Biotempo* **2022**, *19*, 149–164. [CrossRef]
89. Jyothi, P.V.; Sureshkumar, S. Patterns of Vegetation Dynamics across Mild Disturbance Gradient in a Freshwater Wetland System in Southern India. *Wetlands* **2018**, *38*, 807–817. [CrossRef]
90. Mullen, C.; Bertassello, L.E.; Rao, P.S.C.; Müller, M.F. From Wetlands to Wetlandscapes: Remote Sensing Calibration of Process-Based Hydrological Models in Heterogeneous Landscapes. *Hydrol. Process.* **2022**, *36*, e14739. [CrossRef]
91. Mitsch, W.J.; Gosselink, J.G. Wetland Definitions. In *Wetlands*, 5th ed.; Wiley: Hoboken, NJ, USA, 2015; pp. 27–44.
92. Cowardin, L.M.; Carter, V.; Golet, F.C.; LaRoe, E.T. *Classification of Wetlands and Deepwater Habitats of the U.S.*; U.S. Department of the Interior, Fish and Wildlife Service: Washington, DC, USA, 1992.
93. Domínguez-Machín, M.E.; Silva-López, G. Consumo de plantas por el ganado vacuno en una zona de Manglar-Espartal de Alvarado, Veracruz, México. *For. Veracruzana* **2020**, *22*, 7–11.
94. Adams, D.C. Evaluating Modularity in Morphometric Data: Challenges with the RV Coefficient and a New Test Measure. *Methods Ecol. Evol.* **2016**, *7*, 565–572. [CrossRef]

https://doi.org/10.3390/d18040214

95. Lavado-Casimiro, W.S.; Felipe, O.; Silvestre, E.; Bourrel, L. ENSO Impact on Hydrology in Peru. *Adv. Geosci.* **2013**, *33*, 33–39. [CrossRef]
96. Gómez-Sánchez, R.; Cuba, D.; Aponte, H. Sobre la necesidad de descentralización y diversificación de la investigación en humedales costeros peruanos. *Biologist* **2022**, *20*, 121–150. [CrossRef]
97. Rivera, G.; Gonzales, S.; Aponte, H. Wetlands of the South American Pacific Coast: A Bibliometric Analysis. *Wetl. Ecol. Manag.* **2021**, *30*, 869–877. [CrossRef]

https://doi.org/10.3390/d18040214

diversity

MDPI

Article

Impact of Distinct Management Regimes on Wintering Waterbird Communities in China's Coal Mining Subsidence Wetlands

Sen Yang [1,2], Kai Cao [1], Yuanyuan Wang [3], Wenning Shen [4], Tong Lin [2], Ningning Liu [2], Jing Li [1], Lingbo Ji [1], Huiping Chen [1], Yanying Xu [1], Bo Tang [1,*] and Ying Li [1,*]

[1] Zhejiang Key Laboratory of Ecological Environmental Damage Control and Value Transformation, Ecological and Environmental Science and Research Institute of Zhejiang Province, Hangzhou 310007, China; 21110700108@m.fudan.edu.cn (S.Y.); 13210700119@fudan.edu.cn (K.C.); jingli0713@163.com (J.L.); jilingbo19@mails.ucas.ac.cn (L.J.); chenhp1101@163.com (H.C.); 13858046681@163.com (Y.X.)

[2] Ministry of Education Key Laboratory for Biodiversity Science and Ecological Engineering, School of Life Sciences, Fudan University, Shanghai 200438, China; 23210700094@m.fudan.edu.cn (T.L.); nnliu_fudan@163.com (N.L.)

[3] Zhejiang Province Environmental Engineering Co., Ltd., Hangzhou 310012, China; whitewolf.yuan@163.com

[4] Huzhou Municipal Ecology and Environment Bureau Nanxun Branch, Huzhou 313009, China; m13665727124@163.com

* Correspondence: tangbo1984278@163.com (B.T.); liying@zjshky.com.cn (Y.L.); Tel./Fax: +86-0571-87998939 (B.T.)

Abstract

Natural wetland loss constitutes a primary threat to waterbirds worldwide, increasingly forcing them to rely on expanding artificial wetlands. Extensive underground coal mining across the North China Plain has created numerous subsidence wetlands, which could serve as important alternative habitats for migratory and wintering waterbirds. However, the effects of different management regimes on waterbirds in these novel artificial wetlands remain poorly understood, hindering effective strategies for reconciling human development with waterbird conservation. Here, we conducted a long-term field survey (2017–2025) of wintering waterbirds across 15 subsidence wetlands under four distinct human management regimes in the Huaibei coal mining area. We recorded 22,712 waterbirds of 45 species. We found that high-intensity aquaculture and floating photovoltaic systems were associated with reduced waterbird diversity, increased community dissimilarity, altered species composition, and the loss of multiple threatened species from survey records. We also found that ecological aquaculture and unutilized wetlands may serve as favorable habitats as alternatives to natural wetlands. Our findings demonstrate that subsidence wetlands can provide vital wintering habitats when managed sustainably, but intensive development severely compromises their conservation value. Future research should integrate habitat variables and year-round surveys to optimize management strategies for these expanding artificial ecosystems.

Keywords: habitat loss; waterbird conservation; subsidence wetland; human disturbance

1. Introduction

The loss of natural habitats is widely recognized as the foremost threat to global biodiversity [1,2]. Natural wetlands provide critical ecosystem services and support approximately 40% of the world's plant as well as animal species [3,4]. Over the past hundred years, more than half of the global natural wetlands have been lost, and the remaining fragmented wetlands continued to suffer from extensive degradation due to increasing anthropogenic pressure [5–7]. The persistent disappearance and deterioration of natural wetlands adversely affect wetland-dependent wildlife, particularly waterbirds, which represent one of the most sensitive and imperiled groups [8,9]. This has precipitated documented worldwide contractions in numerous waterbird populations, and the remaining species also face multiple threats [2,10].

As the carrying capacity of remaining natural wetlands declines, waterbirds increasingly exploit rapidly expanding networks of artificial wetlands (e.g., aquaculture ponds, constructed lakes, reservoirs, and irrigated lands) as complementary or alternative habitats [1,11]. Evaluating habitat quality, understanding adaptive responses, and identifying associated threats in these artificial wetlands have emerged as key priorities for waterbird conservation [2,12]. Numerous studies demonstrate that properly managed artificial wetlands can effectively substitute for natural habitats in supporting waterbird population [13,14]. This is attributable to the high productivity exhibited by most artificial wetlands, which enables waterbird species to select suitable sites according to their ecological requirements and behaviors [2,15]. Conversely, other research warns that artificial wetlands may be associated with reduced for waterbirds fitness, adversely affecting migration, breeding, foraging, and individual fitness, particularly among less-common species, and this may be attributed to elevated levels of human disturbance [13,16,17]. Despite these ongoing debates, it is undeniable that expanding artificial wetlands have successfully attracted numerous waterbird species [2,12]. Prioritizing and understanding the impacts of these artificial wetlands on waterbirds is crucial for future waterbird conservation, especially given their high sensitivity to rapid habitat alterations [18,19]. While many studies have examined the compensatory role of artificial wetlands for waterbirds, such as aquaculture ponds, constructed lakes, reservoirs, and irrigated lands [2,15], relatively less attention has been directed toward subsidence wetlands formed by extensive and ongoing underground coal mining.

Subsidence wetlands form as a result of land subsidence triggered by subsurface coal mining activities. Combined with abundant rainfall and high groundwater levels, these depressed landforms rapidly evolve into wetland environments over relatively short timeframes [20,21]. This process facilitates a shift from land to wetland ecosystems. By 2017, cumulative land subsidence resulting from underground coal mining had affected an extensive area of approximately 20,000 square kilometers in China, creating contiguous clusters of lakes [21,22]. While this phenomenon has caused various challenging economic and ecological issues, including the loss of arable land, biodiversity decline, and landscape degradation [20,23,24], it has also created opportunities for species that rely on wetland conditions, such as fish and waterbirds [25,26].

Over the past decade, the conservation value of subsidence wetlands in supporting biodiversity has attracted increasing attention from researchers [20,26]. For instance, Li et al. (2018) [20] documented that these wetlands in the coal mining area function as essential habitat for a wide variety of waterbirds, especially migratory waterbirds from the East Asian–Australasian Flyway, including threatened taxa such as the Dunlin (*Calidris alpina*) and Baer's Pochard (*Aythya baeri*). Although subsidence wetlands have been shown to possess important ecological functions [20], they are generally considered unprotected artificial habitats and are predominantly utilized for economic purposes, including the installation of floating photovoltaic (FPV) systems for clean energy generation, aquaculture, and recreational development [24,27]. As the duration and area of subsidence wetlands continue to expand, human utilization intensifies accordingly [23,28]. Given the worldwide decline in natural wetlands, this newly formed subsidence wetlands may provide important alternative habitats where large numbers of waterbirds for foraging, resting, and nesting. However, due to limited understanding of the long-term ecological impacts of

https://doi.org/10.3390/d18030146

human disturbance in these wetlands, mitigation measures to address adverse effects on waterbirds have been largely absent in most regions, resulting in widespread overdevelopment and declining conservation value for waterbirds [20,26]. Long-term monitoring of waterbird community dynamics and human disturbance effects in subsidence wetlands may provide valuable insights for promoting sustainable economic development and optimizing conservation strategies.

The Huaibei coal mining area, situated in eastern China, is renowned for its extensive croplands and thriving coal industry. Over the past 20 years, extensive underground coal mining across the North China Plain has generated numerous subsidence wetlands [21,22]. Against the backdrop that the quality and extent of natural wetlands have declined markedly over the last 30 years [5,6], these novel subsidence wetlands have become important stopover and wintering sites for large numbers of waterbirds, particularly for species using the East Asia–Australasian Flyway [20,23,26]. In recent years, aquaculture and FPV systems have been extensively deployed within these artificial wetlands of the Huaibei coal mining area due to rapid economic development and policy requirements from China's National Energy Administration [26,27]. This may create a long-standing conflict between human development and biodiversity conservation. Nevertheless, the responses of waterbird communities to such prolonged human disturbances in these artificial wetlands remain poorly understood.

Here, we carried out a long-term study to quantify waterbird communities and assess the effects of varying human management regimes in the subsidence wetlands of the Huaibei coal mining area. During the winters (January–February) from 2017 to 2025, we conducted surveys of waterbirds across 15 subsidence wetlands under four distinct management types: unutilized wetland, ecological aquaculture, high-intensity aquaculture, and FPV systems. These four wetland types are subject to different primary land uses, and thus their impacts on waterbird communities may differ. We hypothesized that the different human management regimes would influence both the diversity and community composition of waterbirds in the region, with subsidence wetlands experiencing greater human disturbance supporting lower waterbird diversity. The nine-year duration of this study provides a case study to understand the temporal dynamics of the subsidence wetlands and the impact of management regimes on the waterbird community. Therefore, our study contributes significantly to understanding the ecological consequences of long-term and varied human disturbances in subsidence wetlands, and it provides practical guidance for managing these expanding and extensive artificial wetlands.

2. Materials and Methods

2.1. Survey Area and Sampling Sites

The study was conducted in the Huaibei coal mining area, spanning Suixi and Mengcheng counties in Anhui Province on the North China Plain (33°24′–33°39′ N, 116°31′–116°45′ E; Table A1, Figure 1). This region lies within the East Asian–Australasian Flyway, a major migratory corridor for birds. It features a warm-temperate and semi-humid climate, strongly influenced by the East Asian monsoon. The landscape is dominated by plains and hills interlaced with extensive farmland, making it one of China's principal grain-production zones [20]. The study area, Huaibei, is located in a major coal basin of China [21], where mining-induced subsidence has created extensive wetland habitats for waterbird communities [20,23].

https://doi.org/10.3390/d18030146

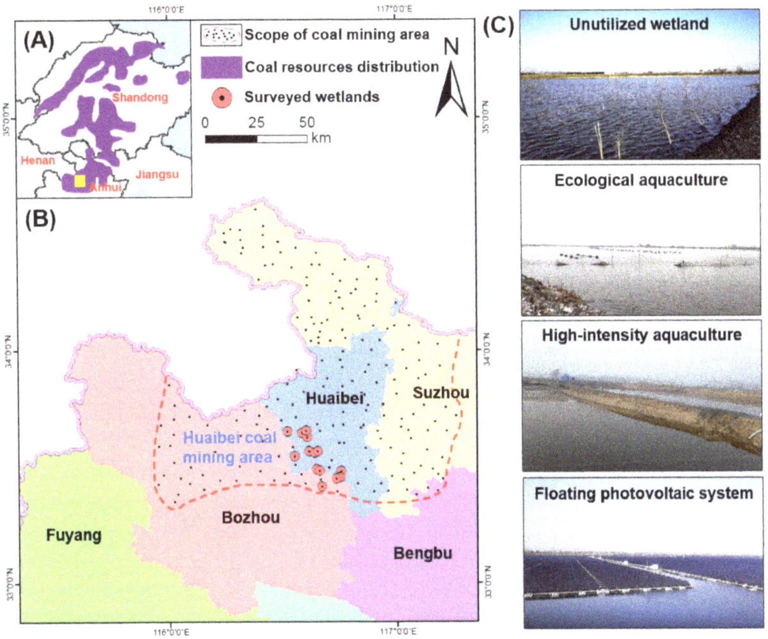

Figure 1. Study area and the distribution of subsidence wetlands surveyed for wintering waterbird communities within Huaibei and Bozhou cities. (**A**) Distribution of coal resources in the North China Plain. (**B**) Locations of the 15 subsidence wetlands in the Huaibei coal mining area. (**C**) Photographs of subsidence wetlands under four management regimes. The yellow square in (**A**) indicates the study area.

In recent years, with the expansion of subsidence wetlands in the Huaibei coal mining area, the intensity of human utilization has grown substantially. Between 2017 and 2025, the primary development forms for subsidence wetlands in this region included ecological aquaculture, intensive aquaculture, and the installation of FPV systems [24,26]. In this study, we classified the subsidence wetlands into four human management regimes based on their primary land use: (1) ecological aquaculture wetland (EA) is defined as utilizing pre-existing water bodies within subsidence wetlands for fish and shrimp farming, integrating aquaculture with constructed wetland features, and without extensive modification of the wetland structure; (2) high-intensity aquaculture wetland (HA) is an aquaculture model conducted within subsidence wetlands, characterized by continuous artificial feeding and frequent use of mechanical devices (e.g., aerators) for water oxygenation, aimed at maintaining high biomass-density production, involving large-scale transformation of subsidence wetlands into aquaculture ponds, deepening banks and dredging sediments, followed by intensive management practices [7]; (3) FPV system wetland (FPVs) refers to a form of water surface utilization where photovoltaic power generation components are installed above subsidence wetland via floating and anchoring systems, primarily for clean energy generation [27]; (4) unutilized wetland (UW) is defined as the subsidence wetlands that have not been subjected to any human use or development. In 2017, all surveyed subsidence wetlands were unutilized except for two under ecological aquaculture. By 2025, the distribution was as follows: five wetlands had FPV systems installed, three were used for high-intensity aquaculture, six for ecological aquaculture, and only one remained unutilized (Table A1). Over the course of the research, the surface size of all 15 subsidence wetlands remained virtually unchanged.

https://doi.org/10.3390/d18030146

2.2. Waterbird Surveys

Annual surveys were carried out across the 15 subsidence wetlands to document the wintering waterbird communities in Suixi and Mengcheng counties in the Huaibei coal mining area from 2017 to 2025, with fieldwork carried out during January and February each year. This approach capitalized on the documented seasonal stability of waterbird communities in these subsidence wetlands during winter (December to February of the following year) [20], providing a reliable basis for comparison across years. Waterbird observation points were established along the boundary of each subsidence wetland, with one to four points per wetland depending on its size, to ensure comprehensive coverage of the waterbird communities [15,20]. Specifically, in wetlands with installed FPV systems or dense vegetation where waterbirds are highly elusive and difficult to observe, we walked around the periphery to ensure all individuals were recorded. Waterbird counts were conducted during morning (8:00–10:30) and afternoon (14:00–17:00) on days with clear weather and calm conditions, by an experienced observer using binoculars (8× magnification, Swarovski) and spotting scope (20–60× magnification, Swarovski) (Swarovski Optik KG, Absam, Tyrol, Austria) [15]. An identical transect route was maintained for every survey within each wetland, and the route characteristics were consistent across all subsidence wetlands. Each waterbird survey at a subsidence wetland lasted between 20 and 40 min. Waterbird species that forage aerially, including gulls and terns, were recorded because their dependence on wetland food resources, whereas individuals that only overflew without using the wetlands were excluded. Additionally, waterbirds observed in the terrestrial habitats (farmlands and trees) surrounding the subsidence wetlands were also recorded, because many waterbirds (such as crakes and egrets) actively forage in subsidence wetlands but frequently use the adjacent terrestrial habitats for resting and wandering. Specifically, we recorded individuals in farmlands within 20 m and trees within 50 m of the subsidence wetland edge. Waterbirds were categorized into two ecological groups based on their primary foraging mode: swimming birds and wading birds [29]. Both ecological groups exhibit strong wetland dependence, wading birds use shallow water zones, whereas swimming birds occupy deeper water zones. The density of each waterbird species was calculated based on the area of the respective subsidence wetland. The primary aim of this study was to assess the effects of structural factors on wintering waterbird communities, not to estimate absolute abundance within each subsidence wetland. Accordingly, analyses utilized raw count data without correcting for detectability.

2.3. Statistical Analysis

For each subsidence wetland on a yearly basis, we assessed four waterbird metrics: species richness, individual density, Shannon-Wiener diversity, and Pielou evenness index [30,31]. For each species, we also calculated the dominance index (Y) across all subsidence wetlands, with species classified as dominant when $Y \geq 0.02$ [32]. Additionally, beta diversity was used to compare community dissimilarities among the 15 subsidence wetlands. Beta diversity was assessed using abundance-based Bray–Curtis dissimilarity metrics with the *betapart* package in R (Version 4.4.2) [33]. The Bray–Curtis metric is robust to occasional missing sites or surveys (i.e., variation in the number or frequency of sampled wetlands), with lower values indicating smaller differences in waterbird composition. Prior to analysis, Shapiro–Wilk tests confirmed that all response variables (species richness, individual density, Shannon-Wiener diversity, and the Pielou evenness index) followed normal distributions. Generalized Linear Mixed Models (GLMMs) used to analyze the effects of four wetland management regimes (ecological aquaculture, high-intensity aquaculture, FPV systems and unutilized wetland) on each of these waterbird community metrics. To address the non-independence of repeated annual surveys at the same site, wetland identity

https://doi.org/10.3390/d18030146

was included as a random intercept, effectively accounting for site-specific baseline differences. To separate management effects from broader temporal trends, year was included as a fixed effect, with management regime as an additional fixed effect.

Non-metric multidimensional scaling (NMDS) and permutational multivariate analysis of variance (PERMANOVA) were employed via the *vegan* package in R (version 4.4.2) to compare waterbirds composition among the four subsidence wetland types [15,33,34]. NMDS was applied to examine differences in species and group composition of waterbird communities across the four wetland types. Based on waterbird individual density, a species community similarity matrix was constructed for the four subsidence wetland types based on the Bray–Curtis distance [33]. Stress values below 0.20 indicate a reliable ordination solution. Standard ellipses representing 95% confidence intervals were used to illustrate compositional variation within each wetland type [15]. Significant differences among the four subsidence wetland types were tested using PERMANOVA, employing a Bray–Curtis distance matrix based on waterbird individual density data [34]. To ensure that observed differences in community composition were not confounded by variations in within-group heterogeneity, we conducted a PERMDISP test (free number of permutations: 999) to evaluate the homogeneity of multivariate dispersions before performing the PERMANOVA. The results of the PERMDISP analysis showed no significant differences in dispersion among management types ($F = 1.40$, $p = 0.256$), thereby confirming that the assumption of homogeneity was met. All permutation tests were performed with the strata argument applied to restrict permutations within individual wetlands, thereby accounting for the lack of independence due to repeated measures taken at the same sites. Statistical analyses were conducted in R (RStudio version 2024.12.1, R 4.4.2; R Core Team, Vienna, Austria, 2024). Results are presented as means ± standard error (SE).

3. Results

3.1. Waterbird Community

The areas of the 15 sampled subsidence wetlands ranged from 9.4 to 231 hectares (mean ± SE: 81.6 ± 16.7 hectares) (Table A1). In this study, a total of 22,712 waterbird individuals belonging to 45 species, 12 families, and nine orders were recorded across the 15 subsidence wetlands during the 2017 to 2025 wintering season (Table A2). This included 8470 waterbirds of 36 species (41 counts) in unutilized wetlands; 12,544 individuals of 40 species (58 counts) in ecological aquaculture wetlands; 1460 individuals of 21 species (25 counts) in high-intensity aquaculture wetlands; and 238 individuals of 11 species (11 counts) in FPV system wetlands. Over 86% of the recorded waterbirds were long-distance migratory species along the East Asian–Australasian Flyway. Swimming birds constituted the vast majority, accounting for 55.6% of all waterbirds (25 of 45 species) and 95.0% of all individuals (21,576 of 22,712 individuals), and all 13 numerically dominant species were swimming birds. Among all waterbirds recorded, the most dominant species were the Common Coot (*Fulica atra*, $Y = 0.40$, 10,094 individuals), Eastern Spot-billed Duck (*Anas zonorhyncha*, $Y = 0.04$, 1410 individuals) and Little Grebe (*Tachybaptus ruficollis*, $Y = 0.04$, 988 individuals), and the combined number of these three species accounts for 55.0% of all waterbirds. The dominant species differed across the four wetland types, in ecological aquaculture and unutilized wetlands, the most common species were the Common Coot, Mallard (*Anas platyrhynchos*), Green-winged Teal (*Anas crecca*), and Eastern Spot-billed Duck, whereas in high-intensity aquaculture and FPV system wetlands, the most dominant species were the Common Coot, Little Grebe, and Common Moorhen (*Gallinula chloropus*), these three waterbirds were highly human-tolerant.

Following the International Union for Conservation of Nature (IUCN) Red List classifications, a total of 922 individuals belonging to six species among the recorded waterbirds

https://doi.org/10.3390/d18030146

were classified as globally threatened: Baer's Pochard (*Aythya baeri*, 80 individuals) is listed as Critically Endangered, Swan Goose (*Anser cygnoid*, 72 individuals) as Endangered; Common Pochard (*Aythya ferina*, 518 individuals) as Vulnerable; Dunlin (*Calidris alpina*, 37 individuals), Northern Lapwing (*Vanellus vanellus*, 199 individuals), and Ferruginous Duck (*Aythya nyroca*, 16 individuals) as Near Threatened. In our study, the period following the installation of FPV systems in subsidence wetlands coincided with no records of threatened waterbird species. In the high-intensity aquaculture wetlands, only the Common Pochard (38 individuals) and Northern Lapwing (2 individuals) were recorded (Table A2).

3.2. Diversity and Composition of Bird Communities

The results of the GLMMs indicated that wetland management regime had a significant effect on all measured waterbird community metrics ($p < 0.05$ for all; Figure A1), the survey year only showed a significant effect on individual density ($F = 13.33$, $p < 0.01$; Figure A1). Species richness, individual density and Shannon-Wiener diversity index were significantly higher in ecological aquaculture and unutilized wetlands than in FPV systems and high-intensity aquaculture ($p < 0.05$ for all; Figure 2A–C), while Pielou evenness index did not differ significantly among the four wetland types ($F_{3,131} = 1.41$, $p = 0.24$, Figure 2D). And there were no significant differences in these four indices between ecological aquaculture and unutilized wetlands, or between high-intensity aquaculture and FPV system wetlands (Figure 2).

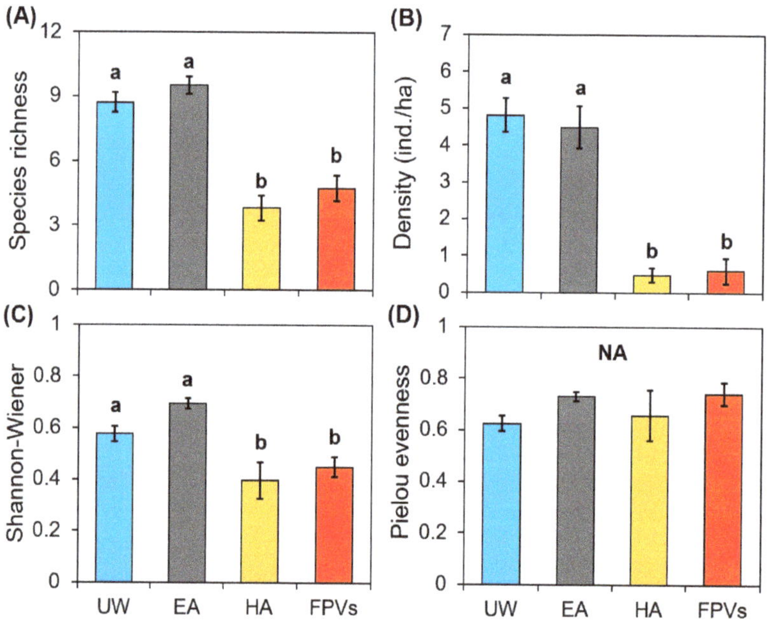

Figure 2. Comparison of waterbird community indices across the four types of subsidence wetlands. (**A**) Species richness. (**B**) Individual density. (**C**) Shannon-Wiener diversity index. (**D**) Pielou evenness index. Different letters (a, b) show significant differences among the four types of subsidence wetlands ($p < 0.05$), bars sharing the same letter are not significantly different, NA: not significant. UW: unutilized wetland, EA: ecological aquaculture, HA: high-intensity aquaculture, FPVs: floating photovoltaic system. Values represent means ± SE.

NMDS analysis revealed that waterbird community structures differed significantly among the ecological aquaculture, high-intensity aquaculture, FPV systems and unutilized

https://doi.org/10.3390/d18030146

wetlands (stress = 0.122, Figure 3A). Samples from high-intensity aquaculture and FPV systems showed greater separation in the NMDS plot. In contrast, samples from ecological aquaculture and unutilized wetlands exhibited moderate overlap but remained statistically distinct (Table 1). This was also supported by PERMANOVA, which confirmed that community composition differed significantly across wetland types ($R^2 = 0.207$, $p < 0.001$, Figure 3A), except between high-intensity aquaculture and FPV systems, where no significant difference was detected ($R^2 = 0.03$, $p = 0.37$, Table 1). Furthermore, beta diversity was significantly higher in high-intensity aquaculture and FPV systems than in ecological aquaculture and unutilized wetlands ($F_{3,2808} = 12.93$, $p < 0.001$, Figure 3B), indicating greater within-type compositional variation in the former. In terms of species and family distributions, the majority of waterbirds (all recorded species and 93% of individuals) showed a preference for ecological aquaculture and unutilized wetlands. In contrast, many families were not recorded across surveys in high-intensity aquaculture (58% of families) and FPV systems (42% of families) (Figure 3C). Collectively, these results indicate that high-intensity aquaculture and FPV systems were associated with substantial alterations in waterbird community composition and a reduction in their diversity.

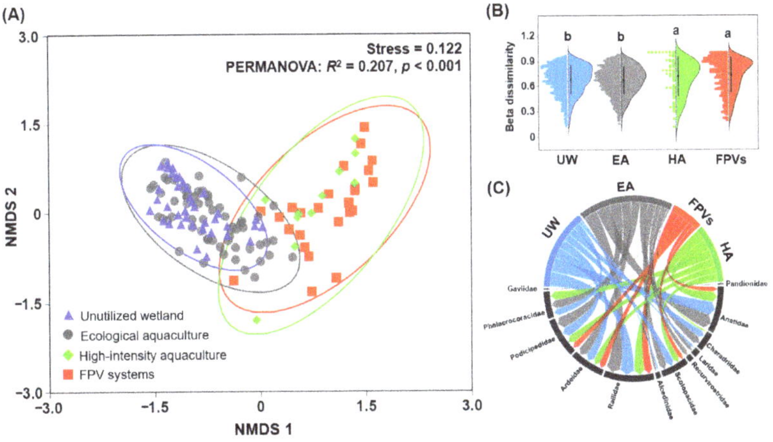

Figure 3. Effects of the four subsidence wetland types on waterbird communities. (**A**) Nonmetric multidimensional scaling (NMDS) plot depicting the waterbirds across the four subsidence wetland types. In the plot, the ellipses indicate the 95% confidence intervals, while the light points depict the raw data. (**B**) Violin plot showing the beta dissimilarity index of waterbird communities among the four subsidence wetland types. Different letters (a, b) denote significant differences among the four types of subsidence wetlands sampled ($p < 0.05$), bars sharing the same letter are not significantly different. (**C**) Subsidence wetland preference of waterbird families across the four subsidence wetland types (abundance-based, data were standardized using log transformation). UW: unutilized wetland, EA: ecological aquaculture, HA: high-intensity aquaculture, FPVs: floating photovoltaic system.

Table 1. Permutational multivariate analysis of variance (PERMANOVA) of waterbird communities among the ecological aquaculture, high-intensity aquaculture, FPV systems and unutilized wetland.

Area Type	R^2	p
Unutilized wetland—High-intensity aquaculture	0.13	0.001
Unutilized wetland—Ecological aquaculture	0.02	0.019
Unutilized wetland—FPV systems	0.10	0.001
High-intensity aquaculture—Ecological aquaculture	0.13	0.001
High-intensity aquaculture—FPV systems	0.03	0.365
Ecological aquaculture—FPV systems	0.11	0.001

https://doi.org/10.3390/d18030146

4. Discussion

Our nine-year study deepens the understanding of waterbird community use of subsidence wetlands and its implications for population trends. Annual surveys across 15 subsidence wetlands revealed substantial wintering waterbird communities (45 species in total and a peak density of 19 individuals per hectare), demonstrating that these novel habitats may provide important alternative stopover and wintering sites along the East Asian–Australasian Flyway. However, increasing human utilization leads to a decline in habitat quality and may exert sustained pressure on wildlife populations [2,12]. Our research indicates that high-intensity aquaculture and FPV systems were associated with significantly lower species richness, individual density, and Shannon–Wiener diversity relative to unutilized wetlands, whereas ecological aquaculture hosted a greater number of waterbird species than the other subsidence wetland types. These findings support our hypothesis and highlight the necessity of assessing the impacts of human activities on the biodiversity of subsidence wetlands. Community composition differed among the four management regimes, with high-intensity aquaculture and FPV systems exhibiting elevated beta dissimilarity, indicating that waterbird assemblages were more heterogeneous. In particular, almost all threatened waterbird species were absent from survey records in subsidence wetlands with high-intensity aquaculture and FPV systems. Our results demonstrate that intensive human development was associated with lower waterbird biodiversity in subsidence wetlands, while ecological aquaculture and unutilized wetlands may provide functional alternatives to natural wetlands in this context.

Artificial wetlands are globally widespread, and their capacity to support waterbirds remains controversial despite growing recognition of their value as alternative habitats in the face of natural wetland loss [2,7,17]. Our research has revealed that subsidence wetlands support a large number of overwintering waterbird communities, including 13% of species and 4.1% of individuals classified as threatened waterbirds, such as the Common Pochard and Baer's Pochard. Given the severe destruction and disappearance of natural wetlands both in China [5,17,35] and worldwide [1], these findings indicate that subsidence wetlands have become crucial wintering and stopover habitats for waterbird communities. These habitats may help support the conservation of threatened species and could play a role in buffering against the loss of natural habitats [2,20,23]. This pattern may reflect two key factors. Firstly, most waterbirds recorded in this study are habitat generalists, which exhibit a high degree of adaptability and tend to benefit more from the conditions provided by subsidence wetlands, such as the Common Coot and Grey Heron (*Ardea cinerea*) [20,26]. Secondly, some carnivorous waterbirds (fish- or shrimp-eating) are able to access reliable sources of animal food or additional foraging opportunities from artificial wetlands, especially those associated with aquaculture production wetlands, which availability of food resources helps sustain their populations, such as herons and Little Grebe [15,36].

Our observational data showed that declines in waterbird species richness, individual density, and Shannon-Wiener diversity were associated with the establishment of FPV systems (since 2018) and high-intensity aquaculture (since 2019) in subsidence wetlands, which is consistent with previous studies [23,24,26]. These changes may stem from habitat loss and landscape alteration caused by human management disturbances and the presence of photovoltaic modules, both of which may increase survival pressures on waterbirds [20,23,26]. High-intensity aquaculture and FPV systems are widespread in the artificial wetlands of the middle and lower reaches of the Yangtze River [24]. These modifications alter aquatic biodiversity and shrink the area of open water, and thus may have decreased the degradation of foraging environments and restricted movement ranges for waterbirds [14,24,37,38]. In addition, we also found that ecological aquaculture wetlands supported the highest waterbird species richness and Shannon–Wiener diversity, although not significantly greater than in unutilized wetlands. This pattern may be because ecological aquaculture wetlands might provide more animal food (such as economic benthos and fish) [2,14,24]. As a result, ecological aquaculture wetlands may provide additional opportunities for wintering waterbirds [14,19,36]. Furthermore, the Pielou evenness index was higher in high-intensity aquaculture and FPV systems than in other wetland types. This may result from the adverse habitat conditions in these systems suppressing populations of otherwise abundant waterbird species (e.g., from families Scolopacidae and Anatidae), thereby increasing abundance-based evenness [20,24]. These findings demonstrate that properly managed subsidence wetlands can provide essential supplementary wintering, foraging, and resting habitats for waterbird communities, particularly for migratory species along the East Asian–Australasian Flyway [10,14,23].

Human disturbance can influence the structure of waterbird communities in artificial wetlands [14,15]. Our findings indicate that the intensity of human disturbance was strongly associated with differences in waterbird community composition within these subsidence wetlands. Among these waterbirds, herbivorous guilds (e.g., Common Coot and Common Moorhen) exhibit pronounced adaptability to human-altered environments; they are more resilient and tend to benefit from the conditions provided by artificial wetlands. Conversely, carnivorous guilds (e.g., ducks and waders) generally show a preference for larger wetlands with minimal human disturbance. Human management practices homogenize habitat structure, which may impede the adaptation and survival of more specialized waterbird groups [2,20]. We also documented markedly lower occurrence of waterbirds in FPV system wetlands, with threatened species detected even less frequently in these habitats, a pattern that may reflect their limited adaptive capacity [26]. This may be because ecological characteristics and landscape structural changes exert a greater influence on waterbird distribution than direct human activity alone [24,39,40]. Such differential species responses ultimately drive community reassembly [2,20,24]. While large subsidence wetlands attract substantial waterbird communities and some species display a degree of disturbance tolerance, if poorly managed, these wetlands may ultimately function as ecological traps for many attracted waterbirds. Therefore, understanding how human management activities affect waterbirds is essential for mitigating future negative impacts and improving habitat quality.

The advantages of subsidence wetlands include providing compensatory habitat for waterbirds, supporting aquaculture development, thereby benefiting human society [27,35]. Our nine-year study of wintering waterbird communities provides a substantial long-term dataset; however, several limitations warrant consideration. First, there is some uncertainty in conducting only a single survey of wintering waterbird communities each year. Increasing survey frequency during breeding and migration seasons would help clarify year-round waterbird utilization patterns [40,41]. Second, we did not measure mechanistic covariates such as water depth, food availability, vegetation structure, water quality, disturbance intensity, as well as wetland age [2,20,42]. Since waterbird utilization of wetlands is determined by multiple ecological factors [2,12,24], incorporating surveys of these habitat variables would help clarify the specific ecological and environmental drivers in subsidence wetlands. Third, the region of the study is situated within Suixi and Mengcheng counties in the Huaibei coal mining area, and the potential effects of unmeasured temporal change in varying meteorological variables (e.g., precipitation, hydrology, solar radiation, and temperature), regional population dynamics and disturbance context, as well as broader landscape contexts (e.g., plains vs. mountains) on waterbird communities remains uncertain [42,43]. Finally, there are methodological limitations in that the detectability and the effective sampling domain differed systematically between open-water and FPV/vegetated

https://doi.org/10.3390/d18030146

contexts. These differences constitute a key alternative explanation for the recorded variation in species observations, which underscores the need for caution in interpreting the observed associations. Accurately evaluating the wide range of environmental and human-induced factors, as well as their respective impacts on waterbird communities is essential for predicting future ecological and environmental outcomes.

5. Conclusions

Extensive underground coal mining has generated extensive artificial wetlands throughout the Huaibei coal mining area, transforming former terrestrial ecosystems (e.g., farmland or villages) into aquatic habitats that effectively compensate for natural wetland loss. This study provides empirical evidence that these novel wetlands serve as important complementary habitats for wintering waterbird communities, especially for those species that are long-distance migrants within the East Asian–Australasian Flyway, including numerous threatened species. However, intensive human development and resource exploitation may be associated with a reduction in the suitability and accessibility of subsidence wetlands for wintering waterbirds, and species with limited adaptive capacity may face particular difficulties in coping with the altered and often poorly understood environmental conditions. Notably, ecological aquaculture might create more beneficial habitats for wintering waterbirds, highlighting that ecological management which preserves natural wetland characteristics may be essential for biodiversity benefits. Future research should integrate broader habitat variables, conduct year-round monitoring, and expand geographical coverage to better elucidate the dependence of waterbirds on these expanding artificial ecosystems.

Author Contributions: Conceptualization: S.Y. and B.T.; methodology, S.Y., T.L. and N.L.; formal analysis, S.Y., K.C., Y.W. and B.T.; investigation, S.Y.; resources, S.Y. and B.T.; data curation, S.Y. and B.T.; writing—original draft, S.Y., K.C., Y.W., W.S., J.L. and B.T.; writing—review and editing, S.Y., K.C., Y.W., W.S., J.L., L.J., H.C., Y.X., Y.L. and B.T.; project administration, B.T. and Y.L.; B.T. and Y.L. contributed equally to this work. Every author made intellectual contributions throughout the research design and manuscript preparation phases, leveraging their individual areas of expertise. All authors have read and agreed to the published version of the manuscript.

Funding: This research was funded by the Young Scientists Fund of Ecological and Environmental Science and Research Institute of Zhejiang Province (grant number 2025QN08).

Institutional Review Board Statement: Not applicable.

Data Availability Statement: The original contributions presented in this study are included in the article. Further inquiries can be directed to the corresponding authors.

Acknowledgments: We express our sincere gratitude to Leilei Ren and Xinyuan Yang for help with facilitating fieldwork and insightful suggestions during the study. We thank Dongfan Xu for producing Figure 1A.

Conflicts of Interest: Yuanyuan Wang is employed by Zhejiang Province Environmental Engineering Co., Ltd. The remaining authors declare that the research was conducted in the absence of any commercial or financial relationships that could be construed as a potential conflict of interest.

Abbreviations

The following abbreviations are used in this manuscript:

UW	unutilized wetland
EA	ecological aquaculture
HA	high-intensity aquaculture
FPVs	floating photovoltaic system

https://doi.org/10.3390/d18030146

Appendix A

Table A1. Locations and details of the 15 subsidence wetlands included in this study. FPV system: floating photovoltaic system.

County	Site Name	Latitude (N)	Longitude (E)	Area (ha)	Human Management Regime
Mengcheng	Zhujiazhuang	33°27′17.86″	116°44′7.35″	9.4	Unutilized wetland
	Qianrenlou	33°27′31.48″	116°44′48.07″	14	High-intensity aquaculture at 2024
	Qizhuang	33°24′53.21″	116°42′7.09″	91.1	Ecological aquaculture at 2018
	Houzhujia	33°27′37.60″	116°45′13.59″	187	FPV systems at November 2017
Suixi	Zhangmiao	33°29′9.95″	116°45′26.10″	152	FPV systems at May 2017
	Xiaoyangjia	33°37′21.00″	116°36′25.55″	231.1	FPV systems at August 2018
	Sunchangling	33°32′43.83″	116°38′58.06″	37.9	Ecological aquaculture at 2019
	Luozhuang	33°38′58.58″	116°31′10.89″	37.4	FPV systems at May 2024
	Ludongcun	33°32′43.92″	116°36′44.47″	51.2	Ecological aquaculture at 2019
	Lujia	33°38′31.65″	116°35′0.39″	29	Ecological aquaculture 2020
	Luomozhangjia	33°29′22.88″	116°39′43.22″	101	Ecological aquaculture at 2019
	Gaohu	33°39′8.09″	116°36′7.71″	32.2	High-intensity aquaculture at 2024
	Dingduzhou	33°33′7.36″	116°33′49.53″	130	FPV systems at August 2023
	Darenjia	33°32′26.06″	116°38′12.60″	37.1	Ecological aquaculture at 2018
	Zengjia	33°29′47.49″	116°38′55.26″	77	High-intensity aquaculture at 2019

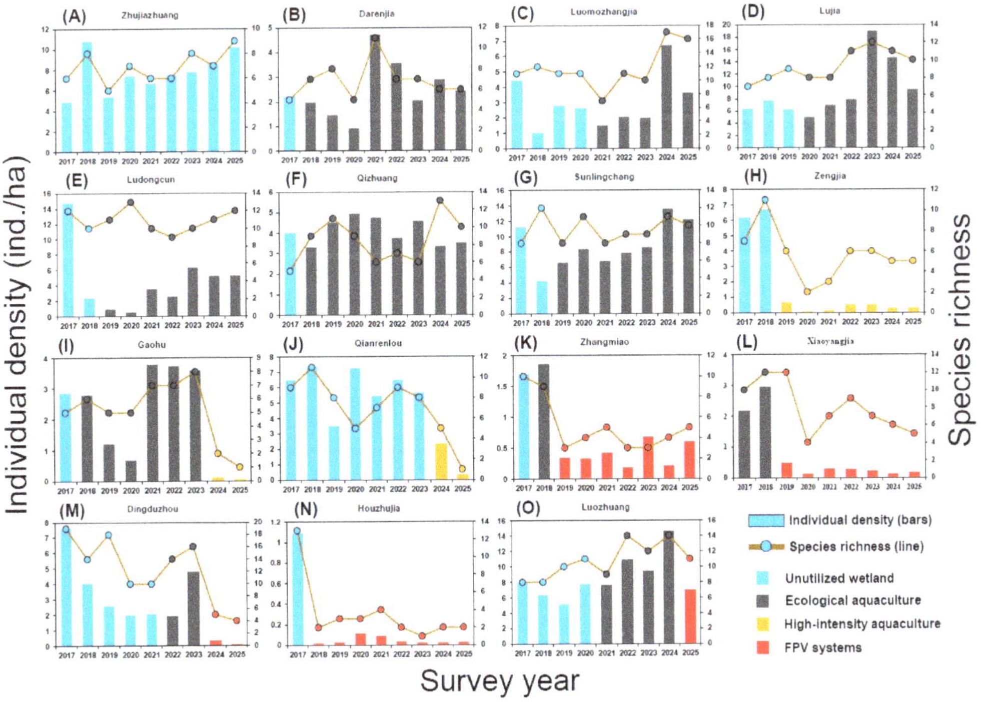

Figure A1. The species richness (line) and individual density (bars) of wintering waterbirds in 15 subsidence wetlands of the Huaibei mining area, and their annual fluctuations.

https://doi.org/10.3390/d18030146

Table A2. Proportion of individuals and McNaughton's dominance index (*Y*) of each waterbird species in this study. The bold values indicate the most dominant waterbirds. CR: Critically Endangered, EN: Endangered, VU: Vulnerable, NT: Near Threatened.

Order	Family	Science Name	English Name	Proportion of Individuals	Frequency	McNaughton's Dominance Index (*Y*)	IUCN
Gaviiformes	Gaviidae	*Gavia arctica*	Black-throated Diver	<1%	1	<1	LC
Accipitriformes	Pandionidae	*Pandion haliaetus*	Osprey	<1%	1	<1	LC
Anseriformes	Anatidae	*Anas crecca*	Green-winged Teal	8%	57	**0.03**	LC
		Anas platyrhynchos	Mallard	6%	72	**0.03**	LC
		Anas zonorhyncha	Eastern Spot-billed Duck	6%	80	**0.04**	LC
		Anser albifrons	Greater White-fronted Goose	<1%	1	<1	LC
		Anser cygnoid	Swan Goose	<1%	2	<1	EN
		Anser fabalis	Bean Goose	5%	10	<1	LC
		Aythya baeri	Baer's Pochard	<1%	3	<1	CR
		Aythya ferina	Common Pochard	2%	28	<1	VU
		Aythya fuligula	Tufted Duck	2%	38	<1	LC
		Aythya nyroca	Ferruginous Duck	<1%	4	<1	NT
		Mareca falcata	Falcated Duck	5%	34	0.01	LC
		Mareca penelope	Eurasian Wigeon	<1%	7	<1	LC
		Mareca strepera	Gadwall	1%	27	<1	LC
		Mergellus albellus	Smew	<1%	6	<1	LC
		Mergus merganser	Common Merganser	<1%	5	<1	LC
		Cygnus columbianus	Tundra Swan	<1%	2	<1	LC
		Sibirionetta formosa	Baikal Teal	<1%	3	<1	LC
		Spatula clypeata	Northern Shoveler	<1%	3	<1	LC
		Anas acuta	Northern Pintail	<1%	5	<1	LC
		Tadorna tadorna	Common Shelduck	<1%	2	<1	LC

https://doi.org/10.3390/d18030146

Table A2. *Cont.*

Order	Family	Science Name	English Name	Proportion of Individuals	Frequency	McNaughton's Dominance Index (*Y*)	IUCN
Charadriiformes	Charadriidae	*Vanellus vanellus*	Northern Lapwing	<1%	21	<1	NT
	Laridae	*Chroicocephalus ridibundus*	Black-headed Gull	<1%	3	<1	LC
	Recurvirostridae	*Himantopus himantopus*	Black-winged Stilt	<1%	2	<1	LC
		Recurvirostra avosetta	Pied Avocet	<1%	1	<1	LC
	Scolopacidae	*Calidris alpina*	Dunlin	<1%	2	<1	NT
		Gallinago gallinago	Common Snipe	<1%	14	<1	LC
		Tringa erythropus	Spotted Redshank	<1%	13	<1	LC
		Tringa nebularia	Common Greenshank	<1%	28	<1	LC
		Tringa ochropus	Green Sandpiper	<1%	4	<1	LC
Coraciiformes	Alcedinidae	*Alcedo atthis*	Common Kingfisher	<1%	2	<1	LC
		Ceryle rudis	Pied Kingfisher	<1%	2	<1	LC
Gruiformes	Rallidae	*Fulica atra*	Common Coot	44%	121	**0.40**	LC
		Gallinula chloropus	Common Moorhen	4%	104	**0.03**	LC
Pelecaniformes	Ardeidae	*Nycticorax nycticorax*	Black-crowned Night Heron	<1%	5	<1	LC
		Ardea alba	Great Egret	<1%	31	<1	LC
		Ardea cinerea	Grey Heron	1%	55	<1	LC
		Ardea intermedia	Intermediate Egret	<1%	1	<1	LC
		Botaurus stellaris	Eurasian Bittern	<1%	5	<1	LC
		Egretta garzetta	Little Egret	<1%	68	<1	LC
		Ixobrychus sinensis	Yellow Bittern	<1%	1	<1	LC
Podicipediformes	Podicipedidae	*Podiceps cristatus*	Great Crested Grebe	3%	60	0.01	LC
		Tachybaptus ruficollis	Little Grebe	4%	110	**0.04**	LC
Suliformes	Phalacrocoracidae	*Phalacrocorax carbo*	Great Cormorant	2%	22	<1	LC

https://doi.org/10.3390/d18030146

References

1. Díaz, S.; Settele, J.; Brondizio, E.S.; Ngo, H.T.; Zayas, C.N. Pervasive human-driven decline of life on Earth points to the need for transformative change. *Science* **2019**, *366*, eaax3100. [CrossRef]
2. Chen, C.; Liu, J.; Ma, Z. Effects of aquaculture on the maintenance of waterbird populations. *Conserv. Biol.* **2022**, *36*, e13913. [CrossRef] [PubMed]
3. Newbold, T.; Hudson, L.N.; Hill, S.L.; Contu, S.; Lysenko, I.; Senior, R.A.; Börger, L.; Bennett, D.J.; Choimes, A.; Collen, B.; et al. Global effects of land use on local terrestrial biodiversity. *Nature* **2015**, *520*, 45–50. [CrossRef] [PubMed]
4. Wood, K.A.; Jupe, L.L.; Aguiar, F.C.; Collins, A.M.; Davidson, S.J.; Freeman, W.; Kirkpatrick, L.; Magalhes, L.D.; Mckinley, E.; Nuno, A. A global systematic review of the cultural ecosystem services provided by wetlands. *Ecosyst. Serv.* **2024**, *10*, 1673. [CrossRef]
5. Niu, Z.; Zhang, H.; Wang, X.; Yao, W.; Zhou, D.; Zhao, K.; Gong, P. Mapping wetland changes in China between 1978 and 2008. *Chin. Sci. Bull.* **2012**, *57*, 1400–1411. [CrossRef]
6. Davidson, N.C. How much wetland has the world lost? Long-term and recent trends in global wetland area. *Mar. Freshwater Res.* **2014**, *65*, 934–941. [CrossRef]
7. Ren, C.; Wang, Z.; Zhang, Y.; Zhang, B.; Chen, L.; Xi, Y.; Xiao, X.; Doughty, R.B.; Liu, M.; Jia, M.; et al. Rapid expansion of coastal aquaculture ponds in China from Landsat observations during 1984–2016. *Int. J. Appl. Earth Obs.* **2019**, *82*, 101902. [CrossRef]
8. Wang, W.; Fraser, J.D.; Chen, J. Wintering waterbirds in the middle and lower Yangtze River floodplain: Changes in abundance and distribution. *Bird Conserv. Int.* **2017**, *27*, 167–186. [CrossRef]
9. Chen, S.; Zhang, Y.; Xu, C.; Cao, L.; Huang, Z.; Li, C.; Chen, B.; Lu, C.; Xu, W.; Song, Y.; et al. Neighbourhood threats: Landscape context and anthropogenic changes can trigger waterbird population collapse. *Landscape Ecol.* **2022**, *37*, 3141–3158. [CrossRef]
10. Wang, X.D.; Kuang, F.L.; Tan, K.; Ma, Z. Population trends, threats, and conservation recommendations for waterbirds in China. *Avian Res.* **2018**, *9*, 14. [CrossRef]
11. Rajpar, M.N.; Ahmad, S.; Zakaria, M.; Ahmad, A.; Guo, X.; Nabi, G.; Wanghe, K. Artificial wetlands as alternative habitat for a wide range of waterbird species. *Ecol. Indic.* **2022**, *138*, 108855. [CrossRef]
12. Zheng, Y.; Chen, Y.; Song, Y.; Wang, G.; Zhang, Y.; Chen, X.; Li, C.; de Boer, W.F. How waterbird communities respond to seasonal and environmental factors in rice fields adjacent to a Ramsar wetland. *Agr. Ecosyst. Environ.* **2025**, *393*, 109778. [CrossRef]
13. Giosa, E.; Mammides, C.; Zotos, S. The importance of artificial wetlands for birds: A case study from Cyprus. *PLoS ONE* **2018**, *13*, e0197286. [CrossRef] [PubMed]
14. Wang, X.; Chen, J.; Zhou, L. Effects of human activities on the diversity of waterbirds wintering in a shallow lake of the Middle and Lower Yangtze River Floodplain, China. *Diversity* **2020**, *12*, 302. [CrossRef]
15. Fan, J.; Wang, X.; Wu, W.; Chen, W.; Ma, Q.; Ma, Z. Function of restored wetlands for waterbird conservation in the Yellow Sea coast. *Sci. Total Environ.* **2021**, *756*, 144061. [CrossRef]
16. Sebastián-González, E.; Green, A.J. Reduction of avian diversity in created versus natural and restored wetlands. *Ecography* **2016**, *39*, 1176–1184. [CrossRef]
17. Chen, Y.; Zhang, J.; Zhang, Y.; Deng, X.; Ao, P.; Gu, D.; Zhang, B.; Damba, I.; Navaandorj, I.; Zhao, Q.; et al. Spatiotemporal diversity patterns and homogenization signs: Ecological responses and conservation strategies for waterbird communities in the Yangtze River floodplain. *Biol. Conserv.* **2025**, *308*, 111244. [CrossRef]
18. Navedo, J.G.; Fernández, G. Use of semi-intensive shrimp farms as alternative foraging areas by migratory shorebird populations in tropical areas. *Bird Conserv. Int.* **2019**, *29*, 263–276. [CrossRef]
19. Wang, X.; Li, X.; Ren, X.; Jackson, M.V.; Fuller, R.A.; Melville, D.S.; Ma, Z. Effects of anthropogenic landscapes on population maintenance of waterbirds. *Conserv Biol.* **2022**, *36*, e13808. [CrossRef]
20. Li, C.; Yang, S.; Zha, D.; Zhang, Y.; de Boer, W.F. Waterbird communities in subsidence wetlands created by underground coal mining in China: Effects of multi-scale environmental and anthropogenic variables. *Environ. Conserv.* **2018**, *46*, 67–75. [CrossRef]
21. Hu, Z.; Yang, G.; Xiao, W.; Li, J.; Yang, Y.; Yu, Y. Farmland damage and its impact on the overlapped areas of cropland and coal resources in the eastern plains of China. *Resour. Conserv. Recy.* **2014**, *86*, 1–8. [CrossRef]
22. Hou, Q.; Luo, P.; Zhou, S. A catchment-based resilient spatial planning framework for coal mining subsidence and hydrological risks. *World Archit.* **2025**, *3*, 104–107. [CrossRef]
23. Liu, K.; Zhou, L. Effects of utilization models of coal mining subsidence wetlands on the diversity of wintering waterbirds. *Chin. J. Ecol.* **2025**, 1–12. Available online: https://link.cnki.net/urlid/21.1148.Q.20251009.1054.002 (accessed on 29 September 2025).
24. Yang, S.; Zhang, Y.; Tian, D.; Liu, Z.; Ma, Z. Water-surface photovoltaic systems have affected water physical and chemical properties and biodiversity. *Commun. Earth Environ.* **2024**, *5*, 632. [CrossRef]
25. Xie, K.; Zhang, Y.; Yi, Q.; Yan, J. Optimal resource utilization and ecological restoration of aquatic zones in the coal mining subsidence areas of the Huaibei Plain in Anhui Province, China. *Desalin. Water Treat.* **2013**, *51*, 4019–4027. [CrossRef]
26. Yang, S.; Lin, T.; Liu, N.; Cao, K.; Zhang, Y. Anthropogenic wetland transformation: Impacts of floating photovoltaic systems on waterbird diversity in subsidence area. *Ecosyst. Health Sustain.* **2025**, *11*, 403. [CrossRef]

https://doi.org/10.3390/d18030146

27. Pouran, H.M. From collapsed coal mines to floating solar farms, why China's new power stations matter. *Energy Policy* **2018**, *123*, 414–420. [CrossRef]
28. Lewin, I.; Spyra, A.; Krodkiewska, M.; Strzelec, M. The importance of the mining subsidence reservoirs located along the trans-regional highway in the conservation of the biodiversity of freshwater molluscs in industrial areas (Upper Silesia, Poland). *Water Air Soil Pollut.* **2015**, *226*, 189. [CrossRef]
29. Ma, Z.; Chen, S. *Seabirds and Waterbirds in China*; Hunan Science and Technology Press: Changsha, China, 2018.
30. Pielou, E.C. The measurement of diversity in different types of biological collections. *J. Theor. Biol.* **1966**, *13*, 131–144. [CrossRef]
31. Ludwig, J.A.; Reynolds, J.F. Statistical ecology: A primer in method and computing. *J. Appl. Ecol.* **1988**, *26*, 1099–1100. [CrossRef]
32. Tsirtsis, G.; Karydis, M. Evaluation of phytoplankton community indices for detecting eutrophic trends in the marine environment. *Environ. Monit. Assess.* **1988**, *50*, 255–269. [CrossRef] [PubMed]
33. Baselga, A.; Orme, C.D.L. Betapart: An R package for the study of beta diversity. *Methods Ecol. Evol.* **2012**, *3*, 808–812. [CrossRef]
34. Anderson, M.J. A new method for non-parametric multivariate analysis of variance. *Austral. Ecol.* **2001**, *26*, 32–46. [CrossRef]
35. Wang, G.; Zhao, J.; Li, W.; Song, S.; Zhang, Y.; Li, C. Responses of breeding waterbird communities to environmental changes in subsidence wetlands in the North China Plain. *Avian Res.* **2023**, *14*, 100110. [CrossRef]
36. Sung, Y.; Tse, I.W.L.; Yu, Y. Population trends of the Blackfaced Spoonbill *Platalea minor*: Analysis of data from international synchronized censuses. *Bird Conserv. Int.* **2018**, *28*, 157–167. [CrossRef]
37. Kloskowski, J. Fish stocking creates an ecological trap for an avian predator via effects on prey availability. *Oikos* **2012**, *121*, 1567–1576. [CrossRef]
38. Zhang, L.; Wang, X.; Zhang, J.; Ouyang, Z.; Chan, S.; Crosby, M.; Watkins, D.; Martinez, J.; Su, L.; Yu, Y. Formulating a list of sites of waterbird conservation significance to contribute to China's Ecological Protection Red Line. *Bird. Conserv. Int.* **2017**, *27*, 153–166. [CrossRef]
39. Yuan, Y.; Zeng, G.; Liang, J.; Li, X.; Li, Z.; Zhang, C.; Huang, L.; Lai, X.; Lu, L.; Wu, H.; et al. Effects of landscape structure, habitat and human disturbance on birds: A case study in East Dongting Lake wetland. *Eco. Eng.* **2014**, *67*, 67–75. [CrossRef]
40. Song, X.; Liu, T.; Wang, G.; Zhang, Y.; Li, C.; Willem, F.D. Floating photovoltaic systems homogenize the waterbird communities across subsidence wetlands in the North China Plain. *J. Environ. Manag.* **2024**, *349*, 119417. [CrossRef]
41. Liu, Q.; Gan, L.; Wu, H.; Liang, L.; Yan, D.; Wang, X.; Li, C.; Sun, T. Water level fluctuations control wetland hydrological connectivity in driving the integrity of wetlands. *J. Hydrol.* **2025**, *657*, 133095. [CrossRef]
42. Zhu, Z.; Song, Z.; Xu, S.; Wang, S.; Chen, X.; Wang, Y.; Zhu, Z. The development of fishery-photovoltaic complementary industry and the studies on its environmental, ecological and economic effects in China: A review. *Energy Nexus* **2024**, *15*, 100316. [CrossRef]
43. Lacoul, P.; Freedman, B. Physical and chemical limnology of 34 lentic waterbodies along a tropical-to-alpine altitudinal gradient in Nepal. *Internat. Rev. Hydrobiol.* **2005**, *90*, 254–276. [CrossRef]

https://doi.org/10.3390/d18030146

diversity

MDPI

Article

Biodiversity and Seasonal Dynamics of Waterbirds in the Danube Wetland North of Kopački Rit

Zlatko Nedić [1,*], Raluca Nicolae [2], Stefan Popescu [2], Vlatko Rožac [3] and Vera Nikolić [4]

1 Competence Centre Ltd. for Research and Development, Glagoljaška 27, 32100 Vinkovci, Croatia
2 Geostud SRL, Sangerului Str., No. 11, 014617 Bucharest, Romania; raluca.nicolae@geostud.ro (R.N.); stefan.popescu@geostud.ro (S.P.)
3 Public Institution Nature Park Kopački Rit, Mali Sakadaš 1, 31327 Kopačevo, Croatia; vlatko.rozac@pp-kopacki-rit.hr
4 Institute of Zoology, Faculty of Biology, University of Belgrade, Studentski trg 16, 11000 Belgrade, Serbia; vera@bio.bg.ac.rs
* Correspondence: nedic@cekom.hr

Abstract

Wetlands play a vital role in conserving waterbirds, particularly along major European river systems such as the Danube River and its tributaries. To promote the importance of wetlands and address the lack of data on bird biodiversity in aquatic ecosystems, this study was conducted to assess the baseline status of a demonstration site within the European project DaWetRest. The research focused on a lesser-studied pilot area, the Danube North of Kopački rit, near the village of Draž, located in the tri-border region of Croatia, Hungary, and Serbia. Systematic ornithological monitoring was carried out monthly from January to December 2024 across three monitoring areas (total transect length: 4200 m). A total of 26 bird taxa were recorded, comprising 2148 individual observations. The most dominant species were the Mallard (*Anas platyrhynchos*), Grey Heron (*Ardea cinerea*), and Great Cormorant (*Phalacrocorax carbo*). Conversely, the species that were the least present were White Stork (*Ciconia ciconia*), Squacco Heron (*Ardeola ralloides*), Great Crested Grebe (*Podiceps cristatus*), Eurasian Teal (*Anas crecca*), and Black Stork (*Ciconia nigra*). This study also examines the seasonal dynamics of waterbirds, providing essential baseline data for evaluating the effectiveness of upcoming restoration measures planned for the area.

Keywords: wetlands; waterbirds; monitoring; Danube River; Kopački rit; DaWetRest

1. Introduction

Wetlands are among the most biologically diverse and ecologically significant ecosystems in Europe, serving as crucial habitats for a wide variety of species, particularly waterbirds [1,2]. They provide essential nesting, feeding, and migratory stopover sites, supporting both common and endangered avian species, and they play a central role in preserving biodiversity on a continental scale [3–5]. The avifauna of countries in Central and Southeastern Europe is strongly dependent on the wetlands of the Danube River Basin, which hold exceptional importance for regional bird populations. One such country is Croatia, a small EU member state [6] that hosts several wetlands critical for ornithological conservation [7]. Some of these wetlands, such as Kopački rit, are widely recognized and protected under the Ramsar Convention [8–10], while many others are part of the Natura 2000 ecological network. Ornithological studies of these wetlands are essential for their

https://doi.org/10.3390/d17100669

sustainable management, as they provide data on habitat conditions, using waterbirds as bioindicators of habitat quality, water quality, and human pressures on ecosystems [11,12].

Croatia's long-standing tradition of ornithological studies, dating back to the late 19th century, has provided a solid foundation for avian research and conservation efforts [7,13,14]. However, smaller, isolated, and lesser-known wetlands have remained largely under-surveyed [15]. This highlights the importance of the DaWetRest project, which was developed to demonstrate that even small wetland areas can be ecologically significant—both at the local and transnational levels—particularly when their geographic location allows them to function as ecological corridors.

Although small and almost isolated, the protected wetland of Draž Municipality is closely connected to Kopački rit, the backbone of wetland biodiversity in Croatia [16]. Situated at the confluence of the Drava and Danube Rivers, Kopački rit is renowned for its rich biodiversity and expansive floodplains. Kopački rit is a hotspot for waterbirds, including herons, egrets, spoonbills, and various duck species [17–20]. Numerous ornithological surveys performed here have confirmed its importance as both a breeding and migratory habitat, underlining the site's transboundary ecological significance [21]. North of Kopački rit, the area of Draž Municipality, with the Topoljski dunavac ("Dunavac" is a term commonly used in Croatia, Serbia, and other countries along the Danube River. It refers to a small side arm or branch of the Danube, i.e., a secondary watercourse that branches off from the main river, often meanders through lowlands, and may eventually rejoin the main channel or become isolated as an oxbow lake or cut-off meander.) oxbow lake and the surrounding Danube canals of Šarkanj, Lorencov dunavac, and Mačkalučki dunavac, further contribute to the regional ornithological richness. Comprehensive studies in the area of Draž Municipality, including all Danube side arms and Topoljski dunavac, have either not been performed or are not available in scientific databases. However, one study has been performed in the ecosystem of Topoljski dunavac that indicates the presence of 58 species (including waterbirds), highlighting the vital role that riverine and wetland ecosystems in the broader Middle Danube region play in sustaining bird populations, particularly water-dependent species [22]. This monitoring area has been chosen as we aimed to assess the baseline condition of waterbird diversity prior to the implementation of active restoration measures in terms of fish repopulation and hydrotechnical interventions that aim to improve the ecosystem of Danube side arms. The aims of this study are to (a) contribute to ornithological knowledge by providing data on waterbird biodiversity from a less-studied area; (b) generate data that can be used to update existing platforms and portals on bird biodiversity; and (c) create a baseline for assessing the long-term effectiveness of the implemented restoration measures on waterbird fauna after the DaWetRest project was completed.

2. Materials and Methods

2.1. Study Area

From June to December 2023, preparatory activities, including transect delineation and initial area assessment, were carried out. The main field study took place from January to December 2024 in the study area of Draž Municipality. This area lies within the boundaries of Podunavlje i donje Podravlje (HR1000016—Special Protection Area under the Birds Directive), which has been part of the Natura 2000 network since 2013 due to its importance for avian conservation [23] (Figures 1 and 2), and it forms part of the protected area Danube North of Kopački rit (45°51′01″ N/18°48′11″ E).

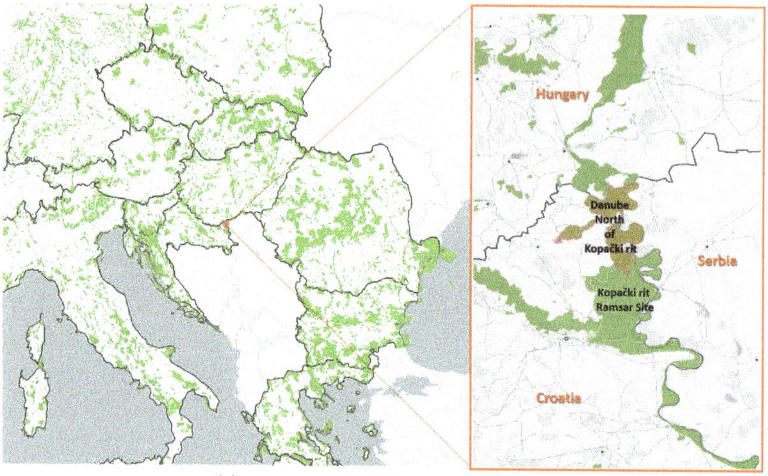

(a) (b)

Figure 1. Danube N of Kopački rit—geographical location: (**a**) Europe; (**b**) regional level. Green color marks areas that belong to the ecological network Natura 2000 (modified by author; source: [16]).

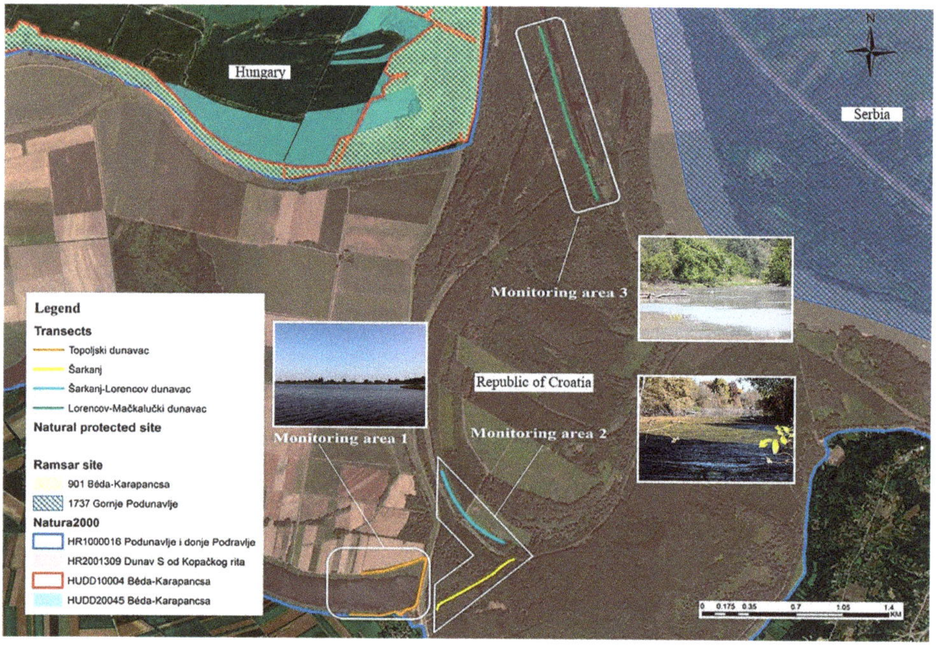

Figure 2. Monitoring areas and overlapping Ramsar and Natura 2000 protected areas in the Danube wetlands near Kopački Rit.

In this area, four transects were delineated:

1. Topoljski dunavac (45°50′55″ N/18°47′45″ E) is a large hydro-accumulation, and it represents the biodiversity core of the entire study area. It was naturally created from an abandoned meander of the old Danube River. Through a system of canals, most notably the Šarkanj canal, it is irrigationally connected to the Danube. The dominant vegetation is reed (*Phragmites australis*), and the surrounding landscape

primarily consists of arable land. The length of the transect is about 1400 m, and this is Monitoring Area 1 (Figure 2).

2. Šarkanj (45°50′52″ N/18°48′17″ E) is predominantly characterized by willow (*Salix* spp.) vegetation and connected via an artificial canal to Topoljski dunavac in the west and the Danube River in the east. The Šarkanj canal is about 4 km long. The transect measures around 700 m and falls within Monitoring Area 2 (Figure 2).
3. Šarkanj-Lorencov dunavac (45°51′11″ N 18°48′16″ E) includes mixed lowland forests, with the Lorencov dunavac canal playing an important ecological role as a biodiversity hotspot. Lorencov dunavac connects further with Mačkalučki dunavac, near the Danube River. This transect is approximately 700 m long and falls within Monitoring Area 2 (Figure 2).
4. Lorencov-Mačkalučki dunavac (45°52′57″ N/18°49′05″ E) is located on the Mačkalučki dunavac side arm, which is close to the Danube River but not connected to it. The surrounding area is dominated by mixed forests, particularly stands of poplar (*Populus* spp.) and willow (*Salix* spp.). This transect measures approximately 1400 m, and this is Monitoring Area 3 (Figure 2).

All of the aforementioned transects were selected in relation to the planned restoration activities and fish repopulation measures defined under the DaWetRest project [24]. The monitoring focused on water-dependent bird species, particularly waterfowl [25] and the Common Kingfisher (*Alcedo atthis*), due to their ecological dependence on aquatic habitats and the role that they play as indicators of fish availability and habitat quality. Using the methodological approach, two smaller and two larger transects were established. Each larger transect defines one monitoring area, while the two smaller transects belong to the same monitoring area (Figure 2). The data obtained through this study will serve as a baseline for evaluating the ecological impacts and effectiveness of the restoration actions proposed within the project framework.

2.2. Survey Implementation

A total of twenty-one field observations were conducted across four transects within three monitoring areas (Figure 2). Surveys took place from January to December 2024, on a monthly basis and covering all seasons, with the following distribution: winter—6 visits; spring—5; summer—4; and autumn—6. No survey was conducted in September, in accordance with the managing authority's recommendation to avoid disturbing the deer during rutting season. Each survey was carried out between 08:30 and 12:30, with standardized observation times: Topoljski Dunavac and Mačkalučki Dunavac were surveyed for 60 min each, and Šarkanj and Lorencov Dunavac were surveyed for 30 min each.

All transects were surveyed on foot by two or three observers. One observer was responsible for direct bird counts, while the other team members collected photographic documentation along the transects to provide additional verification. Photographs were analyzed after each field visit to confirm identifications and detect any individuals that may have been overlooked during the live observation.

Auditory cues and indirect evidence (e.g., feathers, pellets, and eggshells) were not systematically used, as the monitoring focused on water-dependent species typically identified through morphology and behavior in open wetland habitats. Surveys were limited to daytime hours, and no call-playback stimulation was applied to elicit vocal responses. This may have reduced the detectability of cryptic or nocturnal species.

A Canon EOS R6 camera equipped with a Canon RF 600 f/4 I IS USM telephoto lens was used to take photographs during this study. All photographs were geo-tagged at the time of capture. Bird taxa were identified using a standard field identification key [26] and the mobile app "Ptice na dlanu" (from Croatian: "Birds on the palm of your hand") directly on the field and during the analysis of the photographs.

2.3. Data Analysis

Data were initially collected in Microsoft Excel and subsequently transferred to STATISTICA 10 for statistical analysis. The total number of reported individuals per species was calculated from the raw data (Figure 3) as well as relative abundance for species that showed the highest and lowest presence in the entire study period. Frequency of occurrence was calculated based on the species presence per field visits (Tables 1 and 2). To test the distribution of species richness across seasons and monitoring areas a one-way ANOVA was applied, considering the following variables: season—21 cases that represent field observations belonging to a certain season (Figure 4A); species richness per season—the total number of reported species for each of 21 cases (Figure 4A); monitoring areas—63 cases representing field observations assigned to each monitoring area (Figure 4B), and species richness per monitoring area—total number of species reported for each of 63 cases (Figure 4B). Moreover, Tukey's post hoc test with automatically adjusted *p*-value was applied after conducting a one-way ANOVA to identify where significant differences occurred among seasons and monitoring areas. To evaluate bird species richness across the study areas, the exponential Shannon index ($e^{H'}$ also known as Hill Number 1; Table 3) was used as a standard metric for quantifying community heterogeneity [27]. MANOVA (repeated measures) was used to test the statistical significance of species richness across the monitoring areas while also accounting for seasonal effects (Figure 5). The following variables were included: season—21 cases representing field observation per single monitoring area (the cases belong to a specific season according to the number of field visits) and variables Monitoring area 1, Monitoring area 2, and Monitoring area 3 that represent the total number of reported species per each case, separately for each monitoring area.

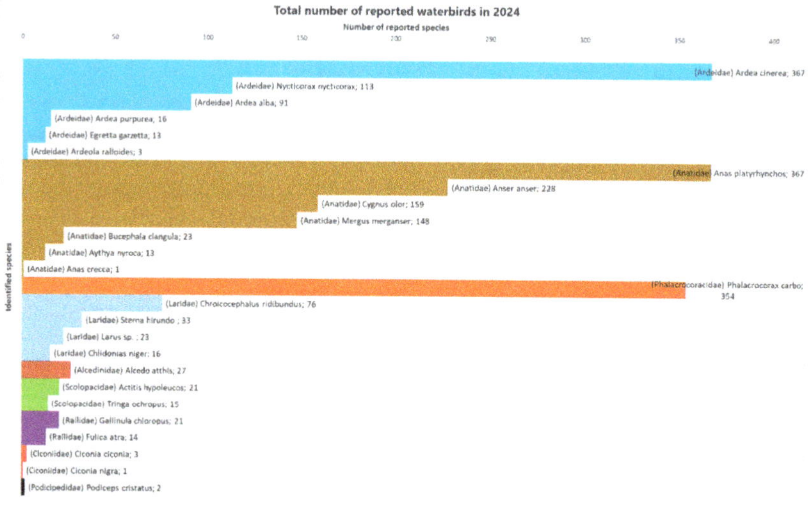

Figure 3. Total number of reported individuals per species and families counted in the entire study area over all monitoring rounds.

Table 1. List of reported waterbirds, their conservation status, and seasonal frequencies: W—winter; SP—spring; SU—summer; A—autumn; Mean—average of reported species per season; SD—standard deviation for seasonal frequency.

Taxa	IUCN [28] /EU Bird Directive [29] /(Bern Convention) [30]	Frequency of Occurence Throughout Season (n = Number of Field Observation Conducted in Certain Season)					
		W (n = 6)	SP (n = 5)	S (n = 4)	S (n = 6)	Mean	SD
ANSERIFORMES							
Anatidae							
Anas crecca	LC/II/(III)	0.17	0.00	0.00	0.00	0.04	0.09
Anas platyrhynchos	LC/II/(III)	22.67	22.40	21.50	8.25	18.71	6.99
Aythya nyroca	LC/I/(III)	0.00	0.00	3.25	0.00	0.81	1.63
Bucephala clangula	LC/II/(III)	3.83	0.00	0.00	0.00	0.96	1.92
Cygnus olor	LC/II/(III)	5.00	3.80	8.75	18.75	9.08	6.79
Mergus merganser	LC/II/(III)	18.00	3.80	0.00	5.25	6.76	7.81
Anser anser	LC/II/(III)	0.00	0.00	2.25	54.75	14.25	27.02
PODICIPEDIFORMES							
Podicipedidae							
Podiceps cristatus	LC(III)	0.00	0.00	0.00	0.50	0.13	0.25
CHARADRIIFORMES							
Scolopacidae							
Actitis hypoleucos	LC(II)	0.00	4.20	0.00	0.00	1.05	2.10
Tringa ochropus	LC(II)	2.17	0.20	0.25	0.00	0.66	1.02
Sternidae							
Chlidonias niger	LC/I/(II)	0.00	0.00	4.00	0.00	1.00	2.00
Sterna hirundo	LC/I/(II)	0.00	6.60	0.00	0.00	1.65	3.30
Laridae							
Chroicocephalus ridibundus	LC/I/(III)	0.00	0.00	0.00	19.00	4.75	9.50
Larus sp.		0.17	0.00	0.50	5.00	1.42	2.4
CICONIIFORMES							
Ardeidae							
Ardea alba	LC/I/(III)	1.67	5.00	12.25	1.75	5.17	4.97
Ardea cinerea	LC(III)	7.17	16.20	39.75	21.00	21.03	13.73
Ardea purpurea	LC/I/(II)	0.00	0.60	1.75	1.50	0.96	0.81
Ardeola ralloides	LC/I/(II)	0.00	0.40	0.00	0.25	0.16	0.20
Egretta garzetta	LC/I/(II)	0.00	2.60	0.00	0.00	0.65	1.30
Nycticorax nycticorax	LC/I/(II)	0.00	6.00	11.75	9.00	6.69	5.04
Ciconiidae							
Ciconia ciconia	LC/I/(II)	0.00	0.20	0.50	0.00	0.18	0.24
Ciconia nigra	LC/I/(II)	0.00	0.20	0.00	0.00	0.05	0.10
Rallidae							
Fulica atra	NT/II/(III)	2.33	0.00	0.00	0.00	0.58	1.17
Gallinula chloropus	LC/II/(III)	0.00	0.00	3.75	1.50	1.31	1.77
PELECANIFORMES							
Phalacrocoracidae							
Phalacrocorax carbo	LC(III)	18.33	5.00	10.50	44.25	19.52	17.37
CORACIIFORMES							
Alcedinidae							
Alcedo atthis	VU/I/(II)	0.17	0.20	3.00	3.25	1.66	1.70

Table 2. Locational frequencies: M1—Monitoring Area 1; M2—Monitoring Area 2; and M3—Monitoring Area 3; Mean—average of reported species per monitoring area; SD—standard deviation for frequency per monitoring areas.

Taxa	Frequency of Occurrence per Monitoring Areas (*n* = Number of Field Observation Conducted in Certain Monitoring Area)				
	M1 (*n* = 21)	M2 (*n* = 21)	M3 (*n* = 21)	Mean	SD
ANSERIFORMES					
Anatidae					
Anas crecca	0.00	0.05	0.00	0.02	0.03
Anas platyrhynchos	10.76	3.48	3.24	5.83	4.27
Aythya nyroca	0.43	0.00	0.19	0.21	0.22
Bucephala clangula	1.10	0.00	0.00	0.37	0.64
Cygnus olor	5.95	0.71	0.90	2.52	2.97
Mergus merganser	2.95	1.81	2.29	2.35	0.57
Anser anser	10.81	0.05	0.00	3.62	6.23
PODICIPEDIFORMES					
Podicipedidae					
Podiceps cristatus	0.00	0.00	0.10	0.03	0.06
CHARADRIIFORMES					
Scolopacidae					
Actitis hypoleucos	1.00	0.00	0.00	0.33	0.58
Tringa ochropus	0.62	0.00	0.10	0.24	0.33
Sternidae					
Chlidonias niger	0.76	0.00	0.00	0.25	0.44
Sterna hirundo	1.57	0.00	0.00	0.52	0.91
Laridae					
Chroicocephalus ridibundus	3.57	0.05	0.00	1.21	2.05
Larus sp.	1.10	0.00	0.00	0.37	0.64
CICONIIFORMES					
Ardeidae					
Ardea alba	2.05	1.57	0.71	1.44	0.68
Ardea cinerea	9.76	5.14	2.57	5.82	3.64
Ardea purpurea	0.76	0.00	0.00	0.25	0.44
Ardeola ralloides	0.14	0.00	0.00	0.05	0.08
Egretta garzetta	0.57	0.05	0.00	0.21	0.32
Nycticorax nycticorax	5.33	0.05	0.00	1.79	3.06
Ciconiidae					
Ciconia ciconia	0.10	0.05	0.00	0.05	0.05
Ciconia nigra	0.00	0.00	0.05	0.02	0.03
Rallidae					
Fulica atra	0.67	0.00	0.00	0.22	0.39
Gallinula chloropus	1.00	0.00	0.00	0.33	0.58
PELECANIFORMES					
Phalacrocoracidae					
Phalacrocorax carbo	12.19	1.71	3.43	5.78	5.62
CORACIIFORMES					
Alcedinidae					
Alcedo atthis	0.67	0.57	0.05	0.43	0.33

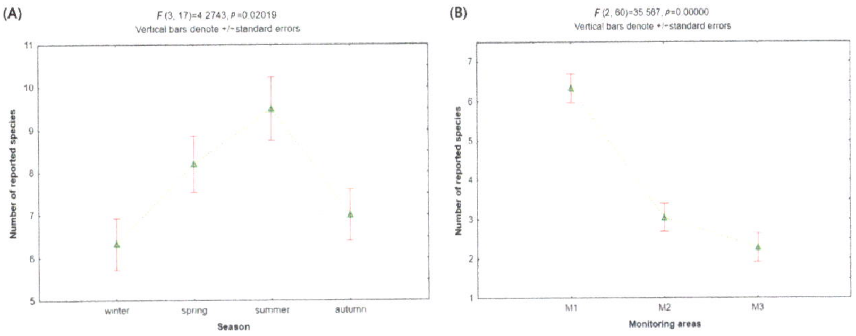

Figure 4. Distribution of the total number of reported species per (**A**) season and (**B**) monitoring area.

Table 3. The exponential Shannon index $e^{H'}$, also known as Hill Number 1 (H): *n* = number of field observations per season; *Min* = lowest value of $e^{H'}$ reported for a given season and monitoring area; *Max* = highest value of $e^{H'}$ reported for a given season and monitoring area; *Mean* = average value of $e^{H'}$ for a given season and monitoring area; *SD* = standard deviation.

Season	*n*	Monitoring Area 1				Monitoring Area 2				Monitoring Area 3			
		Min	Max	Mean	SD	Min	Max	Mean	SD	Min	Max	Mean	SD
Winter	6	3.589	6.376	4.521	1.101	1.738	2.456	1.976	0.326	1.583	4.696	2.735	1.355
Spring	5	3.335	6.823	4.557	1.525	2.090	4.209	3.239	0.947	1.961	3.597	2.845	0.599
Summer	4	4.098	7.365	6.270	1.472	1.419	5.044	2.816	1.950	2.324	2.814	2.547	0.248
Autumn	6	1.606	4.724	3.070	1.041	1.755	3.652	2.766	0.784	1.755	2.219	1.902	0.213
Total	21	1.606	7.365	4.448	1.613	1.419	5.044	2.730	1.034	1.583	4.696	2.526	0.793

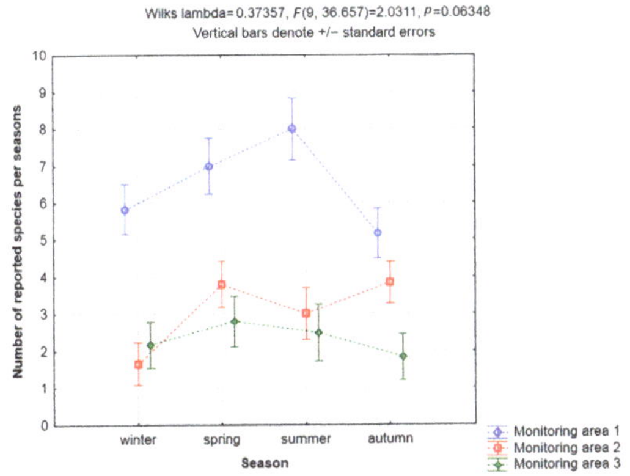

Figure 5. Distribution of the total number of reported species per monitoring area and seasons.

3. Results

During the monitoring activities, a total of 2148 individuals were registered, encompassing 26 taxa from 9 families (Tables 1 and 2, Figure 3). The number of individuals varied widely among species, with some species being highly present while others were relatively rare. The results revealed that two species, Grey Heron (*Ardea cinerea*) and Mallard (*Anas platyrhynchos*), were the most dominant, each accounting for approximately 17.09% of all reported individuals. Following closely was Great Cormorant (*Phalacrocorax carbo*), which made up around 16.48% of the total observations. Together, these three species comprised over 50% of all individuals observed, highlighting their ecological dominance in the study area. In contrast, other species such as Greylag Goose (*Anser anser*) and Mute Swan (*Cygnus olor*) showed notable but lower relative abundances of 10.61% and 7.40%, respectively. At the opposite end of the spectrum, several species demonstrated minimal dominance. For example, Eurasian Teal (*Anas crecca*) and Black Stork (*Ciconia nigra*) each accounted for only 0.05% of all observations, making them the least frequently recorded taxa. They were followed by Great Crested Grebe (*Podiceps cristatus*) with 0.09% and both White Stork (*Ciconia ciconia*) and Squacco Heron (*Ardeola ralloides*), each representing 0.14% of the total abundance. These exceptionally low percentages indicate that these species were rarely encountered and had minimal influence on overall species composition during the monitoring period.

The maximum number of species recorded per season varied significantly ($F = 4.2743$; $p < 0.05$), and more species were reported during the summer compared to the winter ($MS = 2.1843$; df = 17.000; $p < 0.05$; Figure 4A, Table 1). We identified four species that were present across all seasons: *Ardea cinerea*, *Anas platyrhynchos*, *Phalacrocorax carbo*, and *Cygnus olor* (Table 1).

Statistical significance was detected in terms of species richness per monitoring area ($F = 35.667$; $p < 0.05$). The results revealed that Monitoring Area 1 showed a significantly higher mean number of species compared to Monitoring Areas 2 and 3 ($MS = 2.7317$; df = 60; $p < 0.05$). There was no identified statistical significance between Monitoring Areas 2 and 3 ($p > 0.3012$). These findings suggest that Monitoring Area 1 serves as a biodiversity hotspot, whereas Areas 2 and 3 support fewer species and do not differ significantly from each other in terms of richness (Table 2, Figure 4B).

Statistical significance was not detected considering species richness per monitoring area per season ($F = 2.0311$; $p = 0.06348$), but the test results were close to the threshold, highlighting higher species richness in Monitoring Area 1 (Figure 5).

Monitoring Area 1 showed the highest biodiversity, with index values ranging from $H = 1.606$ to $H = 7.365$. Monitoring Area 2 indicated lower values (from $H = 1.419$ to $H = 5.044$), as did Monitoring Area 3 (from $H = 1.583$ to $H = 4.696$). Our analysis clearly identified Monitoring Area 1 as a backbone of waterbird biodiversity for the whole area, underscoring its ecological significance in the context of the broader wetland ecosystem (Table 3).

4. Discussion

Although Croatian ornithology has a long-standing tradition, studies focusing on small biodiversity hotspots remain spatially uneven and fragmented, and they are often concentrated on specific taxa or specific ecological aspects [31]. This study provides the current status of waterbird fauna in Draž (Figures 1 and 2), within the protected Natura 2000 site Danube North of Kopački rit (code HR2001309). One of the most important resources for comparing these findings with previous/ongoing monitoring activities is the platform that aggregates expert-verified data [32]. According to this portal, approximately 45 waterbird taxa have been recorded in the wider Danube North of Kopački rit region. Our study documented 26 taxa, representing over 57% of the area's total recorded waterbird diversity [32]. The presence of these taxa is further corroborated by the eBird platform, which documents observations for many of the focal species over a 40-year period (1985–2024) [33]. The high prevalence of piscivorous species such as the Grey Heron (*Ardea cinerea*) across all seasons and monitoring areas is consistent with previous studies and should be characterized as a common occurrence [33–35]. The favorable habitat conditions of Topoljski dunavac, characterized by heterogeneous feeding grounds and diverse fish availability, further explain the abundance of this species [36]. The Great Cormorant (*Phalacrocorax carbo*) also showed high presence throughout the year in all three areas; however, no nesting activity was recorded for cormorants, which aligns with existing studies identifying Lonjsko polje and Kopački rit as the primary breeding sites for this species [37]. This pattern is consistent with other studies in Croatia, which documented frequent occurrence of non-breeding individuals foraging at fishponds [38].

The high presence of the Anatidae family was largely due to the presence of Greylag Goose (*Anser anser*) and Mallard (*Anas platyrhynchos*). The Croatian Red Book of Birds states that Topoljski Dunavac was considered one of the key nesting sites for Greylag Goose, with up to 30 pairs recorded since 2009 [35,39]. This species was most frequently observed in summer and autumn in Monitoring Area 1, which is surrounded by arable and grassy land that likely satisfies their feeding requirements. Although specific monitoring data for *Anas platyrhynchos* in the study area are lacking, over 170 observations with more than 3500 individuals have been recorded since 2000 within the broader Danube North of Kopački rit [32]. According to open-access biodiversity platforms, Mallards (*Anas platyrhynchos*) were previously recorded in Monitoring Area 3 but not in Areas 1 or 2. This highlights the need for systematic, holistic, and spatially comprehensive monitoring efforts that enable contributions from all interested stakeholders [33,40]. Mallards heavily rely on aquatic invertebrates during brood-rearing, while fish are only a minor dietary component. Studies show that fish-rich waters can reduce duckling survival by depleting invertebrate prey [41,42]. Other studies [43,44] further confirm that fish presence alters invertebrate availability and affects duckling foraging behavior, making fish-dense habitats less suitable. For this reason, it is necessary to determine and recommend future restoration measures that ensure that the habitat remains favorable for all species inhabiting the area.

The number of certain species, such as Common Teal (*Anas crecca*), was lower than in previous studies in areas along the Drava River [45]. The presence of species such as the Black-crowned Night Heron (*Nycticorax nycticorax*), Great White Egret (*Ardea alba*), Mute Swan (*Cygnus olor*), and Goosander (*Mergus merganser*) supports the assumption of generally favorable environmental conditions in the study area, as these species typically require suitable water levels, diverse aquatic vegetation, and sufficient fish resources to thrive [35,39]. Conversely, certain members of the Ardeidae family were less present, such as Purple Heron (*Ardea purpurea*), Little Egret (*Egretta garzetta*), and Squacco Heron (*Ardeola ralloides*). This pattern could be attributed to environmental pressures, including habitat degradation, particularly of reed communities, caused by intensive agriculture, as well as reduced fish availability [39]. Agricultural activities contribute to the reduction in and fragmentation of reedbeds (*Phragmites australis*), which are essential for the foraging and breeding of *A. purpurea* [46,47]. Unlike the Grey Heron (*Ardea cinerea*), an ecological generalist capable of exploiting a wide range of habitats, including open wetlands, fishponds, rivers, and anthropogenically altered landscapes [48,49], *A. purpurea* is strongly dependent on well-preserved reed-dominated habitats for both nesting and food supply [46,47]. This specialization makes the species particularly vulnerable to reedbed degradation. Therefore, while both species may face pressures from reduced fish availability, the added constraint of reedbed loss disproportionately affects *A. purpurea*.

According to data from open-access platforms, the presence of *Ardea purpurea*, *Egretta garzetta*, and *Ardeola ralloides* was lower compared to species from other families, which aligns with our recent findings [32,33]. Similar patterns were observed for Common Goldeneye (*Bucephala clangula*), Ferruginous duck (*Aythya nyroca*), and Common Teal (*Anas crecca*), with the latter two recorded only during winter, consistent with earlier studies [32,45]. Artificially created ecological conditions (caused by dredging and cleaning of Topoljski dunavac) likely contributed to the presence of these species, creating relatively shallow wetlands with well-structured, mosaic-like vegetation and a diversity of microhabitats [39,50].

Among Laridae, the Black-headed Gull (*Larus ridibundus*) was the most commonly recorded species, consistent with studies from similar habitats [22,33]. The Common Tern (*Sterna hirundo*) was observed only once (19 April 2024), likely during the spring migration route to its typical breeding habitats of the Adriatic island region [51,52]. The Black Tern (*Chlidonias niger*) was also detected and, as with *S. hirundo*, its presence is probably linked to seasonal migration [53].

The Common Kingfisher (*Alcedo atthis*) was observed across all monitoring areas. This is likely due to the availability of suitable nesting substrates, such as sandy or loamy embankments formed during canal maintenance [54]. The estimated Croatian population ranges from 700 to 1000 breeding pairs, with over 240 recorded within ecological network sites [35]. Its consistent presence is further supported by recent records, which confirm the stable occurrence of the species in the Draž region over the past two years [33].

The presence of the Common Sandpiper (*Actitis hypoleucos*) in April and May confirms its status as a regular passage migrant from inland Croatia [35]. Its presence was likely facilitated by the artificially constructed sandbanks in Topoljski dunavac. There is no previous evidence of the presence of the mentioned species in the Draž area [32]. In contrast, the Green Sandpiper (*Tringa ochropus*) was reported only twice in Draž in 2023, highlighting the value of continuous monitoring to capture such variations [32].

Two Rallidae species were reported: Common Coot (*Fulica atra*) and Moorhen (*Gallinula chloropus*). Common Coot (*Fulica atra*) is very common in the Draž area, and the available sources suggest that its actual prevalence may be higher than what was observed in this study [32]. The most recent update from 2025 reported more than 600 individuals during feeding events [32]. In our monitoring, the species was observed only during winter, which corresponds with studies indicating its seasonal presence at fishponds during both winter and summer periods [55]. *Gallinula chloropus* was reported only in the summer and autumn, exclusively in the Topoljski dunavac transect. Its presence may be linked to the creation of shallow water habitats through dredging and regular maintenance activities, which led to the formation of artificial accumulations suitable for this species [56].

From Ciconiidae, two species with low abundance were reported: White Stork (*Ciconia ciconia*), with three individuals from May to July in flight, and Black Stork (*Ciconia nigra*), with one individual in April in flight. The available data suggest that breeding populations of *Ciconia ciconia* in Croatia are unknown [39], and past sightings in the area are typically of birds in flight [32]. Although fish represent a substantial component of storks' diet [57], this species was less frequently observed in the monitoring areas. The reasons for this most likely result from environmental pressures and threats, such as pesticide use [39]. The other reason is that breeding pairs nest within Draž village, close to the monitoring areas, and may forage at different times compared with those covered by the survey, resulting in their absence during fieldwork [58]. The presence of *Ciconia nigra* was also low. According to existing data, the breeding population in nearby regions (Gornje Podunavlje and Donje Podravlje) ranges from 35 to 55 pairs [39]. Since 2014, this species has been reported only 19 times in the Danube North of Kopački rit, based on observations by experts and citizen scientists, confirming its rarity in the area [32].

Finally, the Great Crested Grebe (*Podiceps cristatus*), the only representative of the Podicipedidae family observed, showed the lowest abundance, consistent with previous observations indicating just six sightings between 2017 and 2022 [32].

Our results stress the need to view such wetland habitats not in isolation and rather part of a larger ecological network essential to the lifecycles of water-dependent birds. Monitoring Area 1 demonstrated significantly greater species richness compared to Monitoring Areas 2 and 3, likely due to Area 1's optimal microhabitat configuration and permanent hydrological connectivity to the Danube River, which make this area particularly valuable from a biodiversity perspective [12,35,39].

Seasonal fluctuations in the biodiversity index indicate that species diversity is not uniform throughout the year. This is probably due to the large number of taxa that occur across all seasons and species-specific biological characteristics, especially in relation to their feeding habits and migratory patterns [59]. Considering the relatively high number of species registered in the study area, and taking into consideration the Shannon index values alongside the findings from comparable wetland studies, our results reveal a similar trend, namely Draž's slightly higher Shannon index. This is likely attributable to the habitat complexity and ecological richness of Topoljski dunavac, which supported a greater number of reported taxa and higher diversity overall [60].

Our monitoring results indicate that the Draž wetland currently supports fewer waterbird species than expected given its habitat potential. For instance, comparable European wetlands of similar habitat complexity, such as Kopački Rit in Croatia and Neusiedler See—Seewinkel on the Austrian-Hungarian border, regularly support much richer assemblages: Kopački Rit harbors nearly 300 bird species and more than 20,000 waterbirds in winter [9,16], while Neusiedler See—Seewinkel sustains tens of thousands of migratory waterbirds annually, including herons, egrets, and large flocks of geese and ducks [61,62]. Compared to these sites, the Draž wetland appears underutilized by waterbirds despite offering suitable habitat conditions.

This limitation arises primarily from the reduced hydrological connectivity of the lateral canals with the Danube River, which restricts water exchange and fish migration. Hydrological connectivity and flood pulses are widely recognized as central processes supporting biodiversity, nutrient exchange, and fish recruitment in rivers–floodplain systems [63–65]. Planned restoration measures are therefore essential to re-establish stable water levels and enhance habitat suitability for water-dependent birds [66,67]. Beyond biodiversity gains, such interventions would strengthen ecosystem services, including nutrient cycling, carbon sequestration, and flood regulation [68,69], while also supporting local socioeconomic development through ecotourism and nature-based innovation [70]. By restoring ecological processes and improving habitat quality, the site can contribute more effectively to the broader wetland network of the Danube region [71].

This study has some limitations. Transect-based monitoring may under-detect cryptic, nocturnal, or highly mobile species, and the single-year dataset restricts the ability to capture interannual variability. In addition, surveys were limited to daytime hours and did not employ call–playback techniques, which likely reduced the detectability of secretive or nocturnal taxa such as rallids, *Ixobrychus minutus*, *Nycticorax nycticorax*, and *Botaurus stellaris*. Therefore, in future research, monitoring should be extended across multiple years, hydrological and fish population data should be integrated, and expert-led surveys with citizen-science platforms should be combined. Such approaches will ensure a more robust evaluation of restoration outcomes and strengthen adaptive management for wetlands across the wider Danube basin.

5. Conclusions

This study provides the first year-round assessment of waterbird diversity in the Draž Municipality, within the Natura 2000 site "Danube North of Kopački rit". By documenting 26 taxa and over 2100 individuals, we establish a crucial baseline ahead of planned restoration efforts under the DaWetRest project. Importantly, Topoljski dunavac (Monitoring Area 1) was identified as a biodiversity hotspot, underscoring the role that this area plays in terms of hydrological connectivity and habitat diversity to sustain high species richness and community stability.

Our findings contribute to current knowledge by demonstrating the ecological value of small, semi-isolated wetlands, which function as key nodes in regional ecological networks. They also highlight management priorities: restoration measures, particularly hydrotechnical and habitat interventions that should be ecologically calibrated to preserve trophic balance and support sensitive life stages such as breeding and brood-rearing. Moreover, the presence of habitat specialists such as *Ardea purpurea* emphasizes the need to safeguard reedbeds and shallow-water mosaics as critical microhabitats.

Although this study covers only a single year, the results may provide valuable baseline for future restoration and monitoring at Draž. Building on this foundation, subsequent efforts should integrate longer-term ecological observations and expanded methodological approaches. In this way, restoration actions would not only safeguard local biodiversity but also strengthen the role of Draž within the wider Danube wetland network, enhancing both ecological resilience and conservation outcomes.

Author Contributions: Conceptualization Z.N. and V.N.; methodology, Z.N. and R.N.; software, ZN.; validation, V.R. and S.P.; formal analysis, Z.N.; investigation, Z.N.; resources, V.R.; data curation, V.R.; writing—original draft preparation, Z.N. and V.N.; writing—review and editing, V.R. and R.N.; visualization, V.N. and S.P.; supervision, V.R.; project administration, Z.N.; funding acquisition, R.N. All authors have read and agreed to the published version of the manuscript.

Funding: This work is a part of the Danube Wetlands and flood plains Restoration through systemic, community engaged and sustainable innovative actions (DaWetRest) project, which is co-funded by the European Union Horizon Europe programme (HORIZON-MISS-2022-OCEAN-O1-02), under Grant Agreement no. 101113015. Views and opinions expressed are however those of the author(s) only and do not necessarily reflect those of the European Union or CINEA. Neither the European Union nor the granting authority can be held responsible for them.

Data Availability Statement: The original contributions presented in this study are included in the article. Further inquiries can be directed to the corresponding author.

Acknowledgments: Authors express their gratitude to Alma Mikuška from Josip Juraj Strossmayer University of Osijek, Department of Biology, for her support during the preparation of this paper and to Filip Županić from the Competence Centre Ltd. for research and development for technical support during the field observation. This research was also supported by Ministry of Science, Technological Development and Innovation of the Republic of Serbia (Contract No. 451-03-137/2025-03/200178).

Conflicts of Interest: Authors declare that the research was conducted in the absence of any commercial or financial relationships that could be constructed as potential conflicts of interest.

References

1. Wetlands International Europe. Wetland Biodiversity. Available online: https://europe.wetlands.org/home-2/our-work/wetland-biodiversity/ (accessed on 30 August 2025).
2. Luo, J.; Chen, Y.; Lin, L.; Liu, C.; Lei, G. Bird diversity and waterbird habitat preferences in relation to wetland restoration. *Avian Res.* **2019**, *10*, 21. [CrossRef]
3. Morganti, M.; Manica, M.; Bogliani, G.; Gustin, M.; Luoni, F.; Trotti, P.; Perin, V.; Brambilla, M. Multi-species habitat models highlight the key importance of flooded reedbeds for inland wetland birds: Implications for management and conservation. *Avian Res.* **2019**, *10*, 15. [CrossRef]
4. Bhowmik, S. Ecological and Economic Importance of Wetlands and Their Vulnerability: A Review. In *Research Anthology on Ecosystem Conservation and Preserving Biodiversity*; IGI Global: Hershey, PA, USA, 2022; pp. 11–27. [CrossRef]
5. Lipton, D.; Rubenstein, M.A.; Weiskopf, S.R.; Carter, S.; Peterson, J.; Crozier, L.; Fogarty, M.; Gaichas, S.; Hyde, K.J.W.; Morelli, T.L.; et al. *Ecosystems, Ecosystem Services, and Biodiversity*; U.S. Global Change Research Program: Reston, VA, USA, 2018; pp. 268–321. [CrossRef]
6. Kotarski, K.; Šelo-Šabić, S. A small state finding its way in the EU: Croatia and its approach to Brexit. *J. Contemp. Eur. Stud.* **2023**, *31*, 670–686. [CrossRef]
7. Mužinić, J. The state of bird and nature protection in Croatia. *Environmentalist* **1995**, *15*, 188–195. [CrossRef]

8. Pithart, D.; Petrov Rančić, I.; Kutleša, P.; Duplić, A. Study of freshwater ecosystem services in Croatia. *Res. Rep.* **2014**, *71*. Available online: https://www.researchgate.net/publication/316648208_Study_of_Freshwater_Ecosystem_Services_in_Croatia (accessed on 5 August 2025).
9. Rožac, V.; Baković, A. *RIS for Site No. 583: Nature Park Kopački Rit, Croatia*; The Ramsar Sites Information Service: Gland, Switzerland, 2019. Available online: https://rsis.ramsar.org/ris/583 (accessed on 5 August 2025).
10. Brundic, D.; Barbalić, D.; Omerbegović, V.; Schneider-Jacoby, M.; Tusić, Z. Alluvial wetlands preservation in Croatia: The experience of the Central Sava Basin flood control system. *Period. Biol.* **2000**, *96*, 348–356. Available online: https://www.researchgate.net/publication/237559423_ALLUVIAL_WETLANDS_PRESERVATION_IN_CROATIA_THE_EXPERIENCE_OF_THE_CENTRAL_SAVA_BASIN_FLOOD_CONTROL_SYSTEM (accessed on 5 August 2025).
11. Liu, X.; Ho, L.; De Cock, A.; De Saeyer, N.; Pham, K.; Panique-Casso, D.; Forio, M.A.E.; Goethals, P.L.M. Water Quality and Its Influence on Waterbird Habitat Distribution: A Study Along the Lieve River, Belgium. *Water* **2025**, *17*, 595. [CrossRef]
12. Amat, J.A.; Green, A.J. Waterbirds as bioindicators of environmental conditions. In *Conservation Monitoring in Freshwater Habitats: A Practical Guide and Case Studies*; Hurford, C., Schneider, M., Cowx, I., Eds.; Springer: Dordrecht, The Netherlands, 2010; pp. 45–52. [CrossRef]
13. Kralj, J. Ornitofauna Hrvatske tijekom posljednjih dvjesto godina. *Larus* **1997**, *46*, 1–112.
14. Lukač, G. Popis ptica Hrvatske. *Nat. Croat.* **2007**, *16* (Suppl. 1), 1–148. Available online: https://hrcak.srce.hr/file/49182 (accessed on 5 August 2025).
15. Wang, Y.; Zhao, M.; Pei, W.; Guan, Q.; Liu, J.; Chen, Y.; Liu, J.; Zhang, Q. Research Overview on Isolated Wetlands. *Water* **2025**, *17*, 2013. [CrossRef]
16. ICPDR. Kopački Rit—Wetland Paradise in Limbo Between Ongoing Threats and Future Protection. Danube Watch. Available online: https://www.icpdr.org/publications/kopacki-rit-wetland-paradise-limbo-between-ongoing-threats-and-future-protection (accessed on 30 August 2025).
17. Mikuška, A.; Mikuška, T.; Kordić, M. Promjene u ornitofauni Kopačkog rita tijekom posljednjih 150 godina. *Symp. Kopacki Rit* **2022**, *11*, 124–125. Available online: https://www.croris.hr/crosbi/publikacija/prilog-skup/811126 (accessed on 30 June 2025).
18. Mikuška, J.; Mikuška, T.; Romulić, M. Vodič kroz biološku raznolikost Kopačkog rita—Ptice. *Matica Hrvat. Osijek* **2002**.
19. Mikuška, A.; Mikuška, T. Ornitofauna Biljskog i Kopačkog rita. *Symp. Kopacki Rit.* **2021**, *10*, 138–139.
20. Mikuska, T.; Mikuska, A.; Bjedov, D.; Ledinšćak, J.; Tomik, A.; Šetina, N.; Podravec, D. NATURA 2000 status does not guarantee the protection of Croatian spoonbill breeding population. In Proceedings of the X International Workshop Eurasian Spoonbill International Expert Group, Zadar, Croatia, 3–7 October 2022; pp. 55–63. Available online: https://storkibisspoonbill.org/wp-content/uploads/2023/10/Special_Publication_2023.pdf#page=56 (accessed on 30 June 2025).
21. Vereš, M.; Škoro, M.; Rožac, V.; Damjanović, I.; Romanjek, K.; Kučera, S.; Bolšec, B.; Marušić, M.; Agić, I.; Popijač, A. Preliminary data from the monitoring of ornithofauna in the ecological network site Danube North of Kopački rit in the period from 2020 to 2022. In Proceedings of the 12th Symposium with International Participation Kopački Rit Past, Present, Future 2023, Osijek, Croatia, 28–29 September 2023. Available online: https://www.researchgate.net/publication/374914385 (accessed on 30 June 2025).
22. Damjanović, I.; Teni, M.; Uranjek, N.; Galić, A.; Prskalo, M. Bird fauna of the Middle Danube oxbow lakes: Study site Topoljski Dunavac. In Proceedings of the 12th Symposium with International Participation Kopački Rit Past, Present, Future 2023, Osijek, Croatia, 28–29 September 2023. Available online: https://pp-kopacki-rit.hr/wp-content/uploads/2023/09/Prijelom-jucer-danas-sutra-ZA-WEB.pdf (accessed on 30 June 2025).
23. Natura 2000. 2025. Available online: https://natura2000.eea.europa.eu/?views=Sites_View (accessed on 20 June 2025).
24. DaWetRest—Danube Wetlands Restoration Project. DaWetRest—Danube Wetlands and Flood Plains Restoration Through Systemic, Community Engaged and Sustainable Innovative Actions. 2025. Available online: https://dawetrest.eu/ (accessed on 20 June 2025).
25. Wetlands International. Waterbird Population Estimates, Fifth Edition: Summary Report. 2012. Available online: https://www.wetlands.org/publication/waterbird-populations-estimates-fifth-edition/ (accessed on 5 August 2025).
26. Svensson, L.; Mullarney, K.; Zetterström, D. *Collins Bird Guide: The Most Complete Guide to the Birds of Britain and Europe*; HarperCollins: London, UK, 2010; 448p.
27. Shannon, C.E.; Wiener, W. *The Mathematical Theory of Communication*; University of Illinois Press: Champaign, IL, USA, 1949; 117p.
28. International Union for Conservation of Nature. The IUCN Red List of Threatened Species: Search Results for "Larus". Available online: https://www.iucnredlist.org/search?query=larus&searchType=species (accessed on 20 June 2025).
29. European Parliament. Council of the European Union Directive 2009/147/ECon the Conservation of Wild Birds. *Off. J. Eur. Union* **2009**, *L20*, 7–25. Available online: https://eur-lex.europa.eu/legal-content/EN/TXT/?uri=CELEX%3A32009L0147 (accessed on 5 August 2025).

30. Council of Europe. Convention on the Conservation of European wildlife and Natural Habitats (Bern Convention). *Eur. Treaty Ser.* **1979**, *104*. Available online: https://www.coe.int/en/web/conventions/full-list?module=treaty-detail&treatynum=104 (accessed on 5 August 2025).
31. Radović, A.; Gavrilović, M.; Palcu, T.; Ćalić, J. Utilizing remote sensing data for species distribution modeling of birds in Croatia. *Biodiversity* **2025**, *17*, 399. [CrossRef]
32. Observation.org. Available online: https://hr.observation.org/pages/about/ (accessed on 4 September 2025).
33. eBird—Hrvatska. Available online: https://ebird.org/region/HR (accessed on 22 May 2025).
34. Mikuška, J.; Bogdanović, T.; Mikuška, T.; Mikuška, A.; Šalić, V. Size and distribution of breeding colonies of Grey Heron (*Ardea cinerea*) in lowland Croatia. *Acrocephalus* **2005**, *26*, 37–40. Available online: https://www.researchgate.net/publication/375888083 (accessed on 30 June 2025).
35. Dumbović Mazal, V.; Pintar, V.; Zadravec, M. Prvo Izvješće o Brojnosti i Rasprostranjenosti Ptica u Hrvatskoj Sukladno Odredbama Direktive o Pticama. 2019. 49p. Available online: https://www.haop.hr/sites/default/files/uploads/dokumenti/03_prirodne/izvjesca/DumbovicMazal2019_Prvo_izvjesce_art12_ptice_F%20(002).pdf (accessed on 30 June 2025).
36. Jakubas, D.; Manikowska, B. The response of Grey Herons *Ardea cinerea* to changes in prey abundance. *Bird Study* **2011**, *58*, 487–494. [CrossRef]
37. Piria, M. Utjecaj velikog vranca (*Phalacrocorax carbo sinensis*) na riblji stok—Pregled. *Croat. J. Fish.* **2014**, *72*, 164–173. [CrossRef]
38. Opačak, A.; Florijančić, T.; Horvat, D.; Ozimec, S.; Bodakoš, D. Diet spectrum of great cormorants (*Phalacrocorax carbo sinensis* L.) at the Donji Miholjac carp fishponds in eastern Croatia. *Eur. J. Wildl. Res.* **2004**, *50*, 173–178. [CrossRef]
39. Tutiš, V.; Kralj, J.; Ćiković, D.; Barišić, S. Crvena Knjiga Ptica Hrvatske. 2013, pp. 10–243. Available online: https://www.haop.hr/sites/default/files/uploads/dokumenti/03_prirodne/crvene_knjige_popisi/Crvena_knjiga_ptica_web.pdf (accessed on 5 August 2025).
40. Occurrence Data from GBIF. 2025. Available online: https://doi.org/10.15468/2gbkf0 (accessed on 5 August 2025).
41. Hill, D.; Wright, R.; Street, M. Survival of Mallard ducklings *Anas platyrhynchos* and competition with fish for invertebrates on a flooded gravel quarry in England. *Ibis* **1987**, *129*, 159–167. [CrossRef]
42. Cox, R.R.; Hanson, M.A.; Roy, C.C.; Euliss, N.H., Jr.; Johnson, D.H.; Butler, M.G. Mallard Duckling Growth and Survival in Relation to Aquatic Invertebrates. *J. Wildl. Manag.* **1998**, *62*, 124–133. Available online: https://digitalcommons.unl.edu/usgsnpwrc/224 (accessed on 6 August 2025). [CrossRef]
43. Nummi, P.; Sjöberg, K.; Pöysä, H.; Elmberg, J. Individual foraging behaviour indicates resource limitation: An experiment with Mallard ducklings. *Can. J. Zool.* **2000**, *78*, 1891–1895. [CrossRef]
44. Marteau, B.; Sundell, J.; Pesonen, R.; Nummi, P. Duckling body mass increases with abundant aquatic invertebrates: Experimental approach. *Glob. Ecol. Conserv.* **2025**, *58*, e03490. [CrossRef]
45. Jurčević, I.; Mikuška, J. Wintering Anatidae community structure on the Drava River in Croatia. *Limnol. Rep.* **2004**, *35*, 633–638. Available online: https://www.croris.hr/crosbi/publikacija/prilog-skup/496496 (accessed on 30 June 2025).
46. Barbraud, C.; Mathevet, R. Is commercial reed harvesting compatible with breeding purple herons *Ardea purpurea* in the Camargue, southern France? *Environ. Conserv.* **2000**, *27*, 334–340. [CrossRef]
47. Morganti, M.; Viganò, E.; Berlusconi, A.; Cioccarelli, S.; Fasola, M. Post-fledging habitat selection of a Purple Heron *Ardea purpurea* revealed by GPS/GSM telemetry. *Avocetta* **2021**, *45*, 135–145. [CrossRef]
48. Choi, Y.S.; Kwon, I.K.; Yoo, J.C. Foraging Habitat Preferences of Herons and Egrets. *J. Ecol. Environ.* **2007**, *30*, 237–244. [CrossRef]
49. Cieślińska, K.; Manikowska-Ślepowrońska, B.; Zbyryt, A.; Jakubas, D. Foraging Habitat Availability and the Non-Fish Diet Composition of the Grey Heron (*Ardea cinerea*) at Two Spatial Scales. *Animals* **2024**, *14*, 2461. [CrossRef]
50. Petkov, N. Habitat characteristics assessment of the wetlands with breeding Ferruginous Duck *Aythya nyroca* and Pochard *A. ferina* in Bulgaria. *Acrocephalus* **2012**, *32*, 127–134. [CrossRef]
51. Vodopija, M.; Pavlinec, Ž.; Kralj, J. Factors affecting nest attendance in Common Terns. *Larus* **2024**, *59*, 63–82. [CrossRef]
52. Kralj, J.; Martinović, M.; Jurinović, L.; Szinai, P.; Sütő, S.; Preiszner, B. Geolocator study reveals east African migration route of Central European Common Terns. *Avian Res.* **2020**, *11*, 6. [CrossRef]
53. Alerstam, T.; Bäckman, J.; Grönroos, J.; Olofsson, P.; Strandberg, R.; Sjöberg, S. Migration of black terns *Chlidonias niger* and common terns *Sterna hirundo* between south Sweden and the Atlantic coast of Africa. *J. Avian Biol.* **2025**, *2025*, e03348. [CrossRef]
54. Heneberg, P. Soil particle composition of Eurasian Kingfishers' (*Alcedo atthis*) nest sites. *Acta Zool. Acad. Sci. Hung.* **2004**, *50*, 185–193. Available online: http://actazool.nhmus.hu/50/3/heneberg.pdf (accessed on 30 August 2025).
55. Delić, A.; Grlica, I.D. Birds of the Končanica fish-ponds, Croatia. *Nat. Croat.* **2003**, *12*, 63–91. Available online: https://hrcak.srce.hr/file/17568 (accessed on 30 August 2025).
56. Forman, D.W.; Brain, P.F. Reproductive strategies used by moorhens (*Gallinula chloropus*) colonizing an artificial wetland habitat in south Wales. *J. Nat. Hist.* **2004**, *38*, 389–401. [CrossRef]

57. Kosicki, J.Z.; Profus, P.; Dolata, P.T.; Tobółka, M. Food composition and energy demand of the White Stork *Ciconia ciconia* breeding population. In *The White Stork in Poland: Studies in Biology, Ecology and Conservation*; Tryjanowski, P., Sparks, T.H., Jerzak, L., Eds.; Bogucki Wydawnictwo Naukowe: Poznań, Poland, 2006.
58. Bocheński, M.; Jerzak, L. Behaviour of the White Stork Ciconia ciconia: A review. In *The White Stork in Poland: Studies in Biology, Ecology and Conservation*; Tryjanowski, P., Sparks, T.H., Jerzak, L., Eds.; Bogucki Wydawnictwo Naukowe: Poznań, Poland, 2006; pp. 295–324. Available online: https://www.researchgate.net/publication/303234014 (accessed on 5 August 2025).
59. Battisti, C.; Cento, M.; Circosta, A.; Coppola, M.; Muratore, S. Resurrecting seasonal dynamics in waterbirds after wetland restoration: Before-after monitoring highlights the role of a single dominant species. *Wetl. Ecol. Manag.* **2023**, *31*, 203–211. [CrossRef]
60. Wu, J.; Goodale, E.; Li, W.; Zou, F.; Zhang, Q. Patterns in the global diversity of mixed-species bird flocks in relation to environmental factors. *Biol. Divers.* **2024**, *1*, 44–53. [CrossRef]
61. Festetics, A.; Leisler, B. Ecology of waterfowl in the region of Lake Neusiedl, Austria, particularly in the World Wildlife Fund Seewinkel Reserve. *Wildfowl* **1968**, *19*, 83–95. Available online: https://www.nationalparkneusiedlersee.at/sites/default/files/1968-ecology_of_waterfowl_in_the_region_of_lake_neusiedl_festetics-a.pdf (accessed on 5 August 2025).
62. Ramsar Secretariat. Information Sheet on Ramsar Wetlands (RIS): Neusiedler See–Seewinkel (Site No. 271), Austria. Updated December 2005 (Originally Completed April 2003). Ramsar Sites Information Service. Available online: https://rsis.ramsar.org/ris/271 (accessed on 9 September 2025).
63. Heiler, G.; Hein, T.; Schiemer, F.; Bornette, G. Hydrological connectivity and flood pulses as the central aspects for the integrity of a river-floodplain system. *Regul. Rivers Res. Manag.* **1995**, *11*, 351–361. [CrossRef]
64. Tockner, K.; Stanford, J.A. Riverine flood plains: Present state and future trends. *Environ. Conserv.* **2002**, *29*, 308–330. [CrossRef]
65. Wenskus, F.; Hecht, C.; Horchler, P.; Januschke, K.; Rieland, G.; Scholz, M.; Weber, A.; Hering, D. Unravelling direct and indirect effects of river-floodplain connectivity on biodiversity: Insights from the Elbe River floodplains. *Biodivers. Conserv.* **2025**, *34*, 2829–2850. [CrossRef]
66. Hein, T.; Schwarz, U.; Habersack, H.; Nichersu, I.; Preiner, S.; Willby, N.; Weigelhofer, G. Current status and restoration options for floodplains along the Danube River. *Sci. Total Environ.* **2016**, *543*, 778–790. [CrossRef] [PubMed]
67. Junk, W.J.; An, S.; Finlayson, C.M.; Gopal, B.; Květ, J.; Mitchell, S.A.; Mitsch, W.J.; Robarts, R.D. Current state of knowledge regarding the world's wetlands and their future under global climate change: A synthesis. *Aquat. Sci.* **2013**, *75*, 151–167. [CrossRef]
68. Gopal, B. Biodiversity in wetlands. In *The Wetlands Handbook*; Wiley-Blackwell: Hoboken, NJ, USA, 2009; Volume 2, pp. 65–95. [CrossRef]
69. De Groot, D.; Brander, L.; Finlayson, C.M. Wetland ecosystem services. In *The Wetland Book*; Springer: Dordrecht, The Netherlands, 2018; pp. 323–333. [CrossRef]
70. Siuta, M.; Nedelciu, C.E. *Report on Socio-Economic Benefits of Wetland Restoration in Central and Eastern Europe*; CEEweb for Biodiversity: Budapest, Hungary, 2016. Available online: https://www.ceeweb.org/ducuments/publications/report_on_socio_economic_benefits_of_wetland_restoration_in_CEE.pdf (accessed on 30 August 2025).
71. ICPDR. Danube River Basin Management Plan—Update 2021. Available online: https://www.icpdr.org/tasks-topics/tasks/river-basin-management/danube-river-basin-management-plan-2021 (accessed on 30 August 2025).

diversity

MDPI

Article

Spatio-Temporal Distribution of *Limnomysis benedeni* Czerniavsky, 1882 and Other Invaders in the Danube Floodplain Kopački Rit Nature Park, Croatia

Viktorija Ergović [1], Miran Koh [1], Natalija Vučković [2], Mario Rumišek [2], Dubravka Čerba [3], Barbara Vlaičević [1] and Zlatko Mihaljević [2,*]

[1] Department of Biology, University of Osijek, Cara Hadrijana 8/a, 31000 Osijek, Croatia; viktorija.ergovic@biologija.unios.hr (V.E.); miran.koh@biologija.unios.hr (M.K.); barbara.vlaicevic@biologija.unios.hr (B.V.)

[2] Department of Biology, Faculty of Science University of Zagreb, Horvatovac 102a, 10000 Zagreb, Croatia; natalija.vuckovic@biol.pmf.unizg.hr (N.V.); mario.rumisek@biol.pmf.unizg.hr (M.R.)

[3] National Water Reference Laboratory of Slovakia, Department of Assessment and Aquatic Ecosystems Research, Water Research Institute, Nábr. arm. gen. L. Svobodu 5, 81249 Bratislava, Slovakia; dubravka.cerba@vuvh.sk

* Correspondence: zlatko.mihaljevic@biol.pmf.hr

Abstract

The Danube in Croatia serves as an important transport route but also favors the spread of invasive species, especially in the floodplain areas. Many of them originate from the Ponto-Caspian region and influence European ecosystems with their migrations. One of these species, *Limnomysis benedeni*, a mysid shrimp, thrives in shallow waters and plays a crucial role as a food source for fish. *L. benedeni* was first recorded in Croatia in 2004 in Lake Sakadaš (Kopački Rit). Prior to the study on aquatic macroinvertebrates in Kopački Rit Nature Park, conducted from July 2020 to July 2023, there had been no documented records in recent years. Sampling was carried out seasonally for macroinvertebrates and monthly for environmental parameters at 15 sites within the park or in the immediate vicinity. Samples were collected according to standard AQEM methodology. A total of 21 macroinvertebrate groups (407 taxa), out of which the most diverse were Diptera with 20 families, were identified in this study, including nine allochthonous species in addition to *L. benedeni*. The most abundant populations of *L. benedeni* were found in the Danube, the Petreš channel, and Vemeljski Dunavac channel, which supply the floodplain with water from the Danube. Most individuals were collected in summer and spring, with the highest density being 741 individuals per square meter. Environmental parameters such as water level, type of habitats, pH values, chemical oxygen demand, and phosphorus content, were statistically significant for the distribution of species. The dominant microhabitat for *L. benedeni* in Kopački Rit was argyllal in combination with coarse particulate organic matter and wood debris, and the composition of these microhabitats remained consistent throughout the seasons. *L. benedeni* was the only crustacean species to establish a stable population in the floodplain area, excluding *Asellus aquaticus* (water louse), a cosmopolitan species. The ongoing influence of *L. benedeni* on the native community still remains to be determined.

Keywords: allochthonous species; Mysidae shrimp; protected area; macrozoobenthos

1. Introduction

With a course length of around 140 km, the Danube is an important transport route in Croatia. It forms the border between Croatia and Serbia along almost the entire course, from the village of Batina upstream (1425 rkm) to the town of Ilok downstream (1298 rkm) [1]. However, this advantage of being a transport route also carries the risk of transferring allochthonous species, many of which become invasive in their new environment [2]. The main characteristics of invasiveness, such as a high reproductive rate of species, dominance over habitat, predation, and the spread of disease within the native community [3] emphasize the urgent need to address this issue.

A large number of allochthonous macroinvertebrate species that are widely distributed in European freshwaters belong to crustaceans, which are represented with 30 allochthonous species from class Malacostraca–superorder orders Amphipoda, Mysida, and Isopoda (21, 5, and 4 species, respectively) [4]. The brackish-water fauna of the Ponto-Caspian region has shown remarkable success in expanding its range northwards and westwards through the use of canals, rivers, or human transport. It is estimated that 40 different species have spread from the Ponto-Caspian region across Europe [5]. As many as 70% of the invasive species reported in inland waters of North America originate from this region [6,7]. On both continents, these invasions have been shown to have detrimental ecological and economic implications; in some cases, the newcomers altered entire populations and took over food chains [7–9].

The species *Limnomysis benedeni* Czerniavsky, 1882, commonly known as Caspian slender shrimp, Danube mysid or opossum shrimp, is the only described species of the genus *Limnomysis* from the family Mysidae. The natural habitat of this mysid shrimp is the shallow waters of the Ponto-Caspian region. It is found in the littoral zone of shallow, still waters at depths from 0.5 to 1 m [10,11]. It prefers habitats with densely developed submerged and emergent macrophytes or filamentous algae and wood debris. However, it is also frequently found on sheltered shores, in sandy sediment or in empty mussel shells [11–13]. Such habitat conditions also characterize the Kopački Rit floodplain [14,15].

L. benedeni is an active feeder during the day [12]. Its diet is diverse and consists of small organic particles, phytoplankton, epiphytes, detritus, and biofilm [11,16], and, therefore, it is considered omnivorous species [17,18]. This mysid crustacean plays an important role in the diet of omnivorous and carnivorous fish [10,19], which further emphasizes their ecological importance. *L. benedeni* is considered an invasive species because it can rapidly establish dense populations in newly colonized habitats, often outcompeting native macroinvertebrates (through abundance or biomass, alteration of habitat structure, disruption of food webs). Its ability to tolerate a wide range of environmental conditions [20,21] and to spread via interconnected waterways further increases its invasive potential [22].

Together with several other crustacean species, this mysid shrimp was cultivated in the former Soviet Union to be used as supplementary food in fisheries. Until the mid-1930s, there was no significant spread from the autochthonous area [23]. There are several theories about the possible routes of spread. One theory is that they traveled upstream by ship from the Danube Delta towards the North and Baltic Seas and then spread freely across rivers and lakes in Western Europe [12,22,24]. In the 1950s and 1960s, some specimens were introduced into Lake Balaton in Hungary [25,26], into reservoirs along the Dnieper in Ukraine and Belarus [27], and into the Kaunas Reservoir in Lithuania [28,29] as fish food. From the genome analysis of individuals, Audzijonyte et al. [30] concluded that there are at least three waves of spread across Western Europe via the Danube River. All three waves were triggered by human activities [12,24,31]. The first wave of spread started with the introduction of individuals into Lake Balaton, from where they spread via waterways to Slovakia, Austria [32], and Germany [22]. The population to which the individuals found in

Kopački Rit belong has yet to be investigated, but, according to the aforementioned study of fragments of the mitochondrial COI gene, the population from the Danube sampled in the town of Vukovar (in the immediate vicinity of the sampling area) has lineages consistent with individuals from the Danube and Dniester deltas in NW Black Sea coast [30].

The first studies on the impact of this species on the ecosystem in which it was introduced began in the mid-1980s [8,33,34], leading to the first conclusions about its invasiveness. Each new finding provides valuable data for understanding the distribution and invasive potential of this freshwater invertebrate. In European waters, it is considered a species with a moderately high detrimental impact on the native aquatic community [35].

Its first occurrence in Croatia was recorded in 2004 in Lake Sakadaš in the Kopački Rit Nature Park during a weekly sampling of epiphytic communities on the Eurasian watermilfoil (*Myriophyllum spicatum* Linnaeus, 1753) and common hornwort (*Ceratophyllum demersum* Linnaeus, 1753), in the summer months, using a plastic cylinder. Four individuals of *L. benedeni* were found in the community on *C. demersum*, and one individual was found on *M. spicatum* [36]. According to Lindén [37], *M. spicatum* and *C. demersum* act as a kind of repellent for the population of Mysidae species by secreting toxins. Bogut et al. [36] therefore concluded that a larger population could be expected in association with other macrophytes in Lake Sakadaš; however, no conclusion was drawn about the population size of *L. benedeni*. After recording the first occurrence, monitoring of the macrozoobenthos in Kopački Rit continued through various projects; however, there were no new occurrences recorded until 2020. The lack of detections in this period could be related to the low population density, the limited spatial coverage of the sampling area, or the possibility that the sampling methods used were not optimal for the detection of *L. benedeni*.

This study focuses on the distribution and population dynamics as well as the definition of the environmental conditions under which the invasive shrimp *L. benedeni* is present within Kopački Rit Nature Park floodplain. By linking these patterns to physical and chemical parameters, this research contributes to the assessments of habitat suitability that are of direct relevance to the management of invasive species in the wider Danube basin. The objective is also to identify the route of species introduction into the Kopački Rit area and assess its distribution, estimate the size of the population and determine the limits of its spread within the protected area. This research is part of a broader initiative that aims to manage and protect the biodiversity of the Kopački Rit Nature Park.

2. Materials and Methods

2.1. Study Area

The study area is located in the Kopački Rit, a floodplain created by the interaction of two large rivers, Drava and the Danube (Figure 1). Due to its exceptional natural value, it was included in the Ramsar List of Wetlands of International Importance in 1993. It consists of two permanent lakes, Lake Sakadaš and Kopačko lake, which are connected by the Čonakut channel. Kopačko lake is the largest lake in the floodplain with a depth ranging from 1.5 to 5 m, depending on the water level of the Danube, and surface area ranging from 2 to 2.5 km^2 [38]. The floodplain is an area of high biological productivity, and the flood intensity in the park depends on the water level of the Danube. According to Schwarz [14], floods are classified into three categories: at the Danube water levels between 3 and 3.5 m, the water begins to flow into the Kopački Rit; floods of medium intensity are at water levels between 3.5 and 4 m; and, above a water level of 4 m, at the village of Batina, the largest area of the park is underwater, and seasonal flooding occurs. The Hulovo channel, Vemeljski Dunavac, and Petreš are the main channels that are connecting the Danube with the Kopački Rit floodplain. Kopački Rit acts as a biodiversity hotspot, especially for invertebrate species [39,40]. The hydrological diversity of the park is reflected in the variety of water bodies that have formed in the floodplain, creating a mosaic of habitats that support high species richness [41,42]. However, the same hydrological connectivity, combined with anthropogenic pressures such as river regulation and nutrient input [43], facilitates the introduction and spread of invasive species, that alter native community structures and ecosystem functioning [44]. The anthropogenic impacts in the Kopački Rit floodplain are relatively high. Hydrotechnical activities, especially in the last century, such as the construction of embankments for flood protection [45], have significantly reduced the area of natural habitats. The largest embankment has divided the floodplain so that one part is in a near-natural state and the other part is disconnected from the Danube and dependent on precipitation [46]. In addition, three pumping stations, controlled by an extensive channel network, manage the water regime by diverting water for agricultural use. These interventions have changed the hydrological regime in the floodplain and affected its natural habitats [38,46]. In this study, the samples were taken in both sampling areas, the near-natural sites (sites with minimal human impact and where the water supply is dependent on the Danube water level), and the anthropogenically disturbed sites (e.g., dams and channels built by human intervention where the water supply is dependent on rainfall and agricultural runoff) (Table 1).

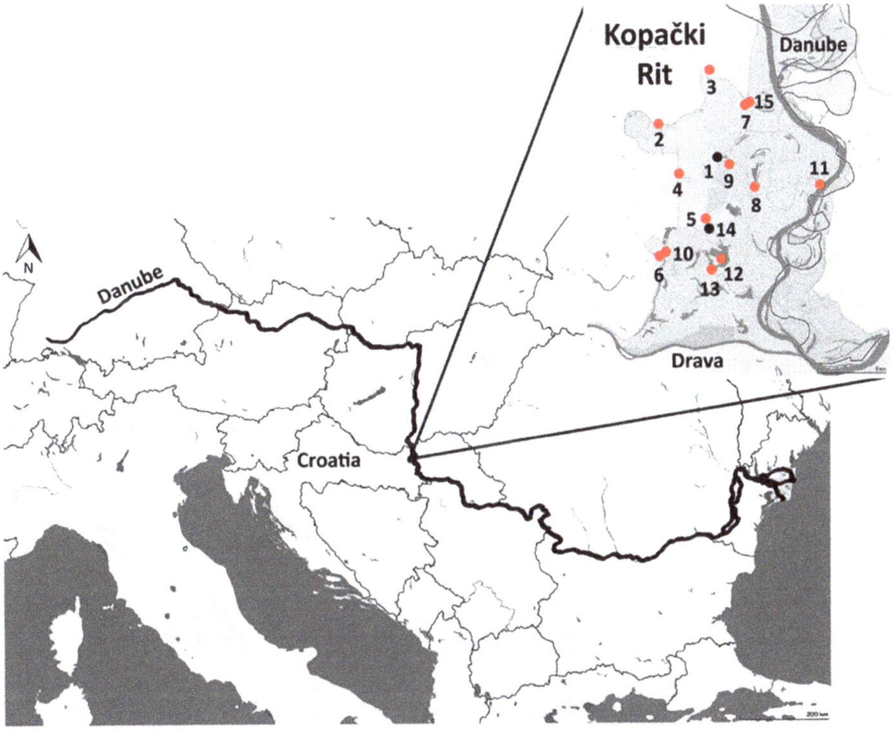

Figure 1. Research area, Kopački Rit Nature Park (locations within the park where *Limnomysis benedeni* was found are marked in red, and locations where *L. benedeni* is absent are marked in black). Sampling sites: 1—Tikveš dam; 2—Šuma Siget; 3—Čarna channel; 4—Podunavlje channel; 5—Podunavlje dam; 6—Kopačevo dam; 7—Zlatna Greda dam; 8—Semenča channel; 9—Vemeljski Dunavac channel; 10—Lake Sakadaš; 11—Danube, channel Petreš; 12—Kopačko lake; 13—Hulovo channel; 14—Četverokut channel; and 15—Zlatna Greda channel.

Table 1. Geographical location of the sampling sites in the Kopački Rit Nature Park, using HTRS 96 coordinates.

Site	Site Number	Site ID	Longitude	Latitude
Anthropogenically disturbed sites				
Tikveš dam	1	Td	683237.31	5061340.25
Šuma Siget	2	SS	678706.31	5063622.38
Čarna channel	3	Cch	682218.06	5067439.25
Podunavlje channel	4	Pch	680277.18	5060124.63
Podunavlje dam	5	Pd	682411.94	5057101.75
Kopačevo dam	6	Kd	679203.19	5054567.25
Zlatna Greda dam	7	ZGd	685015.69	5065153.63
Near-natural sites				
Semenča channel	8	Sch	685830.06	5059549.09
Vemeljski Dunavac channel	9	VDch	683915.42	5061136.08
Lake Sakadaš	10	Sl	679480.15	5054685.72
Danube, channel Petreš	11	DPch	690628.75	5059687.00
Kopačko lake	12	Kl	683582.60	5054400.12
Hulovo channel	13	Hch	682918.69	5053705.88
Četverokut channel	14	Ctch	682516.56	5056586.69
Zlatna Greda channel	15	ZGch	685373.14	5064995.20

2.2. Sampling

This study was conducted from July 2020 to July 2023. Macroinvertebrates were sampled seasonally, while environmental parameters were determined monthly at 14 sampling sites within the park and at one site in the Danube, next to the Petreš channel, which marks the entrance from the Danube to the Kopački Rit area (Table 1, Figure 1). Macroinvertebrate sampling and the collection of environmental parameters of water were usually carried out within the same week. Invertebrate samples were collected at all sites except lakes following the protocols [47]. The sampling method for lakes (Lake Sakadaš and Kopačko lake) is described in detail in Pozojević et al. [48]. Each composite sample consisted of subsamples corresponding to the proportion of different microhabitat types at each site. Sampling was carried out using a hand net with a frame diameter of 25 cm × 25 cm and a mesh size of 500 μm. Each sample was preserved in 96% ethanol. In the laboratory, all macroinvertebrates were isolated and identified. Individuals from the Mysidae family were separated and preserved in 70% ethanol. Isolation and identification were performed using a binocular microscope (Motic SMZ–171 (Motic, Hong Kong)). Each mysid shrimp was photographed with a Moticam 5 MP camera (Motic, Hong Kong) and identified according to the identification key of Eggers et al. [49].

The water level of the Danube indicates the water level in the floodplain. Due to the severe drought in 2022, no samples were taken at the Četverokut channel site during this period. Neither sampling was carried out during the flood phase, and the first sampling date was three weeks after the flood.

2.3. Environmental Conditions

Environmental parameters were measured in situ (water temperature and pH) at the time of invertebrate sampling using an HQ30d multimeter (Hach, Germany), and water transparency was assessed using the Secchi disk. In addition, surface water samples were taken monthly for laboratory analysis of the chemical properties of water and analyzed at the Croatian Institute of Public Health (chemical and biological oxygen demand, alkalinity and water hardness, chlorophyll a, chloride, silicate, ammonium, nitrite, nitrate, total nitrogen, total phosphorous, orthophosphate, total organic carbon, arsenic, chromium, copper and zinc content) using standard analytical procedures [50]. The water level data were obtained by Hrvatske vode, a legal entity for water management in Croatia, at the Batina measuring point.

2.4. Data Analysis

Data analyses were performed using Primer 6.0 [51], CANOCO 4.5 [52], and PAST 4.13 [53]. Both biological and environmental data were log-transformed, and environmental data were normalized prior to analysis. Bray–Curtis similarity was applied to form Resemblance matrices of abundance data, and Euclidean distance was used for the environmental data. Bray–Curtis incorporates species abundance, minimizes artificial similarities, and emphasizes differences between communities. In contrast, Euclidean distance captures absolute differences and reflects straight-line dissimilarities, making it suitable for detecting clear environmental gradients [51]. After transformation of data and creating Resemblance matrices, we applied BEST (Bio-Environmental Similarity Tool) with Spearman's rank correlation analysis [54] to calculate the best subset of the environmental data that correlates with the abundance data of individuals. The strength of Spearman's rank correlation and the statistical significance of the observed correlation were validated by a Monte Carlo permutation test with 999 permutations under the full permutation model.

A hierarchical cluster analysis using the group average linkage method [51] was performed to examine the pattern of similarities between samples of crustacean abundance. The method was used to merge the most similar samples by including percentages.

In addition, redundancy analysis (RDA) and distance-based redundancy analysis (dbRDA) were performed to identify the environmental variables explaining the variation in abundance, followed by a Monte Carlo permutation test with 999 permutations under the full permutation model. The collinearity between the environmental variables was considered by applying forward selection in CANOCO 4.5, which ensures that only non-redundant and statistically significant predictors are retained for the RDA [52].

To analyze the statistical differences between the different study sites and between the seasons, a Normality test was performed, followed by a Kruskal–Wallis H-test and a Dunn's post hoc test, which were performed using the PAST 4.13 software. For all analyses, the significance level was set at $p < 0.050$ [52].

3. Results

3.1. Alien Aquatic Macroinvertebrates

A total of 21 macroinvertebrate groups (407 taxa) were identified in 176 samples collected during the study. Dipterans were the most diverse group, represented with 20 different families. Macroinvertebrate allochthonous taxa, listed in the European Commission list of invasive species in Europe [55], were present in 96 samples. These were Mollusca: *Corbicula fluminea* (O.F. Müller, 1774), *Dreissena polymorpha* (Pallas, 1771), *Ferrissia californica* (Rowell, 1863), *Physella acuta* (Draparnaud, 1805); Crustacea: *Jaera istri* Veuille, 1979, *Chelicorophium curvispinum* (G.O. Sars, 1895), *Dikerogammarus villosus* (Sowinsky, 1894), *L. benedeni*; and Oligochaeta species: *Branchiodrilus hortensis* (Stephenson, 1910) and *Branchiura sowerbyi* Beddard, 1892.

The bivalve *C. fluminea* was present only in the Danube channel Petreš and Vemeljski Dunavac channel during the entire study period. *D. polymorpha* was found in the first sampling period, in the summer of 2020, at several sites, two studied lakes, and the Kopačevo dam, as well as at the Danube channel Petreš, together with the Vemeljski Dunavac channel. From this season until the end of the research, individuals were only found at the Kopačevo dam, with the exception of the Tikveš dam in the summer of 2021.

Most individuals of the species *F. californica* were found in all sampling seasons at the Kopačevo dam and in the Podunavlje channel, both anthropogenically disturbed sites. *P. acuta* was present at almost all sites, with the highest abundance at the Kopačevo dam (Supplement Table S1).

The Oligochaeta species *B. sowerbyi* was found only in the summer period of the first sampling year, at the Hulovo and Semenča channel sites, while *B. hortensis* had a wide distribution, with most individuals found at the Kopačevo and Tikveš dam sites, in spring and summer rather than in winter (Supplement Table S1).

3.2. Spatial and Seasonal Distribution of Crustaceans

The abundance of allochthonous crustacean species individuals was higher at near-natural sites (with minimal human impact) than at the anthropogenically disturbed sites (with man-made facilities) (Figure 2). The allochthonous species *C. curvispinum* and *D. villosus* were detected only once during the winter sampling period in 2022, at the Zlatna Greda dam in the floodplain, while they were present at the Danube channel Petreš and Vemeljski Dunavac channel for almost the entire study period. The species *J. istri* was only present at the Danube channel Petreš site. *Synurella ambulans* (F. Müller, 1846) was found sporadically at several sites but only in the spring and summer sampling periods. *Niphargus* sp. appeared only in spring 2021. Species *Asellus aquaticus* (Linnaeus, 1758) commonly known as water louse, was present at all sites in the floodplain area except Semenča channel and Kopačko lake sites, and no individuals were found at Vemeljski Dunavac channel and at the Danube channel Petreš sites.

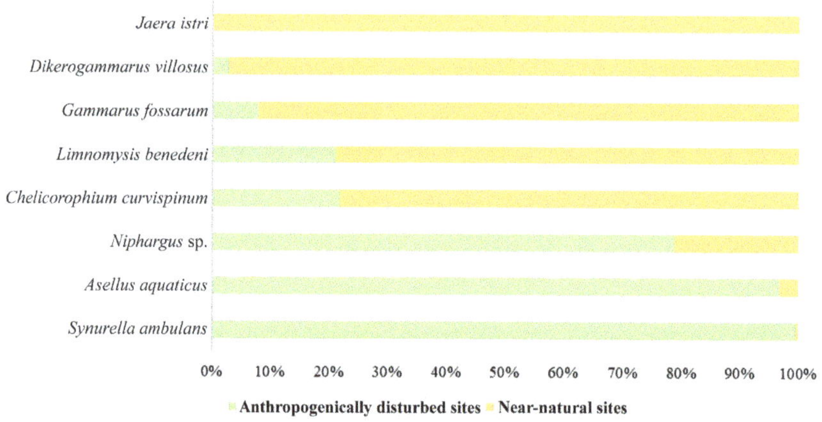

Figure 2. The abundance of recorded crustacean species expressed as a percentage ratio of two types of sites, near-natural and anthropogenically disturbed.

The native species *Gammarus fossarum* Koch, 1863, commonly known as mountain riverine amphipod, was found only three times, at Šuma Siget and Vemeljski Dunavac channel in winter of the first-year sampling (in small numbers) and at the Četverokut channel during autumn sampling of the following year. *Niphargus* sp. and *G. fossarum* are the only crustacean species found in the Četverokut channel.

The results of the hierarchical cluster analysis (Figure 3) revealed three main clusters based on the crustacean abundances, and the sites were grouped on the basis of their similarity. The clusters were defined at approximately 60% similarity ($p < 0.050$). Notably, the Četverokut channel (sampled in the autumn) was separated from all other sites, indicating a distinct community composition. The observed groupings reflect differences in species composition, which are likely due to habitat characteristics, with a clear distinction between anthropogenically disturbed and near-natural sites.

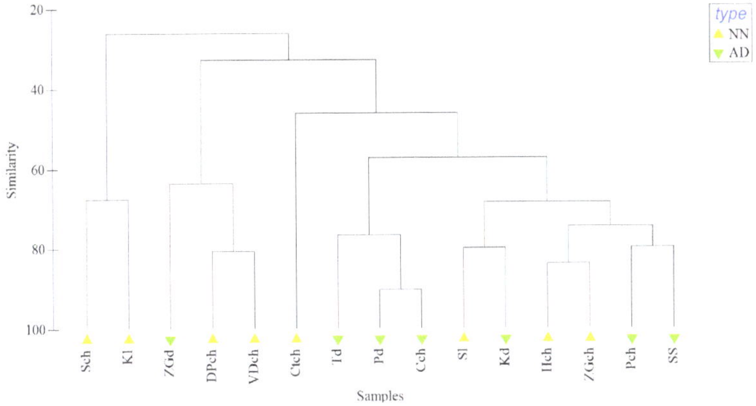

Figure 3. Hierarchical cluster analysis of the group average of crustacean species abundance in two types of sites. Abbreviations: AD—anthropogenically disturbed sites, NN—near-natural sites; site ID given in Table 1.

3.3. Spatial and Seasonal Distribution of *Limnomysis benedeni*

During sampling at 15 sites in the Kopački Rit Nature Park from July 2020 to July 2023, individuals of *L. benedeni* were found in 51 samples at 13 sites. Sites where no individuals were found were the Tikveš dam and Četverokut channel (Figure 1).

The site with the highest number of individuals collected per m^2 was the Vemeljski Dunavac channel with 741 ind./m^2, followed by the Danube channel Petreš with 540 ind./m^2. The lowest number of individuals collected was at Kopačko lake, with only two ind./m^2. *L. benedeni* individuals were most frequently found at the Danube channel Petreš and Vemeljski Dunavac channel sites, both with a frequency of occurrence of 11 times, a total of 12 samplings. At the following sites, *L. benedeni* was sampled only once in different sampling seasons: Čarna channel, Zlatna Greda channel, Zlatna Greda dam, and Podunavlje dam (Table 2).

Table 2. Number of *Limnomysis benedeni* individuals at the sampling sites in Kopački Rit during different seasons (numbers represent the number of individuals per square meter of area). Abbreviations of the sites are listed in Table 1: AD—anthropogenically disturbed sites, NN—near-natural sites. *—Sampling was not conducted.

Site/Season	Summer	Autumn	Winter	Spring	Summer	Autumn	Winter	Spring	Summer	Autumn	Winter	Spring
AD	2020			2021				2022				2023
Td	0	0	0	0	0	0	0	0	0	0	0	0
SS	0	0	10	19	0	0	0	0	0	0	0	0
Cch	0	0	0	38	0	0	0	0	0	0	0	0
Pch	0	0	0	172	7	0	0	26	0	0	0	13
Pd	0	0	0	0	0	0	0	0	28	0	0	0
Kd	250	39	4	0	13	0	0	0	38	0	0	64
ZGd	0	0	0	0	0	0	240	0	0	0	0	0
NN												
Sch	0	0	0	24	20	0	0	0	0	0	0	0
VDch	741	102	16	240	20	32	22	64	64	58	0	179
Sl	82	33	13	0	0	0	0	0	0	0	3	32
DPch	520	62	74	432	34	61	0	138	288	11	13	24
Kl	4	0	2	0	0	0	0	0	0	0	0	0
Hch	0	88	16	48	7	0	0	0	0	0	0	0
Ctch	0	0	0	0	0	0	*	*	*	*	0	0
ZGch	0	0	0	19	0	0	0	0	0	0	0	0

Division of the sampling sites into near-natural and anthropogenically disturbed revealed that 79% of the sampled individuals of *L. benedeni* were found in near-natural sites (Figure 2).

In terms of seasonality, the highest abundance was recorded in summer and spring (2116 and 1532 ind./m^2, respectively), rather than in autumn and winter (486 and 413 ind./m^2, respectively) (Table 2).

3.4. The Relationship Between Environmental Parameters and Crustacean Species

The result of the RDA and Monte Carlo permutation test, habitat type (F = 40.46, p = 0.001), chemical oxygen demand values (F = 8.87, p = 0.001), pH values (F = 8.69, p = 0.011), phosphorus content (F = 3.78, p = 0.016), and total Danube water level (F = 3.61, p = 0.009) were statistically significant environmental parameters for the distribution of crustaceans in the Kopački Rit floodplain. The RDA plot (Figure 4) shows that invasive crustacean species were more abundant at sites with higher total phosphorus content, pH values, and specific habitat type, with the occurrence of *D. villosus* being most strongly influenced by habitat type.

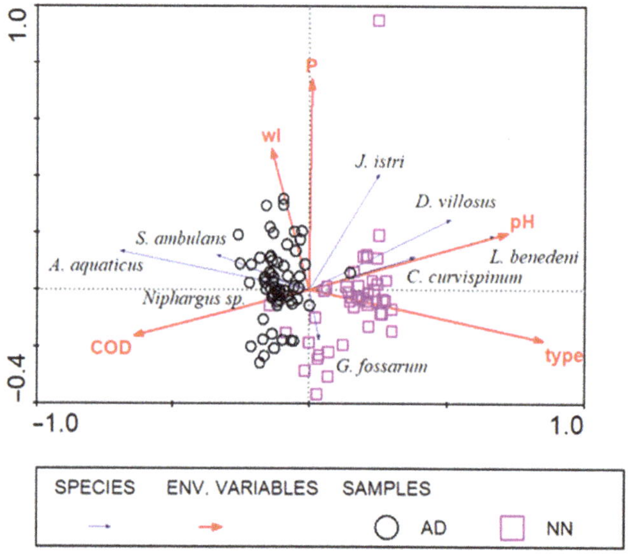

Figure 4. RDA plot indicating statistically significant environmental variables for Crustacea species. Abbreviations: wl—water level of Danube (cm), P—phosphorus content (mg P/L), type—habitat type (anthropogenically disturbed (AD) and near-natural (NN)), and COD—chemical oxygen demand content (mg O_2/L).

The dominant substrate type in the Kopački Rit and the most frequently sampled substrate at the sites where *L. benedeni* was found is argyllal (inorganic particle size of < 6 µm) in combination with CPOM (coarse particulate organic matter), woody debris (dead parts of terrestrial plants), or phytal (algae, mosses, and macrophytes including living parts of terrestrial plants) components. At the study sites Danube channel Petreš, Vemeljski Dunavac channel, and Hulovo channel, the dominant substrates were argyllal and psammal (particle sizes of 6 µm–0.2 cm). The composition of the sampled substrates remained unchanged from year to year.

The highest water level of the Danube measured at the village of Batina was 398 cm at the time of sampling in spring 2023, while the lowest was −18 cm at the time of sampling in autumn 2021. As a result of the Kruskal–Wallis H-test and Dunn's post hoc test, there was no statistical difference between the study sites in terms of the abundance of Crustacea individuals, nor was there a statistically significant difference between the seasons. The Vemeljski Dunavac channel site differed statistically the most from the other sites (Table 3).

Table 3. Values of Kruskal–Wallis H test (H-value) and Dunn's post hoc test among groups of different sites and groups of different seasons. Abbreviation: df—degrees of freedom; *p*-value—statistical significance.

	H-Value	df	*p*-Value	Dunn's Post Hoc Test
sampled sites	9.919	14	0.517	Sch ≠ VDch; Kl ≠ VDch
sampled seasons	0.257	3	0.968	not significant

The Spearman correlation of the best subset of environmental data with the abundance of crustaceans yielded a result of 0.524 ($p < 0.050$), with significant relationships found for habitat type (anthropogenically disturbed or near-natural), pH, total organic carbon (TOC), and nitrate ion (NO_3^-) content. Together, these environmental variables can explain 99.7% of species abundance (Figure 5).

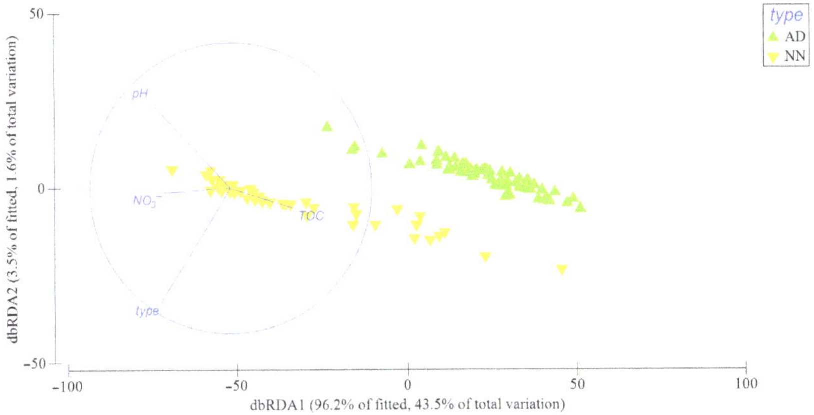

Figure 5. dbRDA of the most strongly correlated environmental variables with the abundance of Crustacea in the entire study. Abbreviations: TOC—total organic carbon, NO_3^-—nitrate ion concentration, type—habitat categories, AD—anthropogenically disturbed sites, and NN—near-natural sites.

4. Discussion

All invasive species recorded in the present study originated from the Ponto-Caspian region [56]. *L. benedeni* was the only crustacean species that could be considered a permanent inhabitant of the studied floodplain, apart from *A. aquaticus*, a cosmopolitan species [57]. *A. aquaticus* was the dominant autochthonous species in the entire study.

Alien species become invasive through a combination of biological characteristics, adaptability, and human impact [58]. Although the first record of *L. benedeni* in the nature park was in 2004, no new data on the populations' size was available, and the ecological consequences of its invasiveness in Kopački Rit are still unclear. It does not necessarily pose a threat to the autochthonous fauna, but, on the other hand, it can become a major problem. The trophic impact of *L. benedeni* could be context-dependent. Mesocosm experiments has shown that it significantly suppresses zooplankton biomass in Cladocera-dominated communities, especially by predation on *Daphnia* sp., while it has little effect on Copepoda [59]. In eutrophic systems, a decrease in zooplankton grazing relaxes top-down control, which typically increases phytoplankton biomass, prolongs the duration of bloom, reduces water clarity, and increases the deposition of labile organic matter that can intensify anoxic conditions [60]. Meanwhile, the introduction of another mysid shrimp, *Mysis diluviana*, Audzijonyte and Väinölä, 2005 [61], to oligotrophic Flathead Lake restructured the food web by reducing large zooplankton grazers, increasing phytoplankton biomass, and altering energy transfer, which facilitated the spread of non-native lake trout (*Salvelinus namaycush* (Walbaum, 1792)) while driving the decline of native species such as zooplanktivorous kokanee salmon (*Oncorhynchus nerka* (Walbaum, 1792)); westslope cutthroat (*Oncorhynchus clarkii lewisi* (Girard, 1856)); and bull trout (*Salvelinus confluentus* (Suckley, 1859)) [61].

From a broader ecological perspective, according to Vuić et al. [62], there are 24 fish species from 10 fish families in the floodplain. The species found in that study [62], such as pumpkinseed sunfish (*Lepomis gibbosus* Linnaeus, 1758), which only occurs in small numbers, and the bleak (*Alburnus alburnus* Linnaeus, 1758), the bream (*Blicca bjoerkna* Linnaeus, 1758), and the Prussian carp (*Carassius gibelio* Bloch, 1782), which occur in large numbers in Kopački Rit, are considered omnivorous and insectivorous fish, and we assume that individuals of *L. benedeni* could be part of their diet. The dense populations of fish species that may feed on *L. benedeni* individuals may result in the fish population keeping the mysid shrimp population at a relatively low abundance [63]. Observed ecological interactions with other invertebrates show what a threat this mysid can pose. For example, in the estuaries of rivers in the Mediterranean area, there is potential competition between *L. benedeni* and species of the genus *Diamysis*, leading to the displacement of native species [16]. In similar studies in the Rhine, it was observed that many native representatives of the macrozoobenthos disappeared with the appearance of invasive invertebrates from the Crustacea subphylum, including *L. benedeni* [64].

The Kopački Rit floodplain is a well-studied protected area [15,40,62,65–68], which has been partly shaped by dams and embankments built for flood protection in the last century. Despite these modifications, part of the Kopački Rit is still connected to the Danube, and regular spring floods occur due to the snowmelt [44,69]. During floods, riverine species have open passages to enter the protected area. Some of these species are alien species that have the potential to spread throughout the nature park via channels. The bivalve *D. polymorpha* has been present in the Čonakut and Hulovo channels since 2003 continuously to the present day [36,70,71]. The diatom *Didymosphenia geminata* (Lyngbye) Mart.Schmidt, 1899, was first observed in 2006 during a major flood in the Kopački Rit. It is an oligotrophic species that is sometimes known to inhabit meso- and eutrophic lowland rivers and reaches the Kopački Rit also via the Danube [72,73]. The invasive cyanobacterium *Raphidiopsis raciborskii* (Wołoszyńska, 1912) Aguilera et al., 2018 (synonym for *Cylindrospermopsis raciborskii* (Wołoszyńska) Seenayya and Subba Raju, 1972) [74], which has been present since 2003, is associated with the global increase in water temperature in lowland lakes, including Lake Sakadaš [75].

It is well known that, when alien species invade a new area, they disturb the substrate. By actively moving, they disturb the upper part of the sandy sediment and make the substrate unfavorable for the native fauna [76]. A high tolerance of alien species to changes in climatic conditions is crucial for their ability to alter the community of native species composition and reduce biodiversity [3]. Their ability to survive in new environments often has a significant negative impact on the environment, biodiversity loss, and the economy, as they disrupt regional ecosystems and displace native species [77,78]. But not every alien species is invasive. High reproduction rates, the absence of natural predators, efficient dispersal mechanisms, competition for available habitats, and even the transmission of diseases to native species [79] are characteristics of invasive species. Some invasive species, such as freshwater gastropods, can serve as intermediate hosts for various parasitic larvae that can infect fish and cause economic problems [80]. Others, such as *C. fluminea* and *D. polymorpha*, can even cause problems by contaminating the water channels of factories and power plants with their overgrowth which has a significant impact on factory revenues [81]. To control and mitigate the impact of invasive species, prevention and control systems need to be improved. In the European Union, EUR 12.5 billion are spent annually to reduce the damage caused by alien plant and animal species [82].

Four pumping stations were built in the study area as a protection against flooding. The purpose of all stations is to pump water from the Kopački Rit into channels in the surrounding area, and only one of them can work into two ways (pumping water in the Kopački Rit area and from the Kopački Rit area into the Danube). All four pumping stations were sampled during this study, and only one of them (Tikveš dam) was found to be free of *L. benedeni*. According to Hrvatske vode, a legal entity for water management, only the pumping station at the Kopačevo dam was opened and closed several times during the sampling period. This information is valuable for the clarification of the spread and definition of the influence of *L. benedeni* in this area. From this information, we can conclude that invasive species *L. benedeni* is present in all areas of Kopački Rit, with the exception of Tikveš. The Tikveš dam is located in the area supplied with water that must first pass-through previous pumping stations (either from the north or from the south), and none of the pumping stations were open during the sampling period. *L. benedeni* is known to spread throughout the Danube [12,22,24] and has entered the floodplain in this way [36]. However, no correlation with the water level of the Danube was found, but the water level was a statistically significant parameter, leading us to assume that *L. benedeni* originated from the Danube and became a permanent inhabitant of the floodplain, invading deeper and deeper into the protected area together with other allochthonous species. Furthermore, *L. benedeni* was detected at most sites and during the entire sampling period, making it the only crustacean species that can be considered a permanent inhabitant of the Kopački Rit floodplain.

The sporadic occurrence of the taxa *S. ambulans* and *Niphargus* sp., only during the spring and summer, is most likely related to the increased water level of the Danube and the subsequent flooding. These species are known as inhabitants of both the groundwater and surface water; therefore, it is likely that shifts in water level are the reason of their presence in the surface water. The native species *G. fossarum* was found only on one sampling occasion in the two near-natural sites, at the Četverokut channel and Semenča channel sites. The Četverokut channel was the shallowest site in the entire study, and it dried out due to a severe drought in the floodplain.

Isopod (*J. istri*) and amphipod species (*D. villosus* and *C. curvispinum*) co-occurred with the mysid *L. benedeni* in the Danube, channel Petreš and the Vemeljski Dunavac channel. These two main channels supply floodplain with water from the Danube, and these species are known to inhabit the Danube littoral zone [83,84].

The correlation between high phosphorus content and invasive crustacean species can be explained by the fact that invasive species thrive better under nutrient-rich conditions, especially Bivalvia species [85,86]. The area surrounding the nature park is partly covered by agricultural fields and drainage channels. The influence of drainage channels on the nutrient intake in the nearest lakes or rivers has been well studied [38,87,88]. Most of the nutrient input in floodplain lakes is due to drainage from agricultural fields, which is why lakes are in a eutrophic state [89]. The RDA plot shows that the invasive crustacean species clustered in the same direction, with habitat type and pH, as well as with increase in phosphorus levels, indicating a positive correlation with nutrient enrichment and specific habitat and chemical conditions. Furthermore, the results of hierarchical clustering supported the hypothesis that invasive species in the area of the floodplain came from the Danube, and in the floodplain are clustered depending on the type of habitats, anthropogenically disturbed or near-natural which is connected with nutrient content. This suggests that such environmental gradients facilitate the establishment and spread of invasive taxa, possibly due to their ecological flexibility and tolerance to eutrophic habitats [42]. According to Vilenica et al. [68] anthropogenically disturbed sites are sites with more densely developed macrophytes compared to near-natural habitats with no or only scarcely developed vegetation. This finding indicates that the presence of vegetation is not necessarily a habitat preference for Mysidae species in new environments.

The seasonal dynamics of Oligochaeta species can be attributed to several ecological factors. Temperature appears to be a key driver, as both *B. sowerbyi* and *B. hortensis* are warm-water adapted species that exhibit accelerated growth and reproduction at elevated temperatures [90,91]. This explains their higher abundance in spring and summer and their near absence in the colder winter months. In addition to thermal dependence, organic matter availability also has a strong influence on their distribution. Oligochaetes typically thrive in fine sediments enriched with detritus, and, in this study, the dominant substrates were argyllal in combination with coarse organic matter, conditions that favor deposit-feeding taxa [91]. Increased nutrient concentrations, particularly phosphorus and total organic carbon, likely increased food availability and promoted higher densities at dam sites where organic accumulation is greater. Despite its tolerance to low-oxygen conditions, *B. sowerbyi* was only detected during the first summer sampling at two sites, suggesting that extreme hypoxia during stratification or decomposition events may have limited its persistence. Hydrological connectivity also plays a role, as channels such as Hulovo and Semenča are directly connected to the Danube and provide pathways for dispersal during high-water periods. However, fluctuating hydrological regimes in subsequent years may have restricted the colonization and prevented *B. sowerbyi* from maintaining stable populations [91].

In the Croatian sections of the Drava river, which is a major tributary of the Danube in Croatia, 10 invasive species were recorded, while, in the Danube, there are 13 invasive species, with the most numerous being the amphipods *D. villosus* and *C. curvispinum* and the most widespread Mollusca being *D. polymorpha* [44,71,92]. These species have been recorded in the current research at the site Danube channel Petreš. In the study by Žganec et al. [44], which was carried out along the course of the Sava River from Slovenia to Serbia at 61 sites, the occurrence of *L. benedeni* was found in the middle and lower sections of the Sava River, along with a recorded presence in the Danube in the city of Vukovar [30]. Due to its invasive potential, species *L. benedeni* can significantly alter the structure of native zooplankton communities, and mysid introductions can affect entire aquatic ecosystems [17].

5. Conclusions

The presence of allochthonous species in the Kopački Rit Nature Park is inevitable due to the Danube regime, and the continued presence of some species indicates successful integration in local communities. This study provides new data on the distribution and populations' size of the allochthonous species *L. benedeni* in the Kopački Rit floodplain, in the eastern part of Croatia. The presence of the species is not seasonally influenced, and we can conclude that this invasive shrimp is now a permanent resident of the Kopački Rit Nature Park. The new data obtained from this study indicate the need for future monitoring of the macrozoobenthos diversity and abundance in order to determine the interactions between *L. benedeni* and native fauna, as well as potential for this species to alter food webs and ecosystem health and functioning in this Danube floodplain.

Supplementary Materials: The following supporting information can be downloaded at https://www.mdpi.com/article/10.3390/d17090630/s1, Table S1: List of allochthonous species in the floodplain Kopački Rit recorded in research.

Author Contributions: Conceptualization, Z.M. and D.Č.; methodology and investigation, V.E., M.K., N.V., M.R., D.Č., B.V. and Z.M.; data curation, V.E., M.K., N.V. and M.R.; writing—original draft preparation, V.E. and Z.M.; writing—review and editing, V.E., M.K., N.V., M.R., D.Č., B.V. and Z.M.; project administration, Z.M. and D.Č. All authors have read and agreed to the published version of the manuscript.

Funding: This study was part of a project called "NATURAVITA—Monitoring of water status, groundwater, recent sedimentation, habitats and fauna; Monitoring of ecological status of surface waters and monitoring of additional biological indicators" funded by Hrvatske vode, a legal entity for water management in Croatia; project number: 22-163/20.

Data Availability Statement: Data are available upon request from the authors.

Acknowledgments: The research was funded by the European Structural and Investment Funds, as part of the Croatian Waters monitoring project and systematic testing of biological quality elements under the Naturavita project in Kopački Rit Nature Park. The authors would like to thank the Croatian Institute of Public Health for providing data on chemical water analysis. Special thanks to colleagues from the Public Institution Kopački Rit Nature Park and the Faculty of Science, Department of Biology, University of Zagreb, for their help in field and laboratory work.

Conflicts of Interest: The authors declare no conflicts of interest.

References

1. Ozimec, S.; Topić, J.; Janauer, G. Inventory of Macrophytes and Habitats along the River Danube in Croatia. In Proceedings of the 38th IAD Conference, Dresden, Germany, 22–25 June 2010; pp. 1–5.
2. Tockner, K.; Zarfl, C.; Robinson, C.T. *Rivers of Europe*, 2nd ed.; Elsevier: Amsterdam, The Netherlands, 2022.
3. Crowl, T.A.; Crist, T.O.; Parmenter, R.R.; Belovsky, G.; Lugo, A.E. The Spread of Invasive Species and Infectious Disease as Drivers of Ecosystem Change. *Front. Ecol. Environ.* **2008**, *6*, 238–246. [CrossRef]
4. Holdich, D.M.; Pöckl, M. Invasive Crustaceans in European Inland Waters. In *Biological Invaders in Inland Waters: Profiles, Distribution, and Threats*; Springer: Dordrecht, The Netherlands, 2007; pp. 29–75. [CrossRef]
5. Leppäkoski, E.; Olenin, S. The Meltdown of Biogeographical Peculiarities of the Baltic Sea: The Interaction of Natural and Man-Made Processes. *Ambio* **2001**, *30*, 202–209. [CrossRef]
6. Chess, D.W.; Stanford, J.A. Comparative Energetics and Life Cycle of the Opossum Shrimp (*Mysis relicta*) in Native and Non-Native Environments. *Freshw. Biol.* **1998**, *40*, 783–794. [CrossRef]
7. Ricciardi, A.; MacIsaac, H.J. Recent Mass Invasion of the North American Great Lakes by Ponto–Caspian Species. *Trends Ecol. Evol.* **2000**, *15*, 62–65. [CrossRef] [PubMed]
8. Ketelaars, H.A.M.; Lambregts-van de Clundert, F.E.; Carpentier, C.J.; Wagenvoort, A.J.; Hoogenboezem, W. Ecological Effects of the Mass Occurrence of the Ponto–Caspian Invader, *Hemimysis anomala* G.O. Sars, 1907 (Crustacea: Mysidacea), in a Freshwater Storage Reservoir in The Netherlands, with Notes on Its Autecology and New Records. *Hydrobiologia* **1999**, *394*, 233–248. [CrossRef]
9. Jazdzewski, K.; Konopacka, A.; Grabowski, M. Recent Drastic Changes in the Gammarid Fauna (Crustacea, Amphipoda) of the Vistula River Deltaic System in Poland Caused by Alien Invaders. *Divers. Distrib.* **2004**, *10*, 81–87. [CrossRef]
10. Kelleher, B.; Van Der Velde, G.; Wittmann, K.J.; Faasse, M.A. Current Status of the Freshwater Mysidae in The Netherlands, with Records of *Limnomysis benedeni* Czerniavsky, 1882, a Pontocaspian Species in Dutch Rhine Branches. *Bull. Zoölogisch Mus.* **1999**, *16*, 89–94.
11. Gergs, R.; Hanselmann, A.J.; Eisele, I.; Rothhaupt, K.O. Autecology of *Limnomysis benedeni* Czerniavsky, 1882 (Crustacea: Mysida) in Lake Constance, Southwestern Germany. *Limnologica* **2008**, *38*, 139–146. [CrossRef]
12. Wittmann, K.J. Zur Einwanderung Potamophiler Malacostraca in Die Obere Donau: *Limnomysis benedeni* (Mysidacea), *Corophium curvispinum* (Amphipoda) und *Atyaephyra desmaresti* (Decapoda). *Lauterbornia* **1995**, *20*, 77–85.
13. van Riel, M.C.; van der Velde, G.; de Vaate, A.B. Dispersal of Invasive Species by Drifting. *Curr. Zool.* **2011**, *57*, 818–827. [CrossRef]
14. Schwarz, U. *Landschaftsökologische Charakterisierung des Kopački rit Unter Besonderer Berücksichtigung von Flusslandschaftsformen Sowie Deren Genese und Typologie*; Universität Wien: Wien, Austria, 2005.

15. Čerba, D.; Koh, M.; Vlaičević, B.; Turković Čakalić, I.; Milošević, D.; Stojković Piperac, M. Diversity of Periphytic Chironomidae on Different Substrate Types in a Floodplain Aquatic Ecosystem. *Diversity* **2022**, *14*, 264. [CrossRef]
16. Wittmann, K.J.; Ariani, A.P. *Limnomysis benedeni*: Mysidace' Ponto-Caspien Nouveau Pour Les Eaux Douces de France (Crustacea, Mysidacea). *Vie Milieu* **2000**, *50*, 117–122.
17. Fink, P.; Kottsieper, A.; Heynen, M.; Borcherding, J. Selective Zooplanktivory of an Invasive Ponto-Caspian Mysid and Possible Consequences for the Zooplankton Community Structure of Invaded Habitats. *Aquat. Sci.* **2012**, *74*, 191–202. [CrossRef]
18. Hanselmann, A.J.; Hodapp, B.; Rothhaupt, K.O. Nutritional Ecology of the Invasive Freshwater Mysid *Limnomysis benedeni*: Field Data and Laboratory Experiments on Food Choice and Juvenile Growth. *Hydrobiologia* **2013**, *705*, 75–86. [CrossRef]
19. Mordukhai-Boltovskoi, P.D. Composition and Distribution of Caspian Fauna in the Light of Modern Data. *Int. Rev. Gesamten Hydrobiol. Hydrogr.* **1979**, *64*, 1–38. [CrossRef]
20. Reid, D.F.; Orlova, M.I. Geological and Evolutionary Underpinnings for the Success of Ponto-Caspian Species Invasions in the Baltic Sea and North American Great Lakes. *Can. J. Fish. Aquat. Sci.* **2002**, *59*, 1144–1158. [CrossRef]
21. Ovčarenko, I.; Audzijonyte, A.; Gasiunaite, Z.R. Tolerance of *Paramysis lacustris* and *Limnomysis benedeni* (Crustacea, Mysida) to Sudden Salinity Changes: Implications for Ballast Water Treatment. *Oceanologia* **2006**, *48*, 231–242.
22. Wittmann, K.J.; Ariani, A.P. Reappraisal and Range Extension of Non-Indigenous Mysidae (Crustacea, Mysida) in Continental and Coastal Waters of Eastern France. *Biol. Invasions* **2009**, *11*, 401–407. [CrossRef]
23. Bacesco, M. Les Mysidacés des eaux Roumaines. *Ann. Sci. Univ. Jassy* **1940**, *26*, 453–804.
24. Wittmann, K.J. Continued Massive Invasion of Mysidae in the Rhine and Danube River Systems, with First Records of the Order Mysidacea (Crustacea: Malacostraca: Peracarida) for Switzerland. *Rev. Suisse Zoologie* **2007**, *114*, 65–86. [CrossRef]
25. Woynárovich, E. Vorkommen Der *Limnomysis benedeni* Czern. Im Ungarischen Donauabschnitt. *Acta Zool. Acad. Sci. Hung.* **1955**, *1*, 178–185.
26. Grigorovich, I.A.; MacIsaac, H.J.; Shadrin, N.V.; Mills, E.L. Patterns and Mechanisms of Aquatic Invertebrate Introductions in the Ponto-Caspian Region. *Can. J. Fish. Aquat. Sci.* **2002**, 1189–1208. [CrossRef]
27. Pligin, Y.V.; Yemelyanova, L.V. Acclimatization of Caspian Invertebrates in the Dnieper Reservoirs. *Hydrobiol. J.* **1989**, *25*, 1–9.
28. Olenin, S.; Leppäkoski, E. Non-Native Animals in the Baltic Sea: Alteration of Benthic Habitats in Coastal Inlets and Lagoons. *Hydrobiologia* **1999**, *393*, 233–243. [CrossRef]
29. Arbaciauskas, K. Ponto-Caspian Amphipods and Mysids in the Inland Waters of Lithuania: History of Introduction, Current Distribution and Relations with Native Malacostracans. In *Invasive Aquatic Species of Europe. Distribution, Impacts and Management*; Leppäkoski, E., Gollasch, S., Olenin, S., Eds.; Springer: Dordrecht, The Netherlands, 2002; pp. 104–115. [CrossRef]
30. Audzijonyte, A.; Wittmann, K.J.; Ovčarenko, I.; Väinölä, R. Invasion Phylogeography of the Ponto-Caspian Crustacean *Limnomysis benedeni* Dispersing across Europe. *Divers. Distrib.* **2009**, *15*, 346–355. [CrossRef]
31. Reinhold, M.; Tittizer, T. *Limnomysis benedeni* Czerniavsky 1882 (Crustacea: Mysidacea), Ein Weiteres Pontokaspisches Neozoon Im Main-Donau-Kanal. *Lauterbornia* **1998**, *33*, 37–40.
32. Weish, P.; Türkay, M. *Limnomysis benedeni* in Austria with Considerations on Its Colonization History (Crustacea: Mysidacea). *Integr. Landsc. Catchment Perspect. Ecol. Manag.* **1975**, *5*, 480–491. [CrossRef]
33. Rieman, B.E.; Falter, C.M. Effects of the Establishment of *Mysis relicta* on the Macrozooplankton of a Large Lake. *Trans. Am. Fish. Soc.* **1981**, *110*, 613–620. [CrossRef]
34. Furst, M.; Hammar, J.; Hill, C. The Introduction of *Mysis relicta* in Sweden: Effects on Fish in Impounded Lakes. In *Habitat Modification and Freshwater Fisheries*; Alabaster, J.S., Ed.; Food and Agriculture Organization of the United Nations: London, UK, 1985; pp. 202–206.
35. Ojaveer, H.; Leppäkoski, E.; Olenin, S.; Ricciardi, A. Ecological Impact of Ponto-Caspian Invaders in the Baltic Sea, European Inland Waters and the Great Lakes: An Inter-Ecosystem Comparison. In *Invasive Aquatic Species of Europe. Distribution, Impacts and Management*; Springer: Dordrecht, The Netherlands, 2002; pp. 412–425. [CrossRef]
36. Bogut, I.; Galir, A.; Čerba, D.; Vidaković, J. The Ponto-Caspian invader, *Limnomysis benedeni* (Czerniavsky, 1882), a new species in the fauna of Croatia. *Crustaceana* **2007**, *80*, 817–826. [CrossRef]
37. Lindén, E. *Antipredator Behaviour of Baltic Planktivores*; Faculty of Biosciences of the University of Helsinki: Helsinki, Finland, 2006; pp. 1–58.
38. Tadić, L.; Bonacci, O.; Dadić, T. Dynamics of the Kopački Rit (Croatia) Wetland Floodplain Water Regime. *Environ. Earth Sci.* **2014**, *71*, 3559–3570. [CrossRef]
39. Sommerwerk, N.; Hein, T.; Schneider-Jacoby, M.; Baumgartner, C.; Ostojić, A.; Siber, R.; Bloesch, J.; Paunović, M.; Tockner, K. The Danube River Basin. In *Rivers of Europe*; Elsevier: Amsterdam, The Netherlands, 2009; pp. 59–112. [CrossRef]
40. Čerba, D.; Vlaičević, B.; Davidović, R.A.; Koh, M.; Ergović, V.; Turković Čakalić, I. Chironomidae in Shallow Water Bodies of a Protected Lowland Freshwater Floodplain Ecosystem. *Sci. Prog.* **2023**, *106*, 1–30. [CrossRef]
41. Museth, J.; Johnsen, S.I.; Walseng, B.; Hanssen, O.; Erikstad, L. Managing Biodiversity of Floodplains in Relation to Climate Change. *Int. J. Clim. Chang. Strategy Manag.* **2011**, *3*, 402–415. [CrossRef]

42. Kao, Y.C.; Adlerstein, S.; Rutherford, E. The Relative Impacts of Nutrient Loads and Invasive Species on a Great Lakes Food Web: An Ecopath with Ecosim Analysis. *J. Great Lakes Res.* **2014**, *40*, 35–52. [CrossRef]
43. Tockner, K.; Stanford, J.A. Riverine Flood Plains: Present State and Future Trends. *Environ. Conserv.* **2002**, *29*, 308–330. [CrossRef]
44. Žganec, K.; Ćuk, R.; Tomović, J.; Lajtner, J.; Gottstein, S.; Kovačević, S.; Hudina, S.; Lucić, A.; Mirt, M.; Simić, V.; et al. The Longitudinal Pattern of Crustacean (Peracarida, Malacostraca) Assemblages in a Large South European River: Bank Reinforcement Structures as Stepping Stones of Invasion. *Ann. Limnol.-Int. J. Limnol.* **2018**, *54*, 15. [CrossRef]
45. Đuroković, Z.; Brnić-Levada, D. Influence of Hydrotechnical Works on Kopački Rit Water Resources. In Proceedings of the 2nd Croatian Conference on Water, Dubrovnik, Croatia, 19–22 May 1999; pp. 661–666.
46. Vidaković Šutić, R.; Šarić, I.; Ričković, V.; Vrcelj, B.; Špoljar, M.; Cvetnić, M.; Hršak, V.; Vuković, N.; Buj, I.; Koller Šarić, K.; et al. *Utvrđivanje Retencijskog Kapaciteta i Nultog Stanja Voda i o Vodama Ovisnih Ekosustava i Izrada Detaljnog Plana Monitoringa i Istraživanja u Svrhu Izrade Studije Revitalizacije Vodenih Ekosustava Poplavnog Područja Parka Prirode Kopački Rit*; Institut za elektroprivredu d.d. and Vitaprojekt: Zagreb, Croatia, 2021; p. 399.
47. AQEM Consortium. Manual for the Application of the AQEM System: A Comprehensive Method to Assess European Streams Using Benthic Macroinvertebrates, Developed for the Purpose of the Water Framework Directive. 2002. Available online: https://www.eu-star.at/pdf/AqemMacroinvertebrateSamplingProtocol.pdf (accessed on 15 April 2020).
48. Pozojević, I.; Dorić, V.; Vučković, N.; Rumišek, M.; Šumanović, M.; Ternjej, I.; Mihaljević, Z. Determining the Ecological Status of the Dinaric Karst Natural Lakes in Croatia: Eight Natural Lakes, Seven Lake Types, and One Very Significant Equation. *Environ. Process.* **2024**, *11*, 45. [CrossRef]
49. Eggers, T.O.; Martens, A.; Grabow, K. *Hemimysis anomala* Sars in the Branch-Canal Salzgitter, Northern Germany (Crustacea: Mysidacea). *Lauterbornia* **1999**, *35*, 43–47.
50. Greenberg, A.E. *Standard Methods for the Examination of Water and Wastewater*, 18th ed.; American Public Health Association (APHA): Washington, WA, USA, 1992.
51. Clarke, K.R.; Gorley, R.N. *Primer V6: User Manual—Tutorial*; PRIMER-E Limited: Plymouth, UK, 2006; p. 190.
52. ter Braak, C.J.F.; Šmilauer, P. *CANOCO Reference Manual and User's Guide to Canoco for Windows: Software for Canonical Community Ordination (Version 4)*; Microcomputer Power: New York, NY, USA, 1998; p. 352.
53. Hammer, Ø.; Harper, D.A.T.; Ryan, P.D. PAST: Paleontological Statistics Software Package for Education and Data Analysis. *Palaeontol. Electron.* **2001**, *4*, 1–9.
54. Clarke, K.R.; Warwick, R.M. *Change in Marine Communities: An Approach to Statistical Analysis and Interpretation*, 2nd ed.; PRIMER-E Limited: Plymouth, UK, 2001; p. 168.
55. Invasive Alien Species. Available online: https://environment.ec.europa.eu/topics/nature-and-biodiversity/invasive-alien-species_en (accessed on 25 August 2025).
56. Bij de Vaate, A.; Jazdzewski, K.; Ketelaars, H.A.M.; Gollasch, S.; Van der Velde, G. Geographical Patterns in Range Extension of Ponto-Caspian Macroinvertebrate Species in Europe. *Can. J. Fish. Aquat. Sci.* **2002**, *59*, 1159–1174. [CrossRef]
57. Sket, B. Distribution of *Asellus aquaticus* (Crustacea: Isopoda: Asellidae) and Its Hypogean Populations at Different Geographic Scales, with a Note on *Proasellus istrianus*. *Hydrobiologia* **1994**, *287*, 39–47. [CrossRef]
58. Kolar, C.S.; Lodge, D.M. Progress in Invasion Biology: Predicting Invaders. *Trends Ecol. Evol.* **2001**, *16*, 199–204. [CrossRef]
59. Patonai, K.; Endrédi, A.; Horváth, Z.; Borza, P.; Pálffy, K.; Dobosy, P.; Vad, C.F. Trophic Impact of an Invasive Mysid Shrimp Depends on Zooplankton Community Composition: A Mesocosm Experiment. *Freshw. Biol.* **2024**, *69*, 623–634. [CrossRef]
60. Rohwer, R.R.; Ladwig, R.; Hanson, P.C.; Walsh, J.R.; Vander Zanden, M.J.; Dugan, H.A. Increased Anoxia Following Species Invasion of a Eutrophic Lake. *Limnol. Oceanogr. Lett.* **2024**, *9*, 33–42. [CrossRef]
61. Devlin, S.P.; Tappenbeck, S.K.; Craft, J.A.; Tappenbeck, T.H.; Chess, D.W.; Whited, D.C.; Ellis, B.K.; Stanford, J.A. Spatial and Temporal Dynamics of Invasive Freshwater Shrimp (*Mysis diluviana*): Long-Term Effects on Ecosystem Properties in a Large Oligotrophic Lake. *Ecosystems* **2017**, *20*, 183–197. [CrossRef]
62. Vuić, N.; Turković Čakalić, I.; Koh, M.; Ergović, V.; Vlaičević, B.; Šag, M.; Rožac, V.; Čerba, D. Floodplains as a Suitable Habitat for Freshwater Fish: The Length–Weight Relationships and Condition Factors of Fish Inhabiting a Danube Floodplain in Croatia. *Sustainability* **2024**, *16*, 7566. [CrossRef]
63. Nurminen, L.; Hellén, N.; Olin, M.; Tiainen, J.; Vinni, M.; Grönroos, M.; Estlander, S.; Horppila, J.; Rask, M.; Lehtonen, H. Fishing-Induced Changes in Predation Pressure by Perch (Perca Fluviatilis) Regulate Littoral Benthic Macroinvertebrate Biomass, Density, and Community Structure. *Aquat. Ecol.* **2018**, *52*, 1–16. [CrossRef]
64. Bernauer, D.; Jansen, W. Recent Invasions of Alien Macroinvertebrates and Loss of Native Species in the Upper Rhine River, Germany. *Aquat. Invasions* **2006**, *1*, 55–71. [CrossRef]
65. Vlaičević, B.; Matoničkin Kepčija, R.; Gulin, V.; Turković Čakalić, I.; Kepec, M.; Čerba, D. Key Drivers Influencing the Colonization of Periphytic Ciliates and Their Functional Role in Hydrologically Dynamic Floodplain Lake Ecosystem. *Knowl. Manag. Aquat. Ecosyst.* **2021**, *422*, 33. [CrossRef]

66. Vlaičević, B.; Gulin, V.; Matoničkin Kepčija, R.; Turković Čakalić, I. Periphytic Ciliate Communities in Lake Ecosystem of Temperate Riverine Floodplain: Variability in Taxonomic and Functional Composition and Diversity with Seasons and Hydrological Changes. *Water* **2022**, *14*, 551. [CrossRef]
67. Žuna Pfeiffer, T.; Špoljarić Maronić, D.; Stević, F.; Galir Balkić, A.; Bek, N.; Martinović, A.; Mandir, T.; Nikolašević, R.; Janjić, D. Plastisphere Development in Relation to the Surrounding Biotic Communities. *Environ. Pollut.* **2022**, *306*, 119380. [CrossRef]
68. Vilenica, M.; Brigić, A.; Ergović, V.; Koh, M.; Alegro, A.; Šegota, V.; Rimac, A.; Rumišek, M.; Mihaljević, Z. Taxonomic and Functional Odonata Assemblage Metrics: Macrophyte–Driven Changes in Anthropogenically Disturbed Floodplain Habitats. *Hydrobiologia* **2024**, *851*, 3787–3807. [CrossRef]
69. Pekarova, P.; Gorbachova, L.; Mitkova, V.B.; Pekar, J.; Miklanek, P. Statistical Analysis of Hydrological Regime of the Danube River at Ceatal Izmail Station. In *IOP Conference Series: Earth and Environmental Science*; Institute of Physics Publishing: Bristol, UK, 2019; p. 221. [CrossRef]
70. Čerba, D.; Bogut, I.; Vidaković, J.; Palijan, G. Invertebrates in *Myriophyllum spicatum* L. Stands in Lake Sakadaš, Croatia. *Ekologia* **2009**, *28*, 94–105. [CrossRef]
71. Stević, F.; Čerba, D.; Turković Čakalić, I.; Žuna Pfeiffer, T.; Vidaković, J.; Mihaljević, M. Interrelations between *Dreissena polymorpha* Colonization and Autotrophic Periphyton Development—A Field Study in a Temperate Floodplain Lake. *Fundam. Appl. Limnol.* **2013**, *183*, 107–119. [CrossRef]
72. Vidaković, J.; Mihaljević, M.; Stević, F.; Špoljarić, D.; Palijan, G.; Čerba, D.; Galir, A.; Cvijanović, V. Invasive Species in the Floodplain Waters of Kopački Rit Nature Park. In *Zbornik Sažetaka 10. Hrvatskog Biološkog kongresa*; Besendorfer, V., Kopjar, N., Vidaković-Cifrek, Ž., Eds.; Hrvatsko Biološko Društvo: Zagreb, Croatia, 2009; pp. 307–308.
73. Mihaljević, M.; Špoljarić, D.; Stević, F.; Cvijanović, V.; Hackenberger Kutuzović, B. The Influence of Extreme Floods from the River Danube in 2006 on Phytoplankton Communities in a Floodplain Lake: Shift to a Clear State. *Limnologica* **2010**, *40*, 260–268. [CrossRef]
74. Aguilera, A.; Gómez, E.B.; Kaštovský, J.; Echenique, R.O.; Salerno, G.L. The Polyphasic Analysis of Two Native Raphidiopsis Isolates Supports the Unification of the Genera *Raphidiopsis* and *Cylindrospermopsis* (Nostocales, Cyanobacteria). *Phycologia* **2018**, *57*, 130–146. [CrossRef]
75. Briand, J.; Leboulanger, C.; Humbert, J.; Bernard, C.; Dufour, P. *Cylindrospermopsis raciborskii* (Cyanobacteria) Invasion At Mid-Latitudes: Selection, Wide Physiological Tolerance, Orglobalwarming? *J. Phycol.* **2004**, *40*, 231–238. [CrossRef]
76. Olenin, S.; Minchin, D.; Daunys, D. Assessment of Biopollution in Aquatic Ecosystems. *Mar. Pollut. Bull.* **2007**, *55*, 379–394. [CrossRef]
77. Gozlan, R.E.; Britton, J.R.; Cowx, I.; Copp, G.H. Current Knowledge on Non-native Freshwater Fish Introductions. *J. Fish. Biol.* **2010**, *76*, 751–786. [CrossRef]
78. Gallardo, B.; Clavero, M.; Sánchez, M.I.; Vilà, M. *Global Ecological Impacts of Invasive Species in Aquatic Ecosystems. Global Change Biology*; Blackwell Publishing Ltd.: Hoboken, NJ, USA, 2016; pp. 151–163. [CrossRef]
79. Maguire, I.; Jelić, M.; Klobučar, G.; Delpy, M.; Delaunay, C.; Grandjean, F. Prevalence of the Pathogen *Aphanomyces astaci* in Freshwater Crayfish Populations in Croatia. *Dis. Aquat. Organ.* **2016**, *118*, 45–53. [CrossRef]
80. Pathak, C.R.; Luitel, H.; Utaaker, K.S.; Khanal, P. One-Health Approach on the Future Application of Snails: A Focus on Snail-Transmitted Parasitic Diseases. *Parasitol. Res.* **2024**, *123*, 28. [CrossRef]
81. Darrigran, G. Potential Impact of Filter-Feeding Invaders on Temperate Inland Freshwater environments. *Biol. Invasions* **2002**, *4*, 145–156. [CrossRef]
82. Kettunen, M.; Genovesi, P.; Gollasch, S.; Pagad, S.; Starfinger, U.; ten Brink, P.; Shine, C. *Technical Support to EU Strategy on Invasive Species (IAS)—Assessment of the Impacts of IAS in Europe and the EU*; Final Report for the European Commission; Institute for European Environmental Policy (IEEP): Brusseles, Belgium, 2009.
83. Bódis, E.; Borza, P.; Potyó, I.; Puky, M.; Weperth, A.; Guri, G. Invasive Mollusc, Crustacean, Fish and Reptile Species Along the Hungarian Stretch of the River Danube and Some Connected Waters. *Acta Zool. Acad. Sci. Hung.* **2012**, *58*, 29–45.
84. Kralj, T.; Ćuk, R.; Valić, D.; Schultz, S.; Žganec, K. The Relationship between Alien Crustaceans and Pollution in Croatian Large Rivers: Implications for Biological Monitoring. *Hydrobiologia* **2022**, *849*, 3315–3334. [CrossRef]
85. Sousa, R.; Novais, A.; Costa, R.; Strayer, D.L. Invasive Bivalves in Fresh Waters: Impacts from Individuals to Ecosystems and Possible Control Strategies. *Hydrobiologia* **2014**, *735*, 233–251. [CrossRef]
86. Schuler, M.S.; Hintz, W.D.; Jones, D.K.; Mattes, B.M.; Stoler, A.B.; Relyea, R.A. The Effects of Nutrient Enrichment and Invasive Mollusks on Freshwater Environments. *Ecosphere* **2020**, *11*, e03196. [CrossRef]
87. Elwaseif, M.; Ismail, A.; Abdalla, M.; Abdel-Rahman, M.; Hafez, M.A. Geophysical and Hydrological Investigations at the West Bank of Nile River (Luxor, Egypt). *Environ. Earth Sci.* **2012**, *67*, 911–921. [CrossRef]
88. Castaldelli, G.; Colombani, N.; Vincenzi, F.; Mastrocicco, M. Linking Dissolved Organic Carbon, Acetate and Denitrification in Agricultural Soils. *Environ. Earth Sci.* **2013**, *68*, 939–945. [CrossRef]

89. Mihaljević, M.; Stević, F. Cyanobacterial Blooms in a Temperate River-Floodplain Ecosystem: The Importance of Hydrological Extremes. *Aquat. Ecol.* **2011**, *45*, 335–349. [CrossRef]
90. Aston, A.R.J.; Sadler, K.; Milner, A.G.P. The Effects of Temperature on the Culture *Branchiura sowerbyi* (Oligochaeta, Tubificidae) on Activated Sludge. *Aquaculture* **1982**, *29*, 137–145. [CrossRef]
91. van Haaren, T.; Soors, M. *Aquatic Oligochaeta of The Netherlands and Belgium: Identification Key to the Oligochaetes*; KNNV Publishing: Zeist, The Netherlands, 2013; 400p.
92. Žganec, K.; Lajtner, J.; Ćuk, R.; Crnčan, P.; Pušić, I.; Atanacković, A.; Kralj, T.; Valić, D.; Jelić, M.; Maguire, I. Alien Macroinvertebrates in Croatian Freshwaters. *Aquat. Invasions* **2020**, *15*, 593–615. [CrossRef]

diversity

MDPI

Article

Relationship Between Groundwater Level and Rodent Community Structure Mediated by Nutrient Composition of Plants in Dongting Lake, China

Tian Huang [1], Yiying He [2,3], Tonglin Yu [4], Meiwen Zhang [3], Chen Zhang [3], Yunlin Zhao [2], Junxiang Lin [4], Xiaoning Nan [4], Zhiyuan Hu [1,*] and Zhenggang Xu [1,2,4,*]

1 Hunan Engineering Research Center of Ecological Environment Intelligent Monitoring and Disaster, Prevention and Mitigation Technology in Dongting Lake Region, College of Information and Electronic, Engineering, Hunan City University, Yiyang 413000, China; huangtian@hncu.edu.cn

2 Hunan Research Center of Engineering Technology for Utilization of Environmental and Resources Plant, Central South University of Forestry and Technology, Changsha 410004, China; hyy123456666662021@163.com (Y.H.); rssq198677@163.com (Y.Z.)

3 Dongting Lake Station for Wetland Ecosystem Research, Institute of Subtropical Agriculture, The Chinese Academy of Sciences, Changsha 410125, China; zhangmw@isa.ac.cn (M.Z.); zhangchenrosetiger@163.com (C.Z.)

4 College of Forestry, Northwest A & F University, Yangling 712100, China; 17527790185@163.com (T.Y.); linjunxiang202506@163.com (J.L.); nxn@nwsuaf.edu.cn (X.N.)

* Correspondence: huzhiyuan@hncu.edu.cn (Z.H.); xuzhenggang@nwafu.edu.cn (Z.X.)

Abstract

The Dongting Lake wetland is an important habitat for rodents. In order to understand the structural characteristics of rodent communities and the effect of groundwater level on them, this study explored the changes in rodent community structure in four different habitats (Carex, Reeds, Carex + Reeds, and Poplar) from 2003 to 2023. Meanwhile, the relationships between groundwater level, nutrient composition of *Carex brevicuspis*, and structural characteristics of rodent communities were analyzed. The results showed that the advantages of rodent species in the four different habitats are different, and the community structure of rodents has undergone significant changes in recent years. A significant correlation between groundwater level and the nutrient composition of *C. brevicuspis* was found. Further analysis shows a significant correlation between the nutritional components of *C. brevicuspis* and the population structure of rodents. Crude protein, total phosphorus, and dry matter were three key plant nutrient indicators that were significantly correlated with both capture rate and the community diversity index ($p < 0.05$). Total phosphorus and crude protein were significantly negatively correlated with capture rate and Simpson's index ($p < 0.05$), but significantly positively correlated with Pielou's index and Shannon–Wiener's index ($p < 0.05$), while the dry matter was completely opposite. The research conclusions suggest that different habitats and groundwater levels affect different characteristics of rodent community structures, and that plant nutrients are likely to be the mediator.

Keywords: diversity index; *Carex brevicuspis*; *Alexandromys fortis*; *Apodemus agrarius*; rodent community succession

1. Introduction

Rodents are a vital part of the ecosystem and can be used as an important indicator for environmental monitoring [1]. Rodent community structures vary across ecosystems, which is the result of the interaction between internal factors of the species and external environmental factors [2]. In the process of the above interactions, information is transmitted, both at the individual scale and at the population scale. Exploring the mechanisms of rodent community formation is not only important for rodent management but also an important basis for evaluating ecosystem health. As the largest mammalian group, rodents have become an important model for the study of animal community mechanisms [3,4]. Research has shown that the number of rodent species increases with succession, and rodent species richness could be considered as a baseline of diversity indices in a variety of grassland habitats [5]. External biological factors, such as prey, are important factors affecting rodent communities. Temporary prey may bring about temporarily mixed animal groups, but current research is more concerned with the structural characteristics of stable mixed animal communities [6]. Research has emphasized that there might be particular value in heterospecifics of the same trophic level, because these species often require similar resources, need to avoid similar predators, and might impose lower competitive costs at close distances than conspecifics [7]. Plants also play an important role in the formation of rodent communities, providing not only a food source but also a hiding place for rodents. Studies have shown that changing plant species within habitats also results in important changes in rodent composition [8]. In addition to biotic factors, abiotic environmental factors such as temperature and rainfall also have important effects on rodent communities [9]. Temperature and maximum precipitation per day were identified as significant drivers of the rodent and shrew presence, which would also affect the already threatened Pannonic root vole [10].

Wetlands are biodiversity hotspots and important habitats for rodents [11]. Current attention to rodent communities is focused on habitats such as forests, agricultural fields, and grasslands, while attention to rodent communities in wetlands needs to be strengthened urgently [12]. Dongting Lake is located in the middle and lower reaches of the Yangtze River and is one of the most important storage lakes of the Yangtze River [13], which is also an important global biodiversity hotspot [14,15]. Mammals in the Dongting Lake area are dominated by rodents and *Alexandromys fortis* (formerly known as *Microtus fortis*) [16] and *Apodemus agrarius* are the most important rodent species in the region [17,18]. Rodents in the Dongting Lake region play a significant role in the local ecosystem. They not only have the potential to transmit diseases but also pose a threat to crops, causing biological disasters [19]. Previous studies have shown that ENSO-driven precipitation [20], vegetation coverage [21], and succession of rodent community structure are closely related. In recent years, research has shown that the structure of rodent communities in the Dongting Lake area has changed dramatically with the changes in environmental conditions, especially following the construction of the Three Gorges Project [22]. Before the construction of the Three Gorges Project, Dongting Lake beaches were dominated by *A. fortis*, which usually lived on the beaches and were forced to move into the farmland only during the high water season, causing disasters to crops [19]. In contrast to *A. fortis*, before the construction of the Three Gorges Project, *A. agrarius* mainly lived in the farmland around Dongting Lake and hardly ever entered the Dongting Lake beaches [17]. After the construction of the Three Gorges Project, the rodent community on the Dongting Lake beach changed significantly, especially characterized by the invasion of *A. agrarius* and the coexistence with *A. fortis* [23]. Different rodent species have different characteristics, and exploring the key driving factors of rodent community succession in the Dongting Lake wetland is a prerequisite for achieving rodent management in the region.

Water is a key environmental factor in wetland ecosystems, playing an important role in all ecological elements such as soil, vegetation, and animals [24]. Dramatic water level changes are one of the typical characteristics of Dongting Lake, and water level is one of the key environmental factors affecting the structure of biological communities in the area, including surface water level and groundwater level [25]. With the construction of the Three Gorges Dam, the hydrological characteristics of Dongting Lake have further changed significantly and have further important impacts on the biological communities [26]. The study showed that the typical vegetation area of Dongting Lake wetland clearly changed on the whole lake scale from 1985 to 2020, and the main hydrological variables affected the change of the area of *Carex brevicuspis* and *Phragmites australis* in Dongting Lake wetland [27]. Serving as a critical wintering habitat for migratory birds along the East Asian–Australasian Flyway, the analysis showed that water level changed the habitat suitability of migratory birds by changing vegetation distribution at different elevations [28]. Until now, the main focus of research on Dongting Lake has been on the impact of surface water level on vegetation communities, and there are no research reports on the relationship between groundwater level and rodent community characteristics in Dongting Lake. Research shows that the groundwater level has a significant impact on vegetation nutrition [29,30] and *C. brevicuspis* is also one of the main foods for both *A. fortis* and *A. agrarius* [31,32]. Based on the principle of material and information flow in ecosystems, this study hypothesizes that groundwater level is an important factor influencing the succession of rodent community structures in Dongting Lake and acts primarily through plants as a mediator.

2. Materials and Methods

2.1. Survey of Rodent Communities in Dongting Lake

The research group has long been concerned about the changes in the structure of the rodent community in Dongting Lake, and 18 typical long-term survey sites have been established around Dongting Lake [21], covering all rodent habitats in the area. Research ensures that the survey sites are evenly distributed along the coast of Dongting Lake, while randomly determining the specific locations of the survey points. Since 2003, when the construction of the Three Gorges Project was completed, four typical habitats (Carex, Reeds, Carex + Reeds, and Poplar) have been monitored to evaluate the succession characteristics of rodent communities in Dongting Lake (Figure S1). More details about the survey sites are shown in Table S1. Snap traps were used in the census, and fresh sunflower seeds were used as bait. Each mouse trap is spaced 5 m apart, with 100 traps placed on each sample line. Set up 3 survey lines at each survey site, with a minimum interval of 20 m between each survey line. Place mouse traps in the evening and retrieve them the next morning to count the species and quantities of rodents captured. The survey is conducted every quarter, unless the survey site is flooded or difficult to reach. Each rodent community survey at each survey site lasts for 3 days. More details of the survey process and survey locations are provided in our previous study [17]. During the survey process, the survey points are not absolutely consistent. In order to ensure the stability of the habitat, we will make acceptable adjustments to the survey points. Then, the associated rodent community indices, including the Simpson index, Shannon–Wiener index, and Pielou index, were calculated according to the following equation [33,34]:

- Capture rate

$$C = \frac{100A}{B}\%, \quad (1)$$

C is the population density (%), A is the number of rodents, and B is the number of effective snap traps.

- Simpson index

$$S = \sum_{i=1}^{s} \left(\frac{N_i}{N}\right)^2, \quad (2)$$

S is the number of rodent species comprising the community, Ni is the number of individuals captured of species i, and N is the total number of individuals captured.

- Shannon–Wiener index

$$H = -\sum_{i=1}^{s} \left(\frac{N_i}{N} \ln \frac{N_I}{N}\right), \quad (3)$$

- Pielou index

$$E = \frac{H}{\ln S'}, \quad (4)$$

2.2. Data Collection on Groundwater Level in Dongting Lake

Day-by-day long-term monitoring of groundwater level data in Dongting Lake was conducted by Dongting Lake Station for Wetland Ecosystem Research, Institute of Subtropical Agriculture, the Chinese Academy of Sciences (29°30′ N, 112°48′ E) during 2019–2022. Low, medium, and high groundwater level monitoring points were set up at distances of 100 m, 1000 m, and 2000 m from the bank, respectively. At the same time, the study obtained the daily water level data of Dongting Lake from the website of Hunan Hydrology Public Service "One Map" (http://yzt.hnswkcj.com, accessed on 18 August 2025) from 2002 to 2023. Based on the above data, a linear regression model was established using the relationship between the groundwater level data and water level data of known years, and all groundwater level data of Dongting Lake were finally obtained.

2.3. Determination of Plant Constituents

Carex brevicuspis is the main plant species on the beaches of Dongting Lake and the most important food source for rodents [19,35]. Five 1 m × 1 m *C. brevicuspis* samples were set up at each groundwater level, with a parallel distance of about 200 m between each sample [35]. The habitat is Carex, and the plant samples were taken every month. The above-ground portion of all *C. brevicuspis* in the sampling was cut, and the collected *C. brevicuspis* samples were placed in large plastic bags. The time and place were recorded. Then, the plant samples were dried and ground for the determination of nutrient content. The dry matter [36], crude ash [37], neutral detergent fiber [38], acid detergent fiber [38], crude protein [39], and total phosphorus [40] of *C. brevicuspis* were determined with reference to standard methods for the determination of plant nutrient content [41].

2.4. Statistical Analysis

The ANOVA and LSD were used to test the differences in nutritional components of *C. brevicuspis* between the different groundwater levels, and the above process was performed using SPSS 23.0 software. Pearson correlation analysis was used to analyze the correlation between the nutritional components of *C. brevicuspis* and community diversity index (Pielou, Shannon, and Simpson). The above correlation analysis process was conducted using Origin 2021. Data results were expressed as mean ± standard deviation (mean ± SD), with $p < 0.05$ considered significant and $p < 0.01$ considered highly significant. The investigation process was authorized by the Dongting Lake Station for Wetland Ecosystem Research, Institute of Subtropical Agriculture, the Chinese Academy of Sciences, and conducted under their supervision.

3. Result and Analysis

3.1. Capture Rate and Rodent Species in Different Habitats of the Dongting Lake Area

Habitat types of the Dongting Lake beach are complex, with a mosaic distribution among different types. Four typical habitats inhabited by rodents in Dongting Lake beach were selected to analyze the characteristics of the rodent community in the study: Carex, Reeds, Carex + Reeds, and Poplar (Figures 1 and S1, Table S1). During the period 2003–2023, a total of 57,921 traps were placed on the beach of Dongting Lake in the study, capturing a total of 5931 rodents. A total of eight species of rodent were captured: *A. agrarius*, *A. fortis*, *Micromys minutus*, *Niviventer confucianus*, *Niviventer fulvescens*, *Rattus losea*, *Rattus norvegicus*, and *Rattus tanezumi*. Among them, *A. agrarius* and *M. fortis* have high capture rates and are the dominant rodent species on the beach of Dongting Lake. A total of six species of rodent were captured in the Carex habitat: *A. agrarius*, *A. fortis*, *M. minutus*, *N. confucianus*, *R. losea*, and *R. norvegicus*. Of these, the proportion of *A. fortis* was higher than that of other rodent species in the Carex habitat from 2005–2009, and after 2011, *A. fortis*, *A. agrarius* were the dominant rodent species (Figure S1). In the Reed habitat, a total of four species of rodent were captured: *A. agrarius*, *A. fortis*, *M. minutus*, and *R. norvegicus*. In 2003, only *A. agrarius* was captured, and *A. agrarius* and *A. fortis* together were the dominant species in the Reeds habitat from 2006 onwards. In the Carex + Reeds habitat, *A. agrarius*, *A. fortis*, *M. minutus*, and *R. norvegicus* were also captured, and *A. agrarius* and *A. fortis* were the dominant species. In the Poplar habitat, a total of seven species of rodents were captured: *A. agrarius*, *A. fortis*, *N. confucianus*, *N. fulvescens*, *R. losea*, *R. norvegicus*, and *R. tanezumi*, with *A. fortis* being the advantage species in 2007, and then *A. agrarius* was the advantage species in 2008 and beyond (Figure S1).

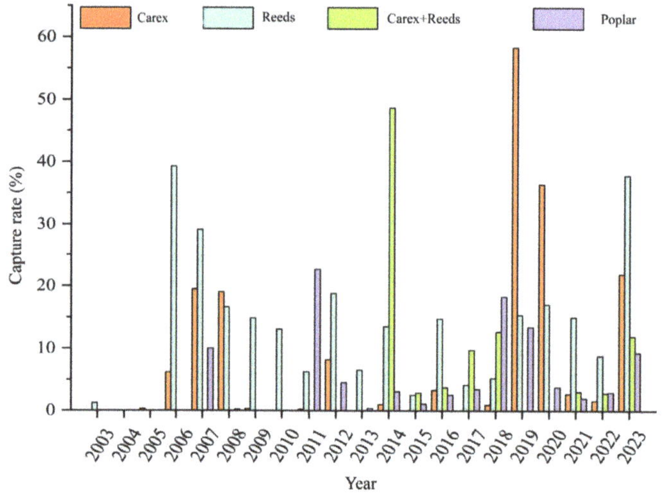

Figure 1. Interannual variation of rodent capture rate in different habitats. Note: Carex, Reeds, Carex + Reeds, and Poplar specifically refer to the four typical habitats selected in the study.

3.2. Diversity Index of Rodent Communities in Different Habitats

Among the four typical habitats of the Dongting Lake beach, the poplar forest (Poplar habitat) was the most species-rich with seven species of rodents, while Reeds and Carex + Reeds had the fewest species (Table 1). In the Carex habitat, Shannon–Wiener index (0.3414) and Pielou index (0.1905) were the lowest, and Simpson index (0.8338) was high. The Reeds habitat was the complete opposite of the Carex habitat, possessing the highest Shannon–Wiener index (0.8315) and Pielou index (0.5998) and the lowest Simpson index (0.4669). Analyzing the diversity index relationship between different habitats (Table S2), the results showed that there was a highly significant correlation between different habitats ($p < 0.05$), except for the non-significant correlation between Carex + Reeds and Poplar habitats. Regression analysis of the three community diversity indices in different habitats showed significant negative correlation between the Simpson index and both the Shannon–Wiener index and the Pielou index (Figure S3, $p < 0.05$). The correlation between Pielou index and Shannon–Wiener index was positive but not significant ($p > 0.05$).

Table 1. Diversity characteristics of rodent communities in different habitats.

Habitat	Species Number	Shannon–Wiener	Pielou	Simpson
Carex	6	0.3414	0.1905	0.8338
Reeds	4	0.8315	0.5998	0.4669
Carex + Reeds	4	0.6189	0.4464	0.6414
Poplar	7	0.6083	0.3126	0.7031

3.3. Relationship Between the Groundwater Level and the Nutrient Composition of Plants

The effect of annual average groundwater level on the various nutrient compositions of *C. brevicuspis* was analyzed, and the results showed that groundwater level had different effects on different nutrients (Figure 2). The content of dry matter, crude ash, neutral detergent fiber, and acid detergent fiber all decreased and then increased with the increase in groundwater level, but the differences were not significant under different groundwater levels ($p > 0.05$). The crude protein content increased with increasing groundwater level, but the difference was not significant either ($p > 0.05$). Total phosphorus content decreased ($p > 0.05$) and then increased significantly ($p < 0.05$) with the increasing groundwater level. Pearson's correlation analysis between annual groundwater level and nutrient composition of *C. brevicuspis* showed that crude protein and total phosphorus were highly significantly positively correlated with annual groundwater level ($p < 0.01$), while dry matter, crude ash, neutral detergent fiber, and acid detergent fiber were all negatively correlated with annual groundwater level ($p < 0.01$, Figure S4).

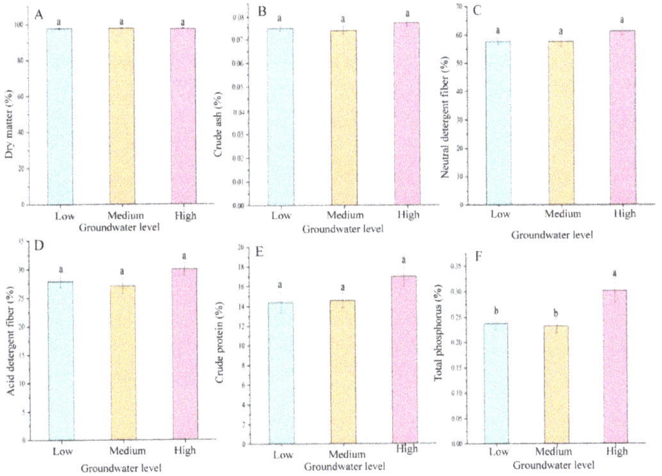

Figure 2. Effects of different annual groundwater levels (low, medium, and high) on nutrient composition of *C. brevicuspis*. (**A**): dry matter; (**B**): crude ash; (**C**): neutral detergent fibers; (**D**): acidic detergent fibers; (**E**): crude protein; (**F**): total phosphorus. Note: The letters represent the results of one-way ANOVA, and completely different letters indicate significant differences ($p < 0.05$).

3.4. Relationship Between the Nutrient Composition of Plant and Rodent Communities

Correlation analysis of the nutrient composition of *C. brevicuspis* with community structure indicators of rodents in different habitats showed that the crude ash was not significantly related to either capture rate or the rodent community diversity index ($p > 0.05$), but other plant nutrients were significantly related to at least one diversity index (Figure 3). Acid detergent fiber and neutral detergent fiber were significantly and positively correlated with rodent capture rate ($p < 0.05$), with correlation coefficients of 0.51 and 0.42, respectively, but there was no significant relationship with community diversity indices ($p > 0.05$). Crude protein, total phosphorus, and dry matter were three key plant nutrient indicators that were significantly correlated with both capture rate and community diversity index ($p < 0.05$). Total phosphorus and crude protein were significantly negatively correlated with capture rate and Simpson's index ($p < 0.0$), but significantly positively correlated with Pielou's index and Shannon–Wiener's index ($p < 0.05$), and the correlation coefficients varied little. Dry matter was completely opposite to total phosphorus and crude protein, and was significantly positively correlated with capture rate and Simpson's index ($p < 0.05$), and significantly negatively correlated with Pielou's index and Shannon–Wiener's index ($p < 0.05$), but the correlation coefficients were all higher than those of total phosphorus and crude protein.

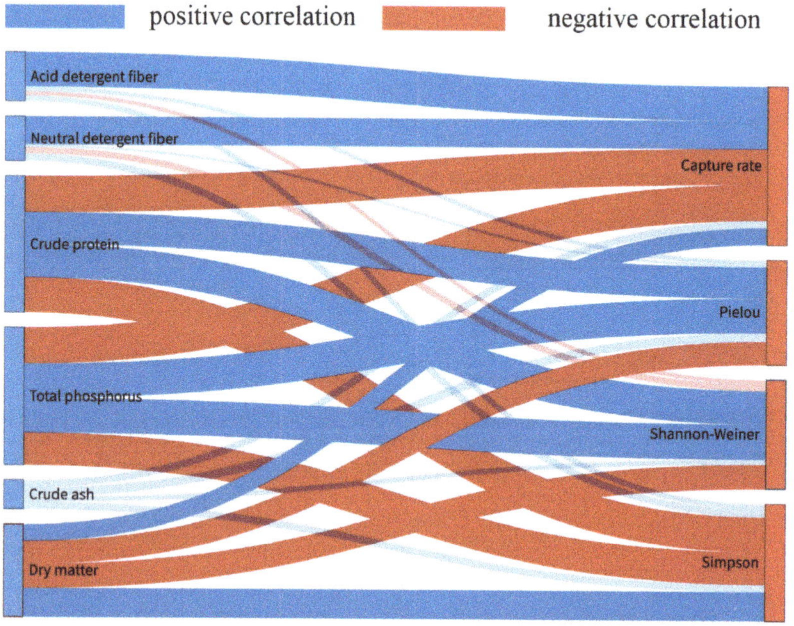

Figure 3. Correlation between the rodent communities and the nutrient composition of *C. brevicuspis* in Dongting Lake. Red color indicates negative correlation, blue color indicates positive correlation, high transparency indicates non-significant ($p > 0.05$), and low transparency indicates significant correlation ($p < 0.05$).

3.5. The Relationship Between Groundwater Level and Rodent Communities

The correlation analysis between groundwater level and rodent community characteristics showed that there was no significant relationship between groundwater level and species number, capture rate, and community diversity index ($p > 0.05$, Figure S5). Further analysis of the impact of different groundwater levels in different habitats on the characteristics of rodent communities revealed that the effects vary across different habitats

(Table 2). In the Poplar habitat, the capture rate, richness, and Shannon–Wiener index of the high groundwater level are the highest, but the Simpson index is the lowest. The capture rate, species number, and Simpson index of high groundwater level are relatively high in both the Carex and Reed habitats, but the Shannon–Wiener index and Pielou index are the smallest. In the Carex + Reeds habitat, the capture rate and Simpson index of high groundwater level are the highest, but the Shannon–Wiener index and Pielou index are the lowest.

Table 2. Diversity characteristics of rodent communities by habitat groundwater level (low, medium, high).

Habitat	Groundwater Level	Capture Rate	Species Number	Shannon–Wiener	Pielou	Simpson
Poplar	low	6.37	4	1.0308	0.7435	0.6459
	medium	4.92	3	0.9997	0.9100	0.5941
	high	6.73	5	1.1594	0.7204	0.6107
Carex	low	4.88	2	0.8631	1.2452	0.6014
	medium	7.29	2	0.7850	1.1325	0.6484
	high	16.86	3	0.3817	0.3474	0.8763
Reeds	low	19.81	3	1.1813	1.0753	0.4870
	medium	17.92	4	1.1997	0.8654	0.5198
	high	28.26	4	1.0966	0.7910	0.5328
Carex + Reeds	low	4.54	4	0.7091	0.5115	0.7692
	medium	6.86	3	0.7965	0.7250	0.6893
	high	8.37	3	0.2026	0.1844	0.9561

4. Discussion

During the research period, significant changes occurred in the rodent communities in different habitats of the Dongting Lake beach. In the Carex habitat, from 2005 to 2009, *A. fortis* was the dominant species. After 2011, *A. agrarius* flooded into the lake beach and became the dominant species of the Carex habitat with *A. fortis*. In 2003, there was only *A. agrarius* in the Reeds habitat, but after 2006, both *A. agrarius* and *A. fortis* became dominant species in the Reeds habitat. In the Carex + Reeds habitat, *A. fortis* was the dominant species only in 2014 and 2021, while *A. agrarius* was the dominant species in other years. In the Poplar habitat, *A. fortis* was the dominant species in 2007, and *A. agrarius* was the absolute dominant species in 2008 and beyond. The changes in the community structure of rodents in different habitats mentioned above are closely related to the Three Gorges Project, which was constructed in 1994 and began storing water in 2003 [42]. The research results are consistent with Zheng Puyang et al. (2020), which suggest that the operation of the Three Gorges Project has changed the hydrological pattern of Dongting Lake. The changes in hydrological characteristics have led to changes in vegetation succession, further causing changes in the dynamics of the rodent communities in Dongting Lake [43]. In the study, the Shannon–Wiener indices of Reeds, Carex + Reeds, and Poplar were higher than those of the Carex habitat, which may be due to higher human interference in the above three habitats, and appropriate human interference can improve species diversity [44]. The changes in rodent communities should be given sufficient attention in order to achieve effective management of rodents. The harm caused to crops by *A. fortis* is stronger than that of *A. agrarius* [19], but the risk of disease transmission by *A. agrarius* is higher [45]. As *A. agrarius* expands between various habitats on the beach, more attention must be paid to the community succession to ensure human health. At the same time, previous studies have shown that due to factors such as environment and food, the reproductive characteristics of *A. agrarius* and *A. fortis* on the Dongting Lake beach are different [23]. Therefore, management measures for these two types of rodents are also different.

Carex spp., especially *C. brevicuspis*, is one of the important plant populations in the wetland ecosystem of Dongting Lake [19]. *Carex* spp. communities are of great significance for water balance and climate environment regulation in the region, and also provide food raw materials and habitats for rare birds, rodents, and fish [46]. In Dongting Lake, the *Carex* spp., which has the characteristics of high biomass and high nutritional value, sprouts and grows rapidly after the water recedes in November each year, and stops growing in May of the following year [47]. In wetland ecosystems, water level is often a key factor restricting vegetation growth and reproductive dynamics, which in turn has a decisive impact on vegetation succession, species richness, distribution patterns, and vegetation composition [48]. Previous studies have shown that the distribution of *C. brevicuspis* in Dongting Lake is closely related to water level, and changes in water level have a significant impact on the growth and clonal reproduction characteristics of *C. brevicuspis* at different elevations [48], but there has been less attention paid to groundwater level. During the dry season, the *Carex* spp. in the Dongting Lake area mainly relies on the supply of groundwater. Therefore, the groundwater level is an important environmental factor that determines the nutritional composition and distribution of *Carex* spp. In this study, the nutritional composition of *C. brevicuspis* varies at different groundwater levels, which can further affect rodents' choices of food and habitat. Different nutrients in plants can have different effects on rodents. Protein is an essential nutrient for rodents to maintain their life, growth, development, and reproduction, and moderate protein in plants can improve the palatability, thereby benefiting population growth [49]. In this study, crude protein, total phosphorus, and the diversity index of rodent communities were significantly positively correlated, indicating that the above elements are beneficial for the maintenance of rodent communities. The research results further suggest that in the process and control of rodents, the community structure and species composition of plants in the ecosystem can be regulated to break the nutritional balance in rodents, ultimately achieving effective regulation of rodent populations [50]. Research suggests that planting crops with low sugar content makes it difficult to meet the nutritional needs of rodents and can reduce the population density of *Lasiopodomys brandtii* [51]. Based on the relationship between plants and rodents, the construction of sustainable agricultural and forestry ecosystems and the control of rodent hazards are increasingly being emphasized [52]. Previous studies have shown that the content of protein, fiber, and tannic acid has significant effects on the feeding behavior and reproduction of *A. fortis* [53]. However, these studies are limited to the laboratory and lack attention to how plant nutrition is transmitted to rodent communities in the wild. In recent years, *C. brevicuspis* in Dongting Lake has been mown in large quantities in winter, further altering the nutritional composition of the plant [54]. Long-term monitoring of changes in the nutrient composition of moss and rodent community structure in Dongting Lake is of great significance. On the other hand, there is also a drawback worth noting, which is that plant samples are not completely synchronized with rodent communities. It is necessary to provide fully synchronized data in subsequent research to obtain more reliable results.

Most rodents have a digging habit, so there is a close relationship between soil and rodents [55,56]. Groundwater can have a series of impacts on the structure of rodent communities through various media, such as soil and plants. This study demonstrates the possible pathways by which groundwater levels ultimately affect rodent communities by altering plant nutrient composition. The groundwater level can also affect the habitat preference of rodents by influencing soil physical and chemical properties. In the study, experiments were designed to explore habitat spaces, selecting for rodents under different water level conditions (Figure S6). Set three groundwater levels, high (−20 cm), medium (−40 cm), and low (−60 cm), respectively, and calculate the proportion of activity duration of *A. fortis* and *A. agrarius* in soil environments with different groundwater levels. The research results show that the activity duration of *A. fortis* and *A. agrarius* in the high groundwater level layer is significantly higher than that in the low groundwater level layer and the medium groundwater level layer (Figure 4, $p < 0.01$). Therefore, the differences in soil selection preferences between *A. fortis* and *A. agrarius* are smaller than the influence of vegetation on the above two mouse species through the groundwater level. It is necessary to pay more attention to the mechanism of groundwater level driving the succession of rodent communities.

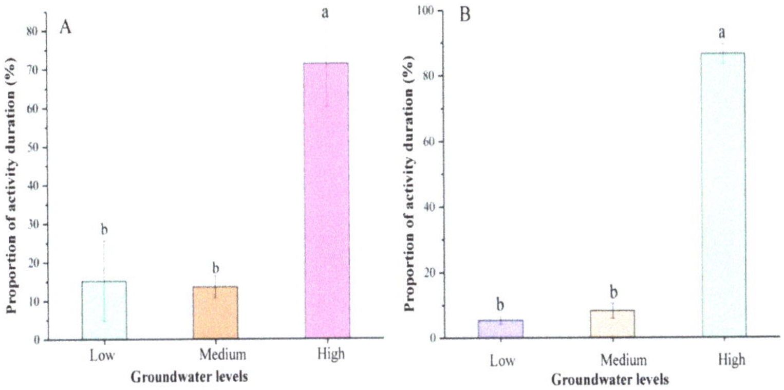

Figure 4. The proportion of activity duration of *A. fortis* and *A. agrarius* in different groundwater levels. (**A**): *A. fortis*; (**B**): *A. agrarius*. Note: The letters represent the results of one-way ANOVA, and completely different letters indicate significant differences ($p < 0.05$).

The rodent community structure is a composite of the number of species, population size, and population distribution [3]. The community diversity index is an important aspect of measuring community structure, but there is currently no diversity index that can independently evaluate community structure, so many diversity indices are calculated [57]. Choosing different diversity indices may yield different results. For example, the Reciprocal Simpson Index is also an important diversity index [58]. If this index is chosen in this study, the relationship between plant nutrients and rodents is opposite to the Simpson index. Therefore, the above research results remind us that it is difficult to generalize the relationship between environmental factors and rodent community structures, and research should focus on specific diversity indices. There are still other challenges to conducting research on the structure of rodent communities, especially in terms of the long-term succession of rodent community structure. These challenges pose difficulties for the effective management of rodents. Rodents are different from plants, and their vigilance makes it possible for any environmental disturbance to affect monitoring results. For example, the temperature during the investigation period can affect the capture rate of rodents.

5. Conclusions

The Dongting Lake wetland is a typical freshwater lake wetland ecosystem in China. In recent years, there have been many outbreaks of rodent damage in the area, which have brought huge economic losses. *Carex brevicuspis* is an important food source for rodents inhabiting the lake beach, and groundwater indirectly affects the growth of *C. brevicuspis* by directly regulating soil. Through long-term monitoring of rodent communities in four typical habitats on the Dongting Lake beach, it has been confirmed that rodent communities vary in different habitats. With the impact of the Three Gorges Project on the hydrological pattern of Dongting Lake, significant changes have occurred in the composition of rodents in all four habitats. Further analysis of the impact of groundwater on the nutritional composition of *C. brevicuspis* revealed that crude protein and total phosphorus showed highly significant positive correlations with annual groundwater levels. Research has shown a significant correlation between plant nutritional components and rodent community structure, particularly in terms of crude protein and total phosphorus content. Different habitats and groundwater levels have different characteristics of rodent community structure, and plant nutrients are likely to be the link between groundwater levels and rodent communities in the Dongting Lake area. The research results further suggest that in the process and control of rodents, the community structure and species composition of plants in the ecosystem can be regulated to break the nutritional balance in rodents, ultimately achieving effective regulation of rodent populations. Exploring the drivers of changes in rodent community structure in Dongting Lake is the theoretical basis for effective management of rodent populations in the region.

Supplementary Materials: The following supporting information can be downloaded at: https://www.mdpi.com/article/10.3390/d17080587/s1, Figure S1. Typical habitat of Dongting Lake beach; Figure S2. The interannual variation of the proportion of rodent species in different habitats; Figure S3. Regression analysis between two indices (A) relationship between Pielou index and Shannon-Wiener index, (B) relationship between Simpson index and Shannon-Wiener index, (C) relationship between Simpson index and Pielou index; Figure S4. Correlation between mean annual groundwater level change and *C. brevicuspis* nutrient composition; Figure S5. Correlation between annual groundwater levels and the community structure of lake beach rodents in Dongting Lake; Figure S6. Experimental design of selection bias of *A. fortis* and *A. agrarius* on soils with different groundwater level; Table S1. The details about the survey sites in the study; Table S2. Comparison of diversity index of rodent communities in different habitats (*t*-test).

Author Contributions: Conceptualization, Y.Z., X.N. and Z.X.; methodology, Y.H., C.Z., and M.Z.; software, Y.H.; formal analysis, Y.H.; investigation, Y.H., C.Z. and M.Z.; writing—original draft preparation, T.H., Y.H., and T.Y.; writing—review and editing, T.H., J.L. and Z.X.; supervision, Y.Z., Z.H. and Z.X.; project administration, Z.H. and Z.X.; funding acquisition, T.H., X.N., M.Z. and Z.X. All authors have read and agreed to the published version of the manuscript.

Funding: This research was funded by the National Key R&D Program of China, grant number 2022YFD1400400; the National Natural Science Foundation of China, grant number U20A20118; the Natural Science Foundation of Hunan Province, grant number 2024JJ5064; the College Students' Innovative Entrepreneurial Training Plan Program of Shaanxi Province, grant number XN2025006031; and the Open Fund of the Hunan Engineering Research Center of Ecological Environment Intelligent Monitoring and Disaster Prevention and Mitigation Technology in Dongting Lake, grant number 2022-DTH-01. The APC was funded by the Natural Science Foundation of Hunan Province.

Institutional Review Board Statement: The animal study protocol was approved by the Ethics Committee of the Institute of Subtropical Agriculture, Chinese Academy of Sciences (protocol code U20A20118, 1 January 2020).

Data Availability Statement: The data are contained within this article.

Conflicts of Interest: The funders had no role in the design of the study; in the collection, analyses, or interpretation of data; in the writing of the manuscript; or in the decision to publish the results.

References

1. Douglass, R.J. Assessment of the use of selected rodents in ecological monitoring. *Environ. Manag.* **1989**, *13*, 355–363. [CrossRef]
2. Minchella, D.J.; Scott, M.E. Parasitism: A cryptic determinant of animal community structure. *Trends Ecol. Evol.* **1991**, *6*, 250–254. [CrossRef] [PubMed]

3. Brown, J.S. Desert Rodent Community Structure: A Test of Four Mechanisms of Coexistence. *Ecol. Monogr.* **1989**, *59*, 1–20. [CrossRef]
4. Hansson, L.; Henttonen, H. Rodent dynamics as community processes. *Trends Ecol. Evol.* **1988**, *3*, 195–200. [CrossRef]
5. Avenant, N.L.; Cavallini, P. Correlating rodent community structure with ecological integrity, Tussen-die-Riviere Nature Reserve, Free State province, South Africa. *Integr. Zool.* **2007**, *2*, 212–219. [CrossRef]
6. Goodale, E.; Sridhar, H.; Sieving, K.E.; Bangal, P.; Colorado, Z.G.J.; Farine, D.R.; Heymann, E.W.; Jones, H.H.; Krams, I.; Martínez, A.E.; et al. Mixed company: A framework for understanding the composition and organization of mixed-species animal groups. *Biol. Rev.* **2020**, *95*, 889–910. [CrossRef]
7. Goodale, E.; Beauchamp, G.; Magrath, R.D.; Nieh, J.C.; Ruxton, G.D. Interspecific information transfer influences animal community structure. *Trends Ecol. Evol.* **2010**, *25*, 354–361. [CrossRef]
8. Larsen, A.L.; Homyack, J.A.; Wigley, T.B.; Miller, D.A.; Kalcounis-Rueppell, M.C. Effects of habitat modification on cotton rat population dynamics and rodent community structure. *For. Ecol. Manag.* **2016**, *376*, 238–246. [CrossRef]
9. Ventura-Rojas, P.D.; González-Romero, A.; Moreno, C.E.; Sosa, V.J. Effect of rainfall, temperature and climate change on the ecology of the rodents of arid zones: A review. *Mammal Rev.* **2025**, *55*, e12372. [CrossRef]
10. Stefke, K.; Landler, L. Long-term monitoring of rodent and shrew communities in a biodiversity hot-spot in Austria using barn owl (Tyto alba) pellets. *Acta Oecologica* **2020**, *109*, 103660. [CrossRef]
11. Noe, K.L.; Rota, C.T.; Frantz, M.W.; Anderson, J.T. Restored Wetland Size and Age Influence Small Mammal Communities in West Virginia, USA. *Wetlands* **2024**, *44*, 48. [CrossRef]
12. Gliwicz, J. Competitive Interactions within a Forest Rodent Community in Central Poland. *Oikos* **1981**, *37*, 353–362. [CrossRef]
13. An, C.; Fang, H.; Zhang, L.; Su, X.; Fu, X.; Huang, H.Q.; Parker, G.; Hassan, M.A.; Meghani, N.A.; Anders, A.M.; et al. Poyang and Dongting Lakes, Yangtze River: Tributary lakes blocked by main-stem aggradation. *Proc. Natl. Acad. Sci. USA* **2022**, *119*, e2101384119. [CrossRef]
14. Zhang, Y.; Huang, D.; Jin, X.; Li, L.; Wang, C.; Wang, Y.; Pellissier, L.; Johnson, A.C.; Wu, F.; Zhang, X. Long-term wetland biomonitoring highlights the differential impact of land use on macroinvertebrate diversity in Dongting Lake in China. *Commun. Earth Environ.* **2024**, *5*, 32. [CrossRef]
15. Zhang, D.; Lin, Y.; Xiao, W.; Wang, D.; Wang, Z.; Jin, X.; Cheng, T.; Zhang, J.; Yi, P. The relationship between landscape pattern and plant species diversity in east Dongting lake wetland based on different Eco-environment. *Environ. Pollut.* **2024**, *355*, 124187. [CrossRef] [PubMed]
16. Lissovsky, A.A.; Petrova, T.V.; Yatsentyuk, S.P.; Golenishchev, F.N.; Putincev, N.I.; Kartavtseva, I.V.; Sheremetyeva, I.N.; Abramson, N.I. Multilocus phylogeny and taxonomy of East Asian voles *Alexandromys* (Rodentia, Arvicolinae). *Zool. Scr.* **2018**, *47*, 9–20. [CrossRef]
17. Zhang, M.; Wang, Y.; Li, B.; Feng, Z.; Zhao, Y.; Xu, Z. Synergistic succession of the small mammal community and herbaceous vegetation after reconverting farmland to seasonally flooded wetlands in the Dongting Lake Region, China. *Mammal Study* **2018**, *43*, 229–243. [CrossRef]
18. Yong, W.; Mei-Wen, Z.; Bo, L.I.; Kai-Rong, W. Rodent community structure and succession in different ecotypic areas in Dongting Lake region. *J. Ecol. Rural Environ.* **2003**, *19*, 13–17.
19. Xu, Z.; Zhao, Y.; Li, B.; Zhang, M.; Shen, G.; Wang, Y. Habitat evaluation for outbreak of Yangtze voles (*Microtus fortis*) and management implications. *Integr. Zool.* **2015**, *10*, 267–281. [CrossRef] [PubMed]
20. Zhang, Z.; Xu, L.; Guo, C.; Wang, Y.; Guo, Y. Effect of ENSO-driven precipitation on population irruptions of the Yangtze vole *Microtus fortis calamorum* in the Dongting Lake region of China. *Integr. Zool.* **2010**, *5*, 176–184. [CrossRef]
21. Huang, T.; Tang, Y.; Sun, Y.; Zhang, M.; Zhang, C.; Zhao, Y.; Nan, X.; Hu, Z.; Xu, Z. Effects of Vegetation Cover on Community Structure of Rodents Based on Long Time Series from Dongting Lake, China. *Biology* **2025**, *14*, 867. [CrossRef] [PubMed]
22. Zhang, M.; Wang, Y.; Li, B.; Guo, C.; Huang, G.; Shen, G.; Zhou, X. Small mammal community succession on the beach of Dongting Lake, China after the Three Gorges Project. *Integr. Zool.* **2014**, *9*, 294–308. [CrossRef]
23. He, J.; Zhang, T.; Yang, G.; Xu, Z.; Zhang, M.; Wang, Y.; Li, B.; Zhou, X.; Huang, H. The reproductive status of *Apodemus agrarius* populations inhabiting the lake beach in Lake Dongting area after Three Gorges Project. *J. Lake Sci.* **2023**, *35*, 2101–2110.
24. Jolly, I.D.; McEwan, K.L.; Holland, K.L. A review of groundwater–surface water interactions in arid/semi-arid wetlands and the consequences of salinity for wetland ecology. *Ecohydrology* **2008**, *1*, 43–58. [CrossRef]
25. Han, Q.; Zhang, S.; Huang, G.; Zhang, R. Analysis of Long-Term Water Level Variation in Dongting Lake, China. *Water* **2016**, *8*, 306. [CrossRef]
26. Lai, X.; Chen, H.; Hou, Y.; Finlayson, B.; Li, M.; Chen, J. Lowering water level of Dongting lake of the Mid-Yangtze River in response to large-scale dam construction: A 60-year analysis. *Geomorphology* **2021**, *391*, 107894. [CrossRef]
27. Liu, Y.; Li, J.; Yan, D.; Chen, L.; Li, M.; Luan, Z. Typical vegetation dynamics and hydrological changes of Dongting Lake wetland from 1985 to 2020. *Ecohydrol. Hydrobiol.* **2024**, *24*, 910–919. [CrossRef]

28. Zhu, Y.; Wang, H.; Guo, W. The impacts of water level fluctuations of East Dongting Lake on habitat suitability of migratory birds. *Ecol. Indic.* **2021**, *132*, 108277. [CrossRef]
29. Takatert, N.; Sanchez-Pérez, J.M.; Trémolières, M. Spatial and temporal variations of nutrient concentration in the groundwater of a floodplain: Effect of hydrology, vegetation and substrate. *Hydrol. Process.* **1999**, *13*, 1511–1526. [CrossRef]
30. Wang, Y.; Wang, J.; Wang, X.; He, Y.; Li, G.; Li, J. Dominant roles but distinct effects of groundwater depth on regulating leaf and fine-root N, P and N:P ratios of plant communities. *J. Plant Ecol.* **2021**, *14*, 1158–1174. [CrossRef]
31. Yong, Z.Y.; Zhang, M.W.; Guo, C.; Wang, Y.; Li, B.; Li, J.X.; Yang, Y.C.; Xu, Z.G. Dietary Habit of *Apodemus agrarius* in Dongting Lake Area. *Chin. J. Zool.* **2012**, *47*, 115–121.
32. Lei, F.; Yunlin, Z.; Meiwen, Z.; Yong, W.; Zhenggang, X.; Jiao, P. Research on feeding of laboratory-bred adult *Microtus fortis calamorum* on carex. *Chin. J. Vector Biol. Control* **2016**, *27*, 546–548.
33. Shannon, C.E. A mathematical theory of communication. *Bell. Syst. Tech. J.* **1948**, *27*, 379–423. [CrossRef]
34. Spellerberg, I.F.; Fedor, P.J. A tribute to Claude Shannon (1916–2001) and a plea for more rigorous use of species richness, species diversity and the 'Shannon–Wiener' Index. *Glob. Ecol. Biogeogr.* **2003**, *12*, 177–179. [CrossRef]
35. Huang, Y.; Chen, X.-S.; Zou, Y.-A.; Zhang, P.-Y.; Li, F.; Hou, Z.-Y.; Li, X.; Zeng, J.; Deng, Z.-M.; Zhong, J.-R.; et al. Exploring the relative contribution of flood regimes and climatic factors to *Carex* phenology in a Yangtze River-connected floodplain wetland. *Sci. Total Environ.* **2022**, *847*, 157568. [CrossRef]
36. Thiex, N.J.; Erem, T.V. Determination of Water (Moisture) and Dry Matter in Animal Feed, Grain, and Forage (Plant Tissue) by Karl Fischer Titration: Collaborative Study. *J. AOAC Int.* **2019**, *85*, 318–327. [CrossRef]
37. Wichmann, H.J.; Tilden, D. Report on Ash in Fruit Products. *J. Assoc. Off. Agric. Chem.* **2020**, *10*, 433–436. [CrossRef]
38. Heckman, M.M.; Lane, S.A. Comparison of Dietary Fiber Methods for Foods. *J. Assoc. Off. Anal. Chem.* **2020**, *64*, 1339–1343. [CrossRef]
39. Thiex, N. Evaluation of Analytical Methods for the Determination of Moisture, Crude Protein, Crude Fat, and Crude Fiber in Distillers Dried Grains with Solubles. *J. AOAC Int.* **2019**, *92*, 61–73. [CrossRef]
40. Jacob, K.D.; Hoffman, W.M. Report on Phosphorus in Fertilizers: I. Preparation of Solution of Sample for Total Phosphorus Determination. *J. Assoc. Off. Agric. Chem.* **2020**, *40*, 690–700. [CrossRef]
41. Zhang, G.; Dai, E.; Luobu, D.; Fu, G. Effects of Climate Change and Fencing on Forage Nutrition Quality of Alpine Grasslands in the Northern Tibet. *Plants* **2023**, *12*, 3182. [CrossRef] [PubMed]
42. Hu, J.-Y.; Xie, Y.-H.; Tang, Y.; Li, F.; Zou, Y.-A. Changes of Vegetation Distribution in the East Dongting Lake After the Operation of the Three Gorges Dam, China. *Front. Plant Sci.* **2018**, *9*, 2018. [CrossRef] [PubMed]
43. Zheng, P.; Zhou, X.; Zhang, M.; Wang, Y.; Zhang, C.; Yang, J. The Situation of Small Mammal Community in Beach of Dongting Lake after the Official Operation of the Three Gorges Reservoir. *Chin. J. Zool.* **2020**, *55*, 141–152.
44. Yang, X.; Gu, H.; Zhao, Q.; Zhu, Y.; Teng, Y.; Li, Y.; Zhang, Z. High seed diversity and availability increase rodent community stability under human disturbance and climate variation. *Front. Plant Sci.* **2022**, *13*, 2022. [CrossRef]
45. Ni, H.; Yang, F.; Li, Y.; Liu, W.; Jiao, S.; Li, Z.; Yi, B.; Chen, Y.; Hou, X.; Hu, F.; et al. *Apodemus agrarius* is a potential natural host of severe fever with thrombocytopenia syndrome (SFTS)—Causing novel bunyavirus. *J. Clin. Virol.* **2015**, *71*, 82–88. [CrossRef]
46. Wu, L.; Zhao, Y.; Zhang, M.; Xu, Z.; He, R.; Wang, S.; Guo, Z. Analysis on Nutritional Content in Different Part of *Carex brevicuspis* in Dongting Lake. *Mod. Agric. Sci. Technol.* **2016**, 305–307.
47. Chen, X.-S.; Xie, Y.-H.; Deng, Z.-M.; Li, F.; Hou, Z.-Y. A change from phalanx to guerrilla growth form is an effective strategy to acclimate to sedimentation in a wetland sedge species *Carex brevicuspis* (Cyperaceae). *Flora-Morphol. Distrib. Funct. Ecol. Plants* **2011**, *206*, 347–350. [CrossRef]
48. Li, Y.; Cheng, X.; Xiang, W.; Xie, Y. Effects of water levels on the growth and reproductive characteristics of *Carex brevicuspis* growing on sites with different elevations. *Acta Ecol. Sin.* **2016**, *36*, 1959–1966.
49. Wang, B.; Chen, J. Effects of Fat and Protein Levels on Foraging Preferences of Tannin in Scatter-Hoarding Rodents. *PLoS ONE* **2012**, *7*, e40640. [CrossRef]
50. Soininen, E.M.; Neby, M. Small rodent population cycles and plants—After 70 years, where do we go? *Biol. Rev.* **2024**, *99*, 265–294. [CrossRef] [PubMed]
51. Zhang, X.Y.; Wang, D.H. Thermogenesis, food intake and serum leptin in cold-exposed lactating Brandt's voles *Lasiopodomys brandtii*. *J. Exp. Biol.* **2007**, *210*, 512–521. [CrossRef]
52. Jurišić, A.; Ćupina, A.I.; Kavran, M.; Potkonjak, A.; Ivanović, I.; Bjelić-Čabrilo, O.; Meseldžija, M.; Dudić, M.; Poljaković-Pajnik, L.; Vasić, V. Surveillance Strategies of Rodents in Agroecosystems, Forestry and Urban Environments. *Sustainability* **2022**, *14*, 9233. [CrossRef]
53. Xing, T.; Wang, Y.; Deng, W.; Zhang, M.; Li, B.; Zhu, J. Effects of protein, fiber, and tannic acid on food intake of *Microtus fortis*. *Acta Ecol. Sin.* **2010**, *30*, 941–948.
54. Wu, T.; Xia, L.; Pan, B.; Zou, Y.; Li, F.; Xie, Y.; Wang, S.; Li, Z. Mowing of *Carex brevicuspis* (Cyperaceae) improves food quality for herbivorous geese in Dongting Lake: The potential mechanisms. *Front. Plant Sci.* **2025**, *16*, 1566808. [CrossRef]

55. Battisti, C.; Dodaro, G.; Di Bagno, E.; Amori, G. Reviewing an eco-biogeographic question at regional scale: The unexpected absence of a ubiquitous mammal species (*Microtus savii*, Rodentia) in coastal Southern Tuscany (central Italy). *Rend. Lincei. Sci. Fis. E Nat.* **2019**, *30*, 715–722. [CrossRef]
56. Battisti, C.; Dodaro, G.; Di Bagno, E.; Amori, G. Small mammal assemblages in land-reclaimed areas: Do historical soil use changes and recent anthropisation affect their dominance structure? *Ethol. Ecol. Evol.* **2020**, *32*, 282–288. [CrossRef]
57. Stevens, R.D.; Tello, J.S. Diversity begets diversity: Relative roles of structural and resource heterogeneity in determining rodent community structure. *J. Mammal.* **2011**, *92*, 387–395. [CrossRef]
58. Gorelick, R. Combining richness and abundance into a single diversity index using matrix analogues of Shannon's and Simpson's indices. *Ecography* **2006**, *29*, 525–530. [CrossRef]

diversity

MDPI

Article

Microplastics Alter Growth and Reproduction Strategy of *Scirpus mariqueter* by Modifying Soil Nutrient Availability

Pengcheng Jiang [1,2], Jingwen Gao [1], Junzhen Li [1], Ming Wu [1], Xuexin Shao [1] and Niu Li [1,*]

[1] Wetland Ecosystem Research Station of Hangzhou Bay, State Key Laboratory of Wetland Conservation and Restoration, Research Institute of Subtropical Forestry, Chinese Academy of Forestry, Hangzhou 311400, China
[2] College of Forestry and Biotechnology, Zhejiang Agriculture and Forestry University, Hangzhou 311300, China
* Correspondence: niuli@caf.ac.cn; Tel./Fax: +86-0571-63342931

Abstract

Microplastic pollution threatens coastal wetland ecosystems, yet its impacts on the dominant plant species and soil properties remain poorly understood. We investigated the effects of four microplastic types (PP, PE, PS, PET) at three concentrations (0.1%, 0.5%, 1% *w*/*w*) on *Scirpus mariqueter*, a keystone species in the coastal wetlands of China, and the associated soil physicochemical properties. In a controlled pot experiment, microplastics significantly altered the plant biomass, vegetative traits, and reproductive strategies, with type-specific and concentration-dependent responses. PET and PE strongly suppressed the belowground and total biomass ($p < 0.05$), with reductions in the belowground biomass of 42.87% and 44.13%, respectively, at a 0.1% concentration. PP promoted seed production, particularly increasing the seed number by 25.23% at a 0.1% concentration ($p < 0.05$). The soil NH_4^+-N, moisture, and EC were key mediators, with NH_4^+-N declines linked to biomass reductions via nitrogen limitation. The Spearman correlations confirmed strong associations between the plant traits and soil properties, particularly nitrogen forms. These findings reveal that microplastics disrupt wetland plant performance and soil environments, potentially impairing carbon sequestration and ecosystem stability. Our study underscores the urgent need for microplastic risk assessments in coastal wetlands and highlights soil–microbe–plant interactions as critical mechanisms for future investigation.

Keywords: microplastics; *Scirpus mariqueter*; coastal wetland; biomass; reproductive strategies; soil physicochemical properties; nitrogen limitation; ecosystem functions

1. Introduction

Microplastics, characterized by their small particle size, high mobility [1], and bioavailability [2], have emerged as a major environmental threat to global ecosystems [3,4]. These particles can accumulate in soils, aquatic environments, and biota through various physical, chemical, and biological processes, thereby exerting profound effects on ecosystem functioning and biodiversity [5,6]. In terrestrial systems, microplastics not only alter soil physicochemical properties but also modulate microbial communities and disrupt nutrient cycling, which can indirectly affect plant growth [7,8].

Coastal wetlands, which serve as crucial interfaces between terrestrial and marine ecosystems, play irreplaceable roles in global carbon cycling [9,10], biodiversity maintenance, and shoreline protection [11–13]. However, these ecosystems are increasingly exposed to land-based microplastic inputs, raising concerns about escalating pollution risks [14]. Previous studies have shown that microplastics can alter the soil structure, reduce water retention by up to 30% in some soils, and significantly suppress microbial biomass carbon by 20–45%, thereby indirectly influencing plant–soil interactions [15,16]. For instance, de Souza Machado et al. (2019) [15] reported that polyethylene microplastics decreased soil bulk density and water holding capacity, while Zhao et al. (2021) found that microplastics significantly reduced microbial respiration in soils [17]. In wetland systems, the limited experimental evidence suggests that microplastics may also inhibit root elongation by 15–25% and reduce nitrogen uptake efficiency in plants such as *Scirpus planiculumis* and *Phragmites australis* [18]. However, most research has focused on simplified settings, with limited attention to the combined effects of microplastic type, concentration, and soil conditions. While previous studies have explored the distribution of microplastics in wetland sediments and their preliminary effects on soil properties [19,20], little is known about how dominant coastal wetland plants—particularly foundation species such as *S. mariqueter*—respond to microplastic stress at both the physiological and ecological levels.

Most existing research has focused on the impacts of microplastics on terrestrial crops or freshwater macrophytes, such as reduced yields and impaired root development [21–23]. In contrast, the synergistic effects of microplastic pollution on plant growth, reproductive strategies, and soil environment dynamics in salt marsh ecosystems remain poorly understood. Moreover, microplastics of different polymer types (e.g., polypropylene (PP), polyethylene (PE), polystyrene (PS), and polyethylene terephthalate (PET)) vary in particle size, density, and surface properties, potentially leading to differential ecological impacts on plant physiology [24,25]. The microplastic types used in this study are among the most commonly produced and discarded polymers globally. They are frequently detected in estuarine and coastal environments due to their extensive use in packaging, textiles, and consumer goods, as well as their persistence and distinct density characteristics that influence their transport and sedimentation dynamics [26,27]. Yet, comparative studies examining these effects are scarce. In particular, the mechanisms through which microplastics influence plant performance through modifications of key soil attributes—such as nitrogen availability, water content, and electrical conductivity—have not been adequately elucidated.

To address these gaps, we conducted a controlled experiment using *S. mariqueter*, a dominant clonal species in coastal wetlands, as a model organism [28]. *S. mariqueter* was chosen because it was the dominant native plant species at the sampling site in Hangzhou Bay, forming large natural populations across the area. It served as a foundation species in the local salt marsh ecosystem and primarily competed with the invasive species *Spartina alterniflora*. We applied four types of microplastics (PP, PE, PS, PET) at varying concentrations (0%, 0.1%, 0.5%, and 1%) to evaluate their effects on plant biomass accumulation, vegetative growth, reproductive traits, and reproductive allocation. We also examined the potential mechanisms through which microplastics modulate plant performance via alterations of soil physicochemical properties. We tested three hypotheses: (1) microplastics of different types and concentrations significantly influence the growth and reproductive performance of *S. mariqueter* exhibiting both polymer-specific and dose-dependent effects; (2) microplastics indirectly regulate plant traits by altering soil water content and nitrogen availability; and (3) under microplastic stress, *S. mariqueter* adjusts its reproductive allocation strategy to better cope with environmental changes. This study aims to improve our understanding of the ecological effects of microplastic pollution and uncover plant response mechanisms in coastal wetlands, thereby informing microplastic risk assessment as well as supporting evidence-based conservation and management strategies for wetland ecosystems.

2. Materials and Methods

2.1. Soil Collection

Soil was collected from a typical coastal wetland in Hangzhou Bay, China (30°17′–30°19′ N, 121°05′–121°07′ E). The site was selected due to its representative soil characteristics and dominance of *S. mariqueter* vegetation. It was a relatively undisturbed natural area with stable hydrology and salinity levels, making it suitable for controlled experiments. In March 2024, surface soil (0–20 cm) was sampled from an area free of industrial or domestic pollution (confirmed by field surveys and historical records, with background microplastic concentration <0.01% *w*/*w*) [29]. Plant debris was removed, and the soil was thoroughly mixed. Samples were stored at 4 °C until subsequent experiments. The background concentration of the microplastics in the collected soil (<0.01% *w*/*w*) was determined using density separation with saturated NaCl solution, followed by stereomicroscopic examination at the Wetland Ecosystem Research Station laboratory. This approach followed the internal protocols established for the long-term monitoring at the site and was used here to confirm that control soils represented near-natural background conditions.

2.2. Microplastic Preparation

Four types of virgin microplastics were used: polyethylene terephthalate (PET), polypropylene (PP), polyethylene (PE), and polystyrene (PS), purchased from Mingyuxing Plastic Materials Co., Ltd., Dongguan, China (purity > 99%, free of plasticizers or colorants). The particle size of the microplastics ranged from 53 to 150 μm. Prior to use, the microplastics were cleaned with 0.1 $mol \cdot L^{-1}$ hydrochloric acid (HCl) to remove potential surface-bound metals. As these microplastics were virgin-grade and free from plasticizers, stabilizers, and dyes, no additional treatment for organic contaminants was deemed necessary.

2.3. Experimental Design

A pot experiment was conducted from March to November 2024 in a greenhouse at wetland ecosystem research station of Hangzhou Bay (temperature 22–28 °C, light intensity 400–600 $\mu mol \cdot m^{-2} \cdot s^{-1}$, 14 h light/10 h dark). The experiment comprised four microplastic types (PET, PP, PE, PS) at three concentrations (0.1%, 0.5%, 1% *w*/*w*), plus a control (0% microplastics, with deionized water added to simulate mixing), totaling 13 treatments, each with three replicates (Figure 1). The selected concentrations (0.1%, 0.5%, and 1% *w*/*w*) were chosen to reflect a gradient from low to high microplastic contamination, corresponding to the levels reported in impacted coastal environments and frequently used in experimental studies to simulate realistic exposure scenarios [30]. Microplastics were weighed according to the specified concentration in each treatment and thoroughly mixed with soil. The mixtures were then placed into sterilized plastic pots (55 cm × 45 cm × 35 cm), each containing 15 kg of soil. In each pot, 20 uniform seedlings of *S. mariqueter* were planted. Throughout the experiment, soil salinity was maintained between 3% and 5%, and no additional nutrients were supplied. Irrigation was applied regularly to keep the soil at saturation. To minimize edge effects, the pots were randomly rearranged and rotated 90° clockwise every week. No fertilizers or pesticides were applied during the experimental period to avoid external interference with soil and plant responses.

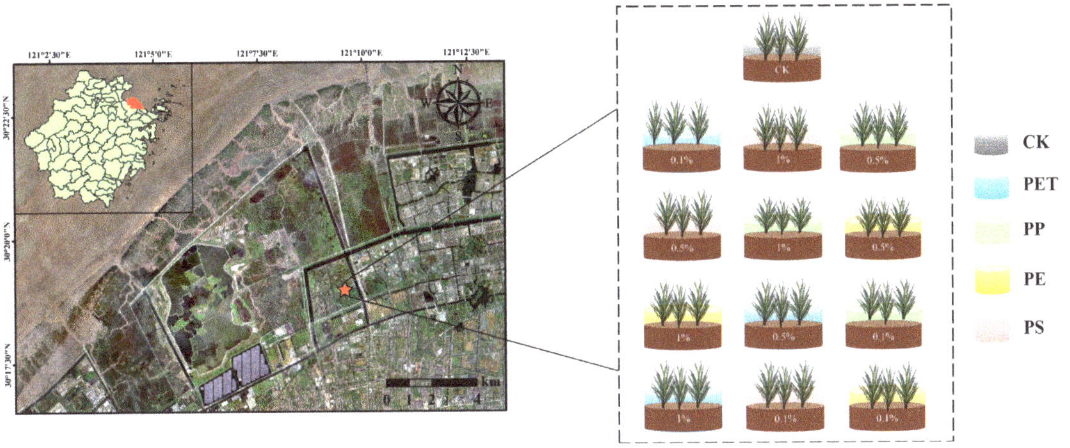

Figure 1. Location of the study area in Hangzhou Bay, China, and schematic of the experimental setup.

2.4. Soil and Plant Measurements

In November 2024, soil samples were harvested. For each replicate pot (biological replicate), five soil cores (0–20 cm, ~200 g each) were collected and composited into a single sample to represent that pot. Plant height was measured from the soil surface to the stem apex using a ruler (precision 0.1 cm) and basal diameter with a vernier caliper (0.01 mm), and ramets, seed, and corm numbers were counted manually. Leaf area was determined using a portable leaf area meter (LI-3000C, LI-COR Biosciences, Lincoln, NE, USA) [31]. Plant samples were oven-dried at 70 °C to constant weight to measure above- and belowground biomass and corm dry weight.

Soil pH and electrical conductivity (EC) were measured in a 1:2.5 (w/v) soil:deionized water suspension using a pH meter (FiveEasy Plus, Mettler Toledo, Greifensee, Switzerland) and a conductivity meter (DDSJ-308A, Shanghai Yidian, Shanghai, China) [32]. Soil moisture was determined by oven-drying at 105 °C for 24 h to constant weight [33]. Soil nitrate nitrogen (NO_3^--N) and ammonium nitrogen (NH_4^+-N) were extracted with 2 mol·L^{-1} KCl (soil:solution 1:10 w/v), shaken for 30 min, filtered, and analyzed using a continuous flow analyzer (AutoAnalyzer 3, SEAL Analytical, Norderstedt, Germany; Mulvaney, 1996) [34]. Available phosphorus (AP) was extracted with 0.5 mol·L^{-1} $NaHCO_3$ and measured at 880 nm using a UV–Vis spectrophotometer (UV-1800, Shimadzu, Kyoto, Japan) [35]. Soil total carbon (TC) was measured using an elemental analyzer (Vario EL III, Elementar, Langenselbold, Germany) [36].

2.5. Statistical Analysis

Data were tested for normality using the Shapiro–Wilk test and, for homogeneity of variances, using Levene's test, with non-normal data log-transformed. Two-way ANOVA was performed in R software (v4.4.3; R Core Team, Vienna, Austria) to evaluate the effects of microplastic type and concentration on plant and soil parameters ($p < 0.05$). Post hoc comparisons were conducted using Tukey's HSD test. Pearson's correlation analysis was used to examine relationships between plant height, biomass, and soil factors (pH, EC, NO_3^--N, etc.; $p < 0.05$). Figures were generated using Origin 2024 (OriginLab Corp., Northampton, MA, USA).

3. Results

3.1. *Effects of Microplastics on Biomass and Vegetative Traits of S. mariqueter*

The microplastic type and concentration affected the belowground, aboveground, and total biomass of *S. mariqueter* (Figure 2). For belowground biomass (Figure 2a), at 0.1% concentration, the PET, PP, and PE treatments significantly reduced the belowground biomass ($p < 0.05$), with reductions of 42.87% and 44.13% for PET and PE, respectively, compared to CK; PS slightly increased the belowground biomass, but the effect was not significant ($p > 0.05$). At 0.5% and 1% concentrations, all microplastic treatments significantly decreased the belowground biomass ($p < 0.05$), with the lowest values observed for PP and PE at 0.5% and for PET at 1%. The aboveground biomass (Figure 2b) was reduced by all microplastic treatments across concentrations. The magnitude of the reduction varied: PET showed a decreasing inhibitory effect with increasing concentration, whereas PP, PE, and PS exhibited a pattern of an initial decline followed by a partial recovery. The total biomass (Figure 2c) was significantly reduced by all treatments except PS at 0.1% ($p < 0.05$). At 0.5%, the PS and PET treatments resulted in significantly higher total biomass than PE and PP ($p < 0.05$). At 1%, PET caused the greatest reduction in the total biomass (31.67%, $p < 0.05$).

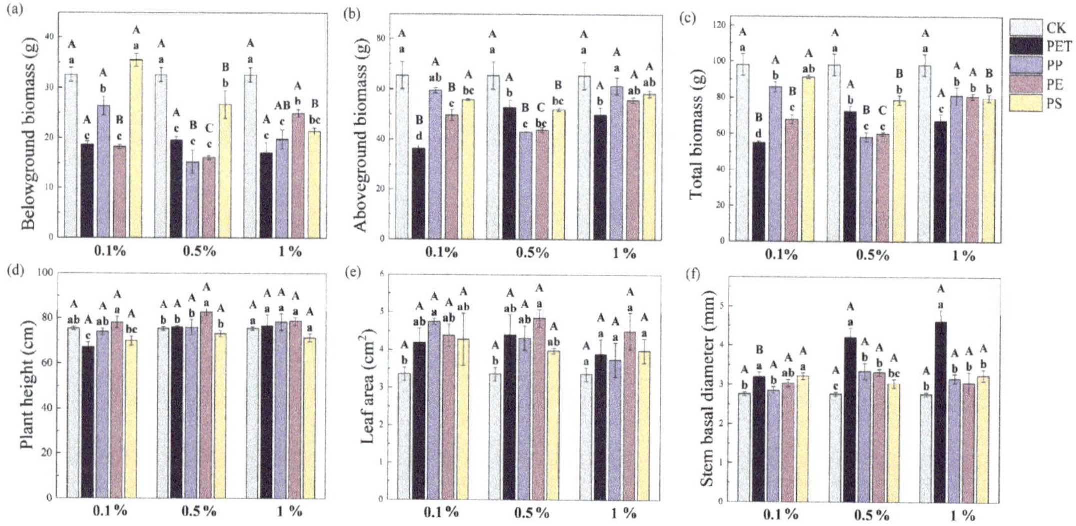

Figure 2. Effects of microplastic type and concentration on *S. mariqueter* biomass and vegetative traits: (**a**) belowground biomass, (**b**) aboveground biomass, (**c**) total biomass, (**d**) plant height, (**e**) leaf area, (**f**) basal diameter. Different lowercase letters indicate significant differences among microplastic types at same concentration ($p < 0.05$); different uppercase letters indicate significant differences among concentrations for same microplastic type ($p < 0.05$).

For the vegetative traits, PET significantly suppressed plant height only at 0.1% ($p < 0.05$), while PE at 0.5% promoted height (Figure 2d, $p < 0.05$). The leaf area (Figure 2e) increased significantly with PP at 0.1% and PE at 0.5% ($p < 0.05$), but the other treatments had no significant effect ($p > 0.05$). The basal diameter (Figure 2f) showed varied responses: at 0.1%, PET and PS significantly increased basal diameter ($p < 0.05$), with PS having the strongest effect; at 0.5% and 1%, PET consistently promoted basal diameter growth ($p < 0.05$).

3.2. Effects of Microplastics on Reproductive Traits and Allocation of *S. mariqueter*

Microplastic addition significantly influenced the asexual and sexual reproduction of *S. mariqueter* (Figure 3a–e). For asexual reproduction, all microplastic types significantly reduced ramet numbers across concentrations ($p < 0.05$, Figure 3a). At 0.1%, the ramet number followed the order CK > PET > PS > PP > PE, with significant differences between PET and PE ($p < 0.05$). The corm number (Figure 3b) increased under PET at 0.1% but was suppressed by PP, PE, and PS, with PE showing the strongest inhibition ($p < 0.05$). The corm biomass (Figure 3c) mirrored the trend in the corm number. For sexual reproduction, the seed number (Figure 3d) responded differently to microplastic types. At 0.1%, PP significantly increased the seed number (25.23%, $p < 0.05$), whereas PE and PS reduced it (>50% decline, $p < 0.05$). At 0.5% and 1%, PP continued to enhance the seed number, while the differences among other treatments were not significant ($p > 0.05$). The seed biomass (Figure 3e) was significantly reduced across all treatments ($p < 0.05$) but remained highest in PP treatments. The reproductive allocation (Figure 3f–h) showed that at 0.1% all microplastic treatments reduced the proportion of seed biomass (Figure 3f); at 0.5% and 1%, PP had the highest seed biomass proportion. The corm biomass proportion (Figure 3g) increased significantly in the PET and PE treatments at 0.5% ($p < 0.05$), but the differences diminished at 1%. The seed-to-corm biomass ratio (Figure 3h) decreased in the PET, PE, and PS treatments but increased in the PP treatments at higher concentrations.

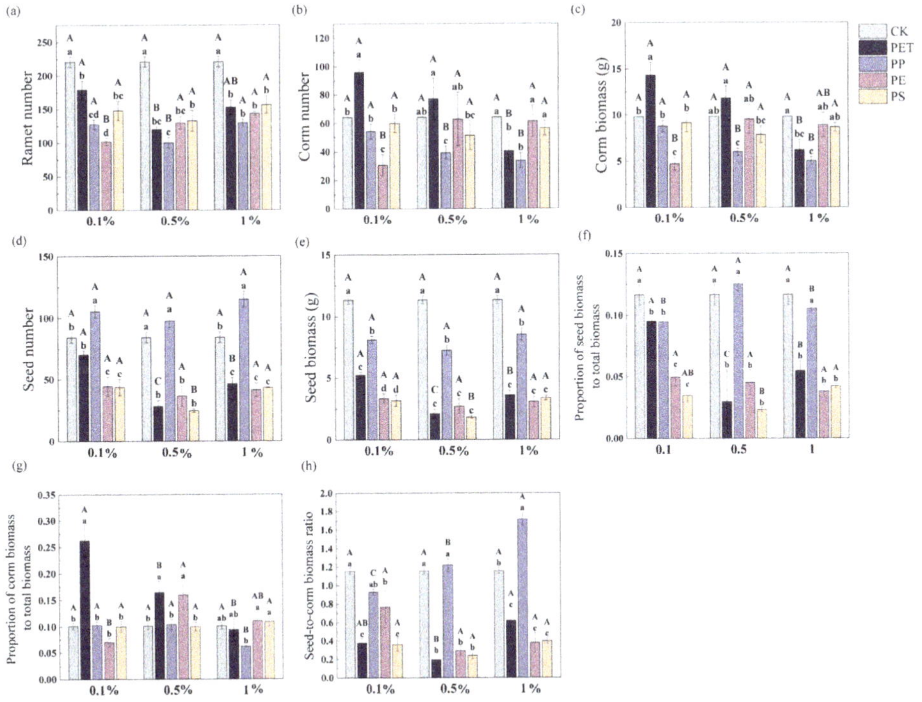

Figure 3. Effects of microplastic type and concentration on asexual reproduction (**a**–**c**), sexual reproduction (**d**,**e**), and reproductive allocation of *S. mariqueter* (**f**–**h**): (**a**) ramet number, (**b**) corm number, (**c**) corm biomass, (**d**) seed number, (**e**) seed biomass, (**f**) proportion of seed biomass to total biomass, (**g**) proportion of corm biomass to total biomass, (**h**) seed-to-corm biomass ratio. Different lowercase letters indicate significant differences among microplastic types at same concentration ($p < 0.05$); different uppercase letters indicate significant differences among concentrations for same microplastic type ($p < 0.05$).

3.3. Effects of Microplastics on Soil Physicochemical Properties

Microplastic addition had complex effects on the physicochemical properties of the coastal wetland soil (Figure 4). For nitrogen, the NH_4^+-N content significantly decreased under PET treatments ($p < 0.05$), with PS showing a concentration-dependent decline, most pronounced at 1% (Figure 4a). Conversely, PE promoted NH_4^+-N accumulation at higher concentrations. The NO_3^--N content (Figure 4b) was highest under low-concentration PS, with a significant increase under PE at 0.5% ($p < 0.05$). The soil moisture (Figure 4c) varied with concentration, peaking under PE at 0.5%, but showed no consistent concentration-dependent trend ($p > 0.05$). The soil pH (Figure 4d) exhibited a slight upward trend but was not significantly affected ($p > 0.05$). The EC and salinity (Figure 4e,f) increased under low-concentration PS but decreased under high-concentration PP. AP (Figure 4g) showed no significant changes ($p > 0.05$), though some treatments displayed fluctuating patterns (e.g., low-high-low or high-low-high). The TC (Figure 4h) varied minimally, with PP showing a slight increase at certain concentrations.

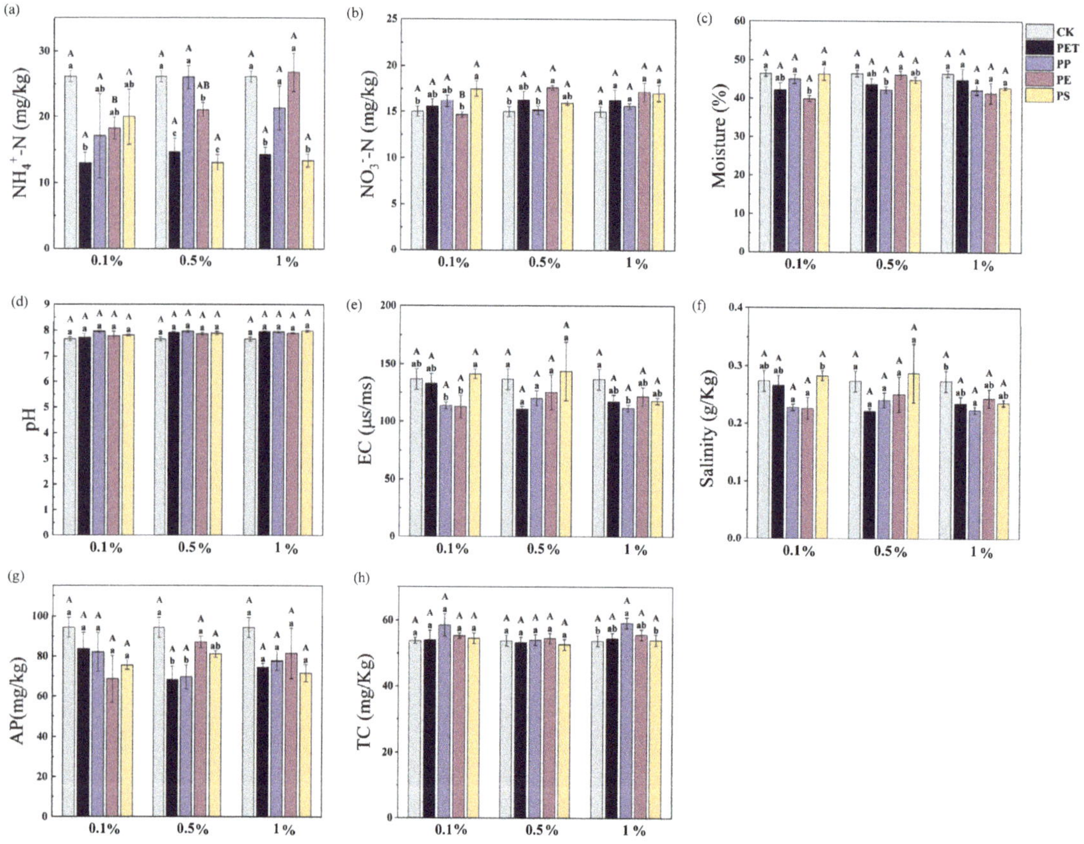

Figure 4. Effects of microplastic type and concentration on soil physicochemical properties: (**a**) NH_4^+-N, (**b**) NO_3^--N, (**c**) soil moisture, (**d**) pH, (**e**) electrical conductivity (EC), (**f**) salinity, (**g**) available phosphorus (AP), (**h**) total carbon (TC). Different lowercase letters indicate significant differences among microplastic types at same concentration ($p < 0.05$); different uppercase letters indicate significant differences among concentrations for same microplastic type ($p < 0.05$).

3.4. Correlation Between Soil Physicochemical Properties and *S. mariqueter* Traits

Spearman correlation analysis (Figure 5) revealed significant relationships between soil properties and *S. mariqueter* traits. The belowground biomass positively correlated with the soil moisture ($p < 0.01$) and salinity ($p < 0.05$), and the total biomass positively correlated with soil moisture ($p < 0.05$). Among the vegetative traits, the plant height strongly positively correlated with the NH_4^+-N content ($p < 0.01$), and the leaf area was strongly positively correlated with the NO_3^--N content ($p < 0.01$). For the reproductive traits, the seed number and seed biomass positively correlated with the NH_4^+-N content and negatively correlated with the NO_3^--N content ($p < 0.05$); the seed number also positively correlated with the TC content ($p < 0.05$). A=the asexual reproductive traits showed no significant correlations with the soil properties ($p > 0.05$).

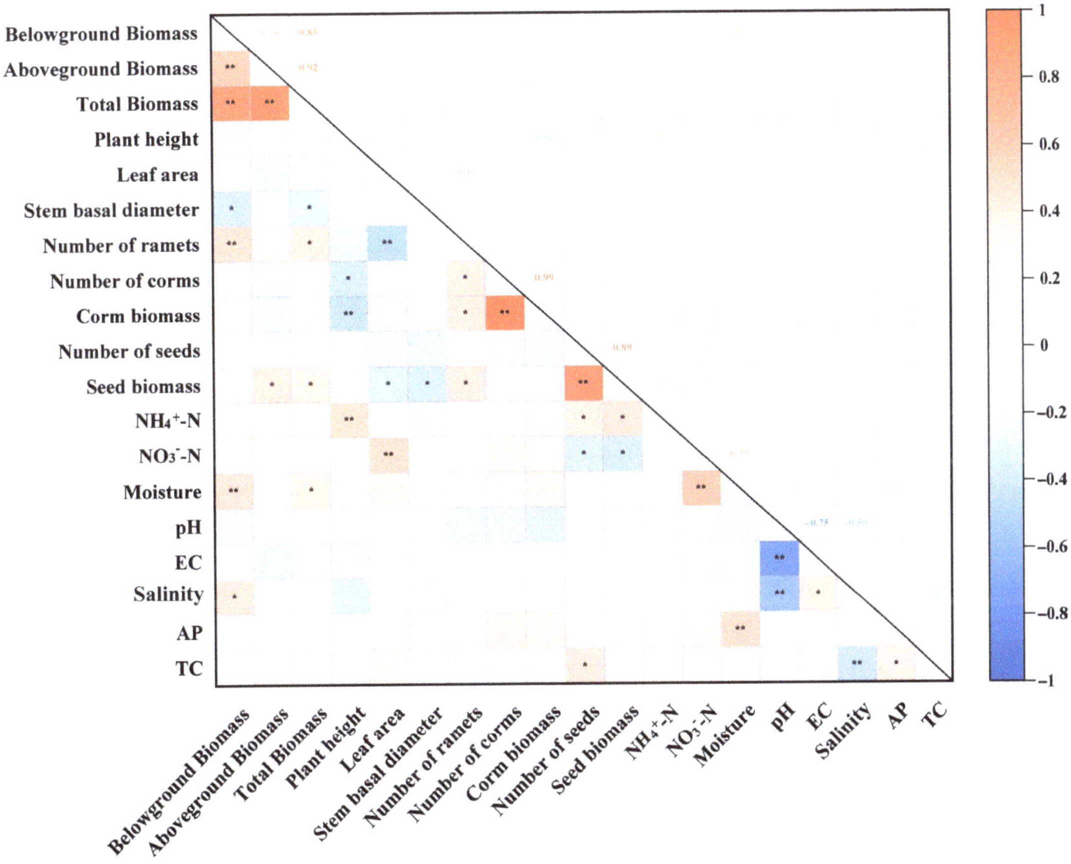

Figure 5. Spearman correlation heatmap of *S. mariqueter* traits and physicochemical soil properties. Color intensity represents correlation coefficient, with significance indicated by asterisks (* $p < 0.05$; ** $p < 0.01$). TC: total carbon; AP: available phosphorus.

4. Discussion

4.1. Microplastics Affect *S. mariqueter* Biomass and Vegetative Growth

Our findings reveal that microplastic type and concentration significantly influence the biomass and vegetative traits of *Scirpus mariqueter*, with PET and PE inducing the strongest suppression and PP showing milder effects (Figures 2 and 3). These negative impacts likely stem from both physical root impediments and altered soil nutrient dynamics, particularly reductions in ammonium nitrogen. Such effects are consistent with terrestrial studies demonstrating microplastic-induced soil compaction, reduced aeration, and disrupted microbial function [37,38]. In salt marsh ecosystems, microplastics may alter soil porosity and water retention [39], reduce NH_4^+-N availability [7] and limit root development [40]. For instance, PET's high density and rigidity may exacerbate soil compaction, hindering root penetration [41,42], while PE's flexible structure could disrupt the fine root architecture [43]. Additionally, changes in plant height, leaf area, and basal diameter suggest that microplastics interfere with morphogenetic processes, such as cell elongation and vascular tissue formation, potentially via altered phytohormone signaling [44]. Beyond these physical and physiological effects, microplastics may also indirectly suppress plant growth by altering soil microbial communities and disrupting nutrient cycling. By altering microbial communities and nutrient cycling, microplastics may influence plant nutrient uptake and overall health [45]. Microplastic contamination has been shown to suppress the beneficial microbial taxa involved in nitrogen fixation and mineralization [30,46], further reducing nutrient availability. Given *S. mariqueter*'s reliance on sediment–root nutrient exchange, such microbial disruptions likely contributed to the observed growth suppression. This pattern, however, is not universally observed across wetland plants. For example, *Lemna minor* exhibited reduced root length but unaffected shoot traits under PE microbead exposure [47], suggesting localized belowground effects. *Phragmites australis* and *Myriophyllum spicatum*, on the other hand, showed no significant changes in biomass or photosynthetic traits under PE or PS treatment in mesocosm settings [48]. These species likely benefit from deeper or more robust root systems or lower dependence on close root–sediment contact. These findings underscore the need to move beyond single-species assessments and consider how functional traits—such as root morphology, sediment dependence, and clonal architecture—influence plant responses to microplastic contamination. As a foundational coastal species, *S. mariqueter* provides a valuable indicator for assessing broader wetland vulnerability to microplastic pollution. Future studies should adopt comparative, trait-based approaches across wetland taxa to better predict ecosystem-level impacts.

4.2. Microplastics Modulate Reproductive Strategies of *S. mariqueter*

Microplastic exposure significantly altered the reproductive traits and allocation strategies of *S. mariqueter*. Treatments with PET, PE, and PS suppressed both sexual (seed) and asexual (corm) reproduction, while PP exposure increased seed number and sexual allocation at specific concentrations (Figure 3b–e). Treatments with PET and PE led to increased investment in corms and reduced seed production (Figure 3f–h), suggesting a shift toward clonal propagation—favoring individual persistence over population expansion under stress [49–51]. This shift mirrors the typical reproductive responses of wetland plants to salinity or heavy metal stress. The contrasting promotion of sexual reproduction under PP exposure highlights a type-specific regulatory effect, likely influenced by microplastic properties such as particle size and surface chemistry [52].

Similar reproductive shifts under microplastic exposure have been reported for other species. Lozano et al. found that film- and fiber-shaped microplastics delayed seed germination in *Daucus carota* while increasing synchrony, implying that particle shape and surface area influence reproductive timing [53]. In *Pisum sativum*, exposure to polystyrene microplastics at higher concentrations (up to 1.0%) and smaller particles ($\leq$50 μm) significantly reduced germination rate, root elongation, and seedling biomass, indicating that both microplastic concentration and particle size critically affect early reproductive development [54]. Bosker et al. further showed that PS beads inhibited seed germination in *Lepidium sativum* by obstructing seed coat pores as well as impairing water and enzyme uptake [55]. Given its shallow, fine roots and clonal growth form, *S. mariqueter* may be particularly sensitive to microplastic-induced soil structure disruptions. PET fragments, being angular and rigid, likely interfere more with nutrient uptake and reproductive resource allocation, while PP's smoother morphology imposes less mechanical and microbial stress [56]. In contrast, recent reviews suggest that plant responses to microplastic stress vary widely depending on root system architecture and particle characteristics. Blanco et al. reported that microplastics can obstruct seed or root pores and impair water and nutrient uptake, but the severity of these effects is strongly influenced by particle size, shape, and plant root morphology [57]. Liu et al. further noted that although microplastics can disrupt the root functions and sediment properties in wetland systems, certain floating or deep-rooted species may maintain growth and reproductive capacity under similar exposures [58]. These findings support the idea that *S. mariqueter*, with its shallow, fine roots, may be more vulnerable to microplastic-induced reproductive suppression. Taken together, our findings indicate that microplastic-induced shifts in plant reproduction result from a combination of polymer traits, stress intensity, and plant-specific functional strategies. As a dominant species in estuarine wetlands, *S. mariqueter*'s reproductive plasticity under microplastic stress may signal broader ecological impacts, underscoring the need for cross-species, trait-based studies in future research.

4.3. *Microplastics Indirectly Affect Plant Growth via Soil Physicochemical Properties*

Microplastics significantly altered soil nitrogen forms, moisture, EC, and salinity, indirectly influencing *S. mariqueter* growth and reproduction. Spearman correlation analysis revealed strong associations between belowground biomass, total biomass, and reproductive traits (seed number and biomass) with soil moisture and nitrogen forms (Figure 5). The decline in NH_4^+-N content closely corresponded to the reductions in above- and belowground biomass, suggesting that limited nitrogen availability is a key mechanism driving growth suppression. This aligns with terrestrial studies where microplastics reduced available nitrogen and impair plant growth [59,60]. Additionally, the microplastic-induced fluctuations in soil moisture and EC likely exacerbated plant growth stress by disrupting soil–plant water relations and the ion balance. Irregular soil moisture can hinder root water uptake, while elevated EC increases osmotic stress, limits nutrient absorption, and disrupts electrochemical gradients at the root–soil interface, negatively affecting plant metabolism and development [61,62]. In our study, soil moisture peaked under PE at 0.5% but did not show a consistent concentration-dependent trend ($p > 0.05$, Figure 4c), indicating that the effects of microplastics on moisture dynamics depend on microplastic type and concentration. Soil pH showed a slight increase, with no significant changes ($p > 0.05$, Figure 4d), and EC and salinity increased under low-concentration PS but decreased under high-concentration PP (Figure 4e,f). These results are consistent with previous studies, although our findings suggest greater variability, indicating that factors such as microplastic type and concentration influence soil properties differently. In addition to directly affecting plant growth through root development disruption, microplastics also alter soil microenvironmental processes. *S. mariqueter* may influence microplastic fate by slowing water flow, enhancing sediment stability, and promoting particle deposition, which could increase microplastic accumulation in the rhizosphere. These results indicate that microplastics exert both direct physical effects (e.g., impeding root development) and indirect effects through changes in soil microenvironmental processes [63], resulting in compounded negative impacts. In addition to directly affecting plant growth through root development disruption, microplastics also alter soil microenvironmental processes.

S. mariqueter may influence microplastic fate by slowing water flow, enhancing sediment stability, and promoting particle deposition, which could increase the microplastic accumulation in the rhizosphere. Recent studies suggest that vegetation can actively mediate microplastic transport and persistence through such physical and biogeochemical interactions [64]. Future research should further explore this bidirectional relationship in coastal wetland systems.

4.4. Limitations and Future Directions

This study focused primarily on the responses of plant biomass and reproductive allocation to microplastic exposure, using a single size range of unaged microplastics (53–150 μm). While this controlled design allowed us to isolate plant-level responses, it limited our ability to assess size-dependent effects or the ecological impacts of aged microplastics. Environmental weathering processes—such as photodegradation, oxidation, and biofilm colonization—can substantially alter the surface properties of microplastics, affecting their interaction with soil, nutrients, and co-contaminants [65,66]. Future studies should include aged particles to better reflect field conditions and improve the environmental relevance of laboratory findings. In addition, although changes in soil nitrogen and moisture were measured in this study, microbial community composition and function—key mediators of nutrient cycling—were not examined. This omission reflects this study's primary focus on plant physiological responses, but we acknowledge that soil microbiota likely play a crucial role in modulating the plant–microplastic interface. Future research should incorporate metagenomic, transcriptomic, or enzyme activity analyses to characterize microbial responses and link them to observed plant-level effects [67]. Moreover, this experiment was conducted under short-term greenhouse conditions. Long-term field studies that integrate variable environmental drivers (e.g., tidal dynamics, seasonal shifts) and multi-trophic interactions (e.g., plant–soil–microbe feedback) will be critical for evaluating how microplastics influence broader ecosystem functions such as carbon storage, nutrient retention, and wetland biodiversity. Addressing these knowledge gaps will help develop more holistic frameworks for predicting and managing microplastic risks in coastal wetland ecosystems.

5. Conclusions

This study revealed that microplastic pollution poses a significant threat to *S. mariqueter*, a keystone species in coastal wetlands, with type-specific and concentration-dependent effects on growth, reproduction, and physicochemical soil properties. PET and PE strongly suppressed belowground biomass, total biomass, and seed production, while PP promoted certain growth and reproductive traits, accompanied by alterations in soil moisture, $NH4^+$-N, and EC. Our results suggest that these effects are linked to changes in soil conditions and potential soil–microbe interactions. While our study identified these patterns, more in-depth experimental studies are needed to clarify the underlying mechanisms and confirm the causal pathways involved. By impairing *S. mariqueter* performance, microplastics may destabilize wetland carbon sinks, biodiversity, and ecosystem services, underscoring the urgent need for risk assessments in coastal ecosystems. Future research should build upon these findings by employing long-term field studies, multi-factorial pollution scenarios, and integrated analyses of plant–soil–microbe dynamics to deepen our understanding of microplastics' impacts and support effective wetland management strategies.

Author Contributions: Conceptualization: N.L., X.S. and M.W.; methodology: J.G., P.J. and J.L.; investigation: P.J., J.G. and J.L.; formal analysis: J.G. and P.J.; data curation: J.L.; writing—original draft: P.J.; writing—review and editing: N.L., X.S. and M.W.; funding acquisition: N.L., X.S. and M.W.; supervision: N.L. and X.S.; project administration: N.L. All authors have read and agreed to the published version of this manuscript.

Funding: This work was funded by Pioneer and Leading Goose R&D Program of Zhejiang (grant No. 2025C02050); Special Fund for Basic Scientific Research Operations of Central Public Welfare Research Institutes (grant No. CAFYBB2024MA029); the Zhejiang Provincial Science and Technology Program (grant No. 2023C03120); the Zhejiang Province Commonwealth Projects (grant No. LTGS24C160001).

Institutional Review Board Statement: Not applicable.

Data Availability Statement: The datasets used and analyzed during the current study are available from the corresponding author.

Acknowledgments: We sincerely thank Xiaohong Zhu from the Wetland Ecosystem Research Station of Hangzhou Bay and Chuanliang Li for their invaluable assistance with plant cultivation and final sampling. Their dedicated efforts in managing the experimental setups and collecting samples were critical to the success of this study. We are also grateful to the anonymous peer reviewers for their insightful and constructive feedback, which greatly improved the quality of this manuscript.

Conflicts of Interest: There are no conflicts of interest to declare.

References

1. Tirkey, A.; Upadhyay, L.S.B. Microplastics: An overview on separation, identification and characterization of microplastics. *Mar. Pollut. Bull.* **2021**, *170*, 112604. [CrossRef]
2. Lin, H.; Yuan, Y.; Jiang, X.; Zou, J.-P.; Xia, X.; Luo, S. Bioavailability quantification and uptake mechanisms of pyrene associated with different-sized microplastics to *Daphnia magna*. *Sci. Total Environ.* **2021**, *797*, 149201. [CrossRef] [PubMed]
3. Lin, Z.; Jin, T.; Zou, T.; Xu, L.; Xi, B.; Xu, D.; He, J.; Xiong, L.; Tang, C.; Peng, J. Current progress on plastic/microplastic degradation: Fact influences and mechanism. *Environ. Pollut.* **2022**, *304*, 119159. [CrossRef] [PubMed]
4. Hendrickson, E.; Minor, E.C.; Schreiner, K. Microplastic abundance and composition in western Lake Superior as determined via microscopy, Pyr-GC/MS, and FTIR. *Environ. Sci. Technol.* **2018**, *52*, 1787–1796. [CrossRef]
5. De Souza Machado, A.A.; Kloas, W.; Zarfl, C.; Hempel, S.; Rillig, M.C. Microplastics as an emerging threat to terrestrial ecosystems. *Glob. Change Biol.* **2018**, *24*, 1405–1416. [CrossRef]
6. Kumar, M.; Kumar, R.; Sharma, S.; Mukherjee, S.; Bolan, S.; Shrivastava, A.; Hettiarachchi, M.; Siddique, K.H.; Bolan, N. Microplastics pollution modulating soil biological health–A review. *Soil Use Manag.* **2025**, *41*, e70009. [CrossRef]
7. Yao, Y.; Wang, L.; Pan, S.; Li, G.; Liu, H.; Xiu, W.; Gong, L.; Zhao, J.; Zhang, G.; Yang, D. Can microplastics mediate soil properties, plant growth and carbon/nitrogen turnover in the terrestrial ecosystem? *Ecosyst. Health Sustain.* **2022**, *8*, 2133638. [CrossRef]
8. Liu, S.; Suo, Y.; Wang, J.; Chen, B.; Wang, K.; Yang, X.; Zhu, Y.; Zhang, J.; Lu, M.; Liu, Y. Impact of Polystyrene Microplastics on Soil Properties, Microbial Diversity and *Solanum lycopersicum* L. Growth in Meadow Soils. *Plants* **2025**, *14*, 256. [CrossRef]
9. Perillo, G.; Wolanski, E.; Cahoon, D.R.; Hopkinson, C.S. *Coastal Wetlands: An Integrated Ecosystem Approach*; Elsevier: Amsterdam, The Netherlands, 2018.
10. Ward, N.D.; Megonigal, J.P.; Bond-Lamberty, B.; Bailey, V.L.; Butman, D.; Canuel, E.A.; Diefenderfer, H.; Ganju, N.K.; Goñi, M.A.; Graham, E.B. Representing the function and sensitivity of coastal interfaces in Earth system models. *Nat. Commun.* **2020**, *11*, 2458. [CrossRef]
11. Wang, J.; Yu, G.; Han, L.; Yao, Y.; Sun, M.; Yan, Z. Ecosystem carbon exchange across China's coastal wetlands: Spatial patterns, mechanisms, and magnitudes. *Agric. For. Meteorol.* **2024**, *345*, 109859. [CrossRef]
12. Hansen, V.D.; Nestlerode, J.A. Carbon sequestration in wetland soils of the northern Gulf of Mexico coastal region. *Wetl. Ecol. Manag.* **2014**, *22*, 289–303. [CrossRef]
13. Barr, J.G.; Troxler, T.G.; Najjar, R.G. Understanding Coastal Carbon Cycling by Linking Top-Down and Bottom-Up Approaches. *Eos Trans. Am. Geophys. Union* **2014**, *95*, 315. [CrossRef]
14. Li, N.; Wu, M.; Zhang, Y.; Yuan, W.; Wu, J.; Shao, X. A review on microplastics pollution in coastal wetlands. *Watershed Ecol. Environ.* **2023**, *5*, 24–37. [CrossRef]

15. De Souza Machado, A.A.; Lau, C.W.; Till, J.; Kloas, W.; Lehmann, A.; Becker, R.; Rillig, M.C. Impacts of microplastics on the soil biophysical environment. *Environ. Sci. Technol.* **2018**, *52*, 9656–9665. [CrossRef]
16. Wang, F.; Wang, Q.; Adams, C.A.; Sun, Y.; Zhang, S. Effects of microplastics on soil properties: Current knowledge and future perspectives. *J. Hazard. Mater.* **2022**, *424*, 127531. [CrossRef] [PubMed]
17. Zhao, T.; Lozano, Y.M.; Rillig, M.C. Microplastics increase soil pH and decrease microbial activities as a function of microplastic shape, polymer type, and exposure time. *Front. Environ. Sci.* **2021**, *9*, 675803. [CrossRef]
18. Qi, Y.; Yang, X.; Pelaez, A.M.; Lwanga, E.H.; Beriot, N.; Gertsen, H.; Garbeva, P.; Geissen, V. Macro-and micro-plastics in soil-plant system: Effects of plastic mulch film residues on wheat (*Triticum aestivum*) growth. *Sci. Total Environ.* **2018**, *645*, 1048–1056. [CrossRef] [PubMed]
19. Zhang, X.; Li, C.; Liu, X.; Zhao, Y.; Wang, Y.; Zhang, Y.; Liu, J. Distribution Characteristics and Sources of Microplastics in Inland Wetland Ecosystem Soils. *Water* **2025**, *17*, 231. [CrossRef]
20. Hu, B.; Guo, P.; Han, S.; Jin, Y.; Nan, Y.; Deng, J.; He, J.; Wu, Y.; Chen, S. Distribution characteristics of microplastics in the soil of mangrove restoration wetland and the effects of microplastics on soil characteristics. *Ecotoxicology* **2022**, *31*, 1120–1136. [CrossRef]
21. Mészáros, E.; Bodor, A.; Kovács, E.; Papp, S.; Kovács, K.; Perei, K.; Feigl, G. Impacts of plastics on plant development: Recent advances and future research directions. *Plants* **2023**, *12*, 3282. [CrossRef]
22. Kumari, A.; Rajput, V.D.; Mandzhieva, S.S.; Rajput, S.; Minkina, T.; Kaur, R.; Sushkova, S.; Kumari, P.; Ranjan, A.; Kalinitchenko, V.P. Microplastic pollution: An emerging threat to terrestrial plants and insights into its remediation strategies. *Plants* **2022**, *11*, 340. [CrossRef]
23. Tang, K.H.D. Effects of microplastics on agriculture: A mini-review. *Asian J. Environ. Ecol.* **2020**, *13*, 130170. [CrossRef]
24. Chen, Z.; Carter, L.J.; Banwart, S.A.; Pramanik, D.D.; Kay, P. Multifaceted effects of microplastics on soil-plant systems: Exploring the role of particle type and plant species. *Sci. Total Environ.* **2024**, *954*, 176641. [CrossRef]
25. Muthuvairavasamy, R. Microplastics in the Biotic Systems. In *Microplastics: Footprints on the Earth and Their Environmental Management*; Springer: Berlin/Heidelberg, Germany, 2022; pp. 55–57.
26. Andrady, A.L. Microplastics in the marine environment. *Mar. Pollut. Bull.* **2011**, *62*, 1596–1605. [CrossRef] [PubMed]
27. Chae, Y.; An, Y.-J. Current research trends on plastic pollution and ecological impacts on the soil ecosystem: A review. *Environ. Pollut.* **2018**, *240*, 387–395. [CrossRef]
28. Li, L.; Li, D.; Kong, R.; Ren, Z.; Liu, L.; Zhu, Y.; Sun, Y.; Peng, N.; He, J.; Ji, Y. Strategies in growth and reproduction of the native endangered plant species *Scripus mariqueter* and the driving factors in a coastal salt marsh wetland, eastern China. *Acta Oecol.* **2024**, *122*, 103979. [CrossRef]
29. Wang, T.; Hu, M.; Song, L.; Yu, J.; Liu, R.; Wang, S.; Wang, Z.; Sokolova, I.M.; Huang, W.; Wang, Y. Coastal zone use influences the spatial distribution of microplastics in Hangzhou Bay, China. *Environ. Pollut.* **2020**, *266*, 115137. [CrossRef]
30. Dalvand, M.; Hamidian, A.H. Occurrence and distribution of microplastics in wetlands. *Sci. Total Environ.* **2023**, *862*, 160740. [CrossRef]
31. Flumignan, D.L.; Adami, M.; Faria, R.T.d. Área foliar de folhas íntegras e danificadas de cafeeiro determinada por dimensões foliares e imagem digital. *Cof. Sci.* **2008**, *3*, 1–6.
32. Miller, R.O.; Kissel, D.E. Comparison of soil pH methods on soils of North America. *Soil Sci. Soc. Am. J.* **2010**, *74*, 310–316. [CrossRef]
33. Bittelli, M. Measuring soil water content: A review. *HortTechnology* **2011**, *21*, 293–300. [CrossRef]
34. Sattolo, T.M.S.; Otto, R.; Mariano, E.; Kamogawa, M.Y. Adaptation and validation of colorimetric methods in determining ammonium and nitrate on tropical soils. *Commun. Soil Sci. Plant Anal.* **2016**, *47*, 2547–2557. [CrossRef]
35. Song, S.-h.; Wen, Y.-j.; Zhang, J.-y.; Wang, H. Rapid spectrophotometric measurement with a microplate reader for determining phosphorus in $NaHCO_3$ soil extracts. *Microchem. J.* **2019**, *146*, 210–213. [CrossRef]
36. Lee, J.-H.; Park, K.-S.; Woo, H.-J. Characteristics of total carbon and total organic carbon using elemental analyzer in Hyung-do intertidal zone sediments. *Econ. Environ. Geol.* **2012**, *45*, 673–684. [CrossRef]
37. Xu, H.; Chen, C.; Pang, Z.; Zhang, G.; Zhang, W.; Kan, H. Effects of microplastics concentration on plant root traits and biomass: Experiment and meta-analysis. *Ecotoxicol. Environ. Saf.* **2024**, *285*, 117038. [CrossRef] [PubMed]
38. Zhao, S.; Rillig, M.C.; Bing, H.; Cui, Q.; Qiu, T.; Cui, Y.; Penuelas, J.; Liu, B.; Bian, S.; Monikh, F.A. Microplastic pollution promotes soil respiration: A global-scale meta-analysis. *Glob. Change Biol.* **2024**, *30*, e17415. [CrossRef]
39. Wang, Z.; Li, J.; Li, W.; Yang, W.; Zhao, G.; Lyu, J. Effects of microplastics on the porosity and connectivity with different soil textures: Based on CT scanning. *Res. Sq.* **2023**, *317*, 137762. [CrossRef]
40. Paudel, P.; Kumar, R.; Pandey, M.K.; Paudel, P.; Subedi, M. Exploring the impact of micro-plastics on soil health and ecosystem dynamics: A comprehensive review. *J. Exp. Biol. Agric. Sci.* **2024**, *12*, 163–174. [CrossRef]

41. Frene, J.P.; Pandey, B.K.; Castrillo, G. Under pressure: Elucidating soil compaction and its effect on soil functions. *Plant Soil* **2024**, *502*, 267–278.
42. Wiersum, L. The relationship of the size and structural rigidity of pores to their penetration by roots. *Plant Soil* **1957**, *9*, 75–85. [CrossRef]
43. Farow, D.; Lebel, R.; Crossman, J.; Proctor, C. Root traits of soybeans exposed to polyethylene films, polypropylene fragments, and biosolids. *Environ. Pollut.* **2024**, *363*, 125141. [CrossRef] [PubMed]
44. Ren, F.; Huang, J.; Yang, Y. Unveiling the impact of microplastics and nanoplastics on vascular plants: A cellular metabolomic and transcriptomic review. *Ecotoxicol. Environ. Saf.* **2024**, *279*, 116490. [CrossRef]
45. Hasan, M.M.; Jho, E.H. Effect of microplastics on the germination and growth of terrestrial plants. *J. Korean Soc. Environ. Eng.* **2022**, *44*, 375–382. [CrossRef]
46. Ya, H.; Jiang, B.; Xing, Y.; Zhang, T.; Lv, M.; Wang, X. Recent advances on ecological effects of microplastics on soil environment. *Sci. Total Environ.* **2021**, *798*, 149338. [CrossRef]
47. Di Mauro, R.; Kupchik, M.J.; Benfield, M.C. Abundant plankton-sized microplastic particles in shelf waters of the northern Gulf of Mexico. *Environ. Pollut.* **2017**, *230*, 798–809. [CrossRef] [PubMed]
48. Liu, L.; Codogno, B.; Wei, W.; Zhang, X.; Gao, J.; Dokuchaeva, V.; Ma, L.; Wu, P.; Yu, Q.; Guo, W. Investigating the Potential Effects of Microplastics on the Growth and Functional Traits in Two Aquatic Macrophytes (*Myriophyllum spicatum* and *Phragmites australis*) in Mesocosm Experiments. *Water* **2024**, *17*, 14. [CrossRef]
49. Yang, Y.Y.; Kim, J.G. The optimal balance between sexual and asexual reproduction in variable environments: A systematic review. *J. Ecol. Environ.* **2016**, *40*, 1–18. [CrossRef]
50. Li, F.; Xie, Y.-H.; Qin, Y.-Y. Adaptive strategies of wetland plants in salt stress environment. *Chin. J. Ecol.* **2009**, *28*, 314.
51. Zhang, S.; Yuan, R.; Peng, J.; Liu, R.; Liu, H.; Qu, J. The Mechanism of a Submerged Aquatic Plant to Micro-Nano Plastics Stress in Ecological Constructed Wetland. *Chem. Eng. J.* **2023**, *480*, 147756. [CrossRef]
52. Carvallo, G.O.; Muñoz-Michea, V. Polypropylene fragments block pollen–pistil interactions and reduce seed production in a monkeyflower species. *Environ. Sci. Technol. Lett.* **2024**, *11*, 426–432. [CrossRef]
53. Lozano, Y.M.; Caesaria, P.U.; Rillig, M.C. Microplastics of different shapes increase seed germination synchrony while only films and fibers affect seed germination velocity. *Front. Environ. Sci.* **2022**, *10*, 1017349. [CrossRef]
54. Li, Y.; Yang, G.; Yu, C.; Lei, X.; Xing, X.; Ma, X.; Sun, Y. The Impact of Microplastic Concentration and Particle Size on the Germination and Seedling Growth of *Pisum sativum* L. *Agron. J.* **2024**, *14*, 923. [CrossRef]
55. Bosker, T.; Bouwman, L.J.; Brun, N.R.; Behrens, P.; Vijver, M.G. Microplastics accumulate on pores in seed capsule and delay germination and root growth of the terrestrial vascular plant *Lepidium sativum*. *Chemosphere* **2019**, *226*, 774–781. [CrossRef]
56. Rana, A.K.; Thakur, M.K.; Saini, A.K.; Mokhta, S.K.; Moradi, O.; Rydzkowski, T.; Alsanie, W.F.; Wang, Q.; Grammatikos, S.; Thakur, V.K. Recent developments in microbial degradation of polypropylene: Integrated approaches towards a sustainable environment. *Sci. Total Environ.* **2022**, *826*, 154056. [CrossRef]
57. Blanco, A.; Högy, P.; Zikeli, S.; Pignata, M.L.; Rodriguez, J.H. Assessment of elevated CO_2 concentrations and heat stress episodes in soybean cultivars growing in heavy metal polluted soils: Crop nutritional quality and food safety. *Environ. Pollut.* **2022**, *303*, 119123. [CrossRef] [PubMed]
58. Liu, L.; Hua, Y.; Sun, J.; Ahmad, S.; He, X.; Zhuo, Y.; Tang, J. Carbon Cycling in Wetlands Under the Shadow of Microplastics: Challenges and Prospects. *Toxics* **2025**, *13*, 143. [CrossRef] [PubMed]
59. Liu, Y.; Chen, S.; Zhou, P.; Li, H.; Wan, Q.; Lu, Y.; Li, B. Differential impacts of microplastics on carbon and nitrogen cycling in plant-soil systems: A meta-analysis. *Sci. Total Environ.* **2024**, *948*, 174655. [CrossRef]
60. Fang, X.Z.; Fang, S.Q.; Ding, Y.; Ma, J.W.; Ye, Z.Q.; Liu, D.; Zhao, K.L. Microplastic exposure inhibits nitrate uptake and assimilation in wheat plants. *Environ. Pollut.* **2024**, *360*, 124626. [CrossRef] [PubMed]
61. Jia, L.; Liu, L.; Zhang, Y.; Fu, W.; Liu, X.; Wang, Q.; Tanveer, M.; Huang, L. Microplastic stress in plants: Effects on plant growth and their remediations. *Front. Plant Sci.* **2023**, *14*, 1226484. [CrossRef]
62. Bakhshaee, A.; Babakhani, P.; Ashiq, M.M.; Bell, K.; Salehi, M.; Jazaei, F. Potential impacts of microplastic pollution on soil–water–plant dynamics. *Sci. Rep.* **2025**, *15*, 9784. [CrossRef]
63. Chen, Y.; Li, Y.; Liang, X.; Lu, S.; Ren, J.; Zhang, Y.; Han, Z.; Gao, B.; Sun, K. Effects of microplastics on soil carbon pool and terrestrial plant performance. *Carbon Res.* **2024**, *3*, 37. [CrossRef]
64. Ullah, R.; Tsui, M.T.K.; Chen, H.; Chow, A.; Williams, C.; Ligaba-Osena, A. Microplastics interaction with terrestrial plants and their impacts on agriculture. *J. Environ. Qual.* **2021**, *50*, 1024–1041. [CrossRef] [PubMed]
65. Liu, H.; Yue, L.; Zhao, Y.; Li, J.; Fu, Y.; Deng, H.; Feng, D.; Li, Q.; Yu, H.; Zhang, Y. Changes in bacterial community structures in soil caused by migration and aging of microplastics. *Sci. Total Environ.* **2022**, *848*, 157790. [CrossRef] [PubMed]

66. Gong, K.; Peng, C.; Hu, S.; Xie, W.; Chen, A.; Liu, T.; Zhang, W. Aging of biodegradable microplastics and their effect on soil properties: Control from soil water. *J. Hazard. Mater.* **2024**, *480*, 136053. [CrossRef]
67. Rillig, M.C.; Lehmann, A.; Ryo, M.; Bergmann, J. Shaping up: Toward considering the shape and form of pollutants. *Environ. Sci. Technol.* **2019**, *53*, 7925–7926. [CrossRef]

diversity

MDPI

Article

Population Viability Analysis Revealed the Vulnerability of Yangtze Finless Porpoise (*Neophocaena asiaeorientalis*) in Poyang Lake

Bin Wu [1,2,*], Weiping Wang [3], Yuehua Wang [2] and Zhihong Zhang [4,*,†]

1 Jiangxi Fisheries Research Institute, Nanchang 330000, China
2 College of Life Sciences, Nanjing Normal University, Nanjing 210023, China; 211202081@njnu.edu.cn
3 Jiangxi Provincial Aquatic Biology Protection and Rescue Center, Nanchang 330000, China; wangweiping916@163.com
4 Institute of Biological Resources of Jiangxi Academy of Sciences, Nanchang 330096, China
* Correspondence: wubinjx@163.com (B.W.); zzh04scau@126.com (Z.Z.)
† Dr. Zhihong Zhang is the primary corresponding author responsible for coordinating the research and communication.

Abstract: Poyang Lake in China is the most critical habitat and final refuge for the Yangtze finless porpoise (*Neophocaena asiaeorientalis*), YFP. In 2022, its population reached approximately 492 individuals, an increase of 35 from the 457 individuals recorded in 2017, showing a steady upward trend. The infrequent movement of YFPs between Poyang Lake and the Yangtze River represents a considerable threat to the long-term viability of this population. Additionally, serious water shortages in the lake during the dry season have led the government to consider the establishment of a hydraulic project. Therefore, a reliable risk assessment and quantitative analysis of conservation scenarios are urgently needed for this population. Population viability analysis of the YFP population in Poyang Lake was conducted using the VORTEX software. The baseline model predicted a probability of extinction of 0.241 over the next 100 years, with no probability of extinction in the first 30 years; the genetic diversity would be on a continuous downward trend and decline by 91.5%. The comprehensive protection model predicted a probability of extinction of 0.0028 and that the genetic diversity would be maintained at about 0.996 in 100 years. Breeding rate, sex ratio at birth, mortality rate, and gene flow were the factors that were sensitive to maintaining population viability. The results showed that the population of YFPs in Poyang Lake was at a high risk of extinction due to the decline in genetic diversity and the higher mortality and lower birth rate caused by habitat degradation. A total ban on productive fishing and the rescue and interchange of YFPs are conducive to enhancing the viability of the YFP population in Poyang Lake.

Keywords: *Neophocaena asiaeorientalis*; VORTEX; gene flow; translocating individual; conservation strategies

1. Introduction

The extinction risk of a population is caused by deterministic and stochastic factors [1]. Certain small populations, particularly of endangered species, are vulnerable to inbreeding and loss of genetic diversity [2–5]. Population viability analysis (PVA) can enable the prediction of the dynamics of population size and extinction probability [6,7]. PVA can help identify the factors that contribute to the loss in viability, thus allowing appropriate conservation and management measures to be selected and implemented [8].

Yangtze finless porpoise (*Neophocaena asiaeorientalis*), YFP, is a relatively small, freshwater, toothed cetacean that occurs only in the middle and lower reaches of the Yangtze River and its adjoining lakes (Poyang and Dongting lakes) [9–11]. The maximum weight of the YFP is 78.5 kg, with an average of 39.1 kg. The maximum body length is 1.77 m, and the average is 1.30 m [11]. The species is genetically isolated from other porpoise populations and reveals the genomic signatures of adaptation to its freshwater environment [10]. The YFP has been upgraded to the Grade-1 National Key Protected Animals, and Poyang Lake is the "last refuge" for YFPs. As a generalist apex predator, the YFP consumes semi-migratory and non-migratory fish prey [12]. Its population size is considered to be in decline across its entire range due to habitat degradation, with only about 1000 individuals extant and critically endangered. However, the results of the 2022 scientific expedition show that the population of Yangtze finless porpoises has reached 1249, an increase of 237 compared to 2017 [13]. In Poyang Lake, the population size of porpoises remained relatively stable between 2005 and 2007 (484), 2012 (450), 2017 (457), and 2022 (492) [13,14]. The likely habitats were distributed discontinuously along the main channel during the dry season, with a total area of 112 km^2. During the flood season, the habitats extended into the central areas of the lake and tributaries, with a total area of 628 km^2 [15]. In Poyang Lake, the stranding risk of the critically endangered YFPs is high due to the significant differences in hydrological and landscape conditions between the flood and dry seasons [16].

In contrast to its previous endangered status and compared with other endangered species in the Yangtze River Basin, such as the extinct Chinese paddlefish and the functionally extinct Yangtze river dolphin, the Yangtze finless porpoise (YFP) has been regarded as the "most successfully conserved natural population" with more remarkable conservation achievements. Although the YFP population in Poyang Lake represents the most successful case of natural population protection, numerous threats still persist at the population level. In recent years, the water level in Poyang Lake has been relatively low; the low water period is ahead of schedule, the water level recedes after the flood, and so on. At the same time, the construction of sluice gates in Poyang Lake has become an issue of concern. Therefore, it is of great theoretical and practical significance to study the dynamic characteristics of the Yangtze finless porpoise population in Poyang Lake using a population viability analysis. Here, we used a VORTEX model to simulate the Poyang Lake YFP population dynamics over the next 100 years to determine the key factors affecting its survival through sensitivity analysis. The process we used can serve as an example for developing a framework of conservation priorities at the population level.

2. Materials and Methods

2.1. Study Area

Poyang Lake is located in the northern part of Jiangxi Province and the southern coast of the Yangtze Plain. Its geographical coordinates are between 28°22′–29°45′ N and 115°47′–116°45′ E. The water in the lake is collected from the Gan River, Xiushui, Raohe, Xin River, Fu River and flows northward through Hukou into the Yangtze River. The main lake area of Poyang Lake is generally gourd-shaped, bounded by Songmenshan island, and divided into north and south parts. The northern section comprises the river channel, with a narrow surface and deep water, and the main southern lake has a wide surface and shallow water. The north–south span is more than 170 km, the east–west width is 74 km, and the average width is 16.9 km. Lake Poyang is the largest freshwater lake in China and one of the two remaining lakes that is freely connected with the Yangtze River in the middle reaches of the Yangtze River (Figure 1). Poyang Lake is a "through-flow type" and seasonal freshwater lake whose water level, area, and volume can vary significantly with seasons and years. Factors such as different measurement times, measurement methods, and the scope of the study area may also lead to differences in data on average water depth. Currently, when the water level of Poyang Lake is 21 m (Wusong Datum), the north–south length is 170 km, the maximum east–west width is 74 km, the minimum is 3 km, and the average width is 20.4 km; the maximum water depth is 23.7 m, and the average water depth is 5.1 m; the water surface area is 3960 square kilometers, and the volume is 26 billion cubic meters [17,18]. Natural waters are mainly distributed in the eastern and the northern parts of the lake. Wet meadows are widely scattered in Nanji Town, and the beaches on both sides of river channels lead to Poyang Lake, taking up the largest area among other types of wetlands during low flow periods [17]. Based on the remote sensing image data over the past 20 years (2002–2019), the analysis showed that the overall water area of Poyang Lake presented a decreasing trend and distinct seasonal variations. The monthly change in water area was evident across the whole year, with the largest average area of 3509 km^2 in July and the smallest average area of 1434 km^2 in January [18]. YFPs mainly inhabit waters with a depth of 2 to 9 m. When the water depth exceeds 3 m, the number of YFP decreases exponentially. They prefer relatively slow flow velocities, generally ranging from 0.3 to 1.2 m per second [11,12].

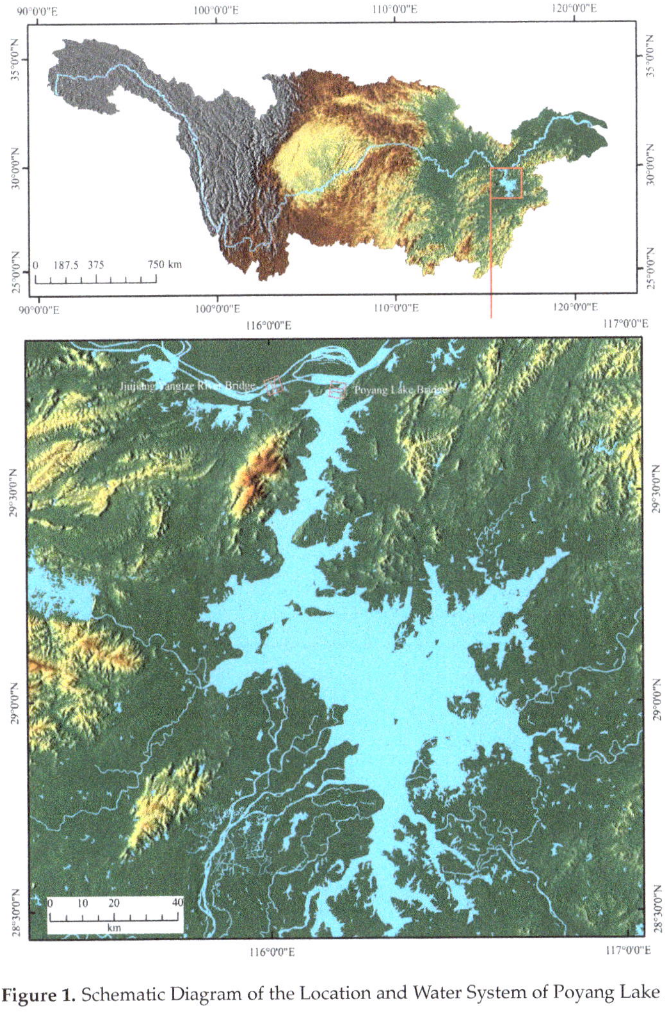

Figure 1. Schematic Diagram of the Location and Water System of Poyang Lake Area.

2.2. VORTEX Simulation Model

VORTEX 10.50 software was used to model the viability of YFP populations in Poyang Lake. VORTEX is an individual-based model that is most appropriate for analyzing extinction dynamics in small populations [19]. The software is developed and recommended by the Conservation Breeding Specialist Group (IUCN/SSC/CBSG), and its owner is the Species Conservation Toolkit Initiative (SCTI) project under the Species360 organization. A VORTEX baseline model was based on three main sources of data: (i) the third Yangtze Freshwater Dolphin Expedition in waters of the Yangtze River, Dongting, and Poyang Lakes, conducted between November and December 2017 [20]; (ii) four capture/release surveys in Poyang Lake conducted in the spring of 2009, 2010, 2011, and 2015, during which all captured individuals were marked with an internal ID and genetic samples were obtained [21]; and (iii) published literature and personal observations.

The YFP population in Poyang Lake was modeled for 100 years so that long-term population trends could be evaluated. The generation length of YFPs was estimated to be 8–9 years [22,23], and the 100-year simulation covered approximately 12 generations. Instead of 1000 replicate simulations that are usually used, we used 10,000 replicate simulations [24]. Extinction was defined as the total removal of at least one sex [25]. The impact of inbreeding was modeled as 3.14 lethal equivalents [26], with 50% of the effect of inbreeding ascribed to recessive lethal alleles.

2.3. Reproductive System and Rates

Polygyny has been shown to be the mating system of YFPs [27]. The mature age of females is 4–6 years, and that of males is 4.5–7 years [27]. However, the female YFPs in Poyang Lake may be able to reproduce at the age of 4 years. The species may still be able to reproduce at the age of 18–19 years [28]. Thus, in this modeling, the first reproductive age of females was 4 years, and that of males was 5 years. The maximum age of reproduction for both male and female YFPs was set as 18 years. Female YFPs do not give birth to more than one young per litter [29]; therefore, the maximum litter size of progeny was set to a value of 1. The birth ratio of YFPs was 1:1 in the Yangtze River Tian-e-zhou ancient channel [29], and the sex ratio from the age of 0 to 1 was 1.1 in Poyang Lake (female to male was 11:10) [28]; therefore, the sex ratio at birth was considered to be 50% female and 50% male.

Assuming that pregnancy and lactation did not overlap, a reproductive cycle that includes both pregnancy and lactation lasted for approximately 18 months. However, taking into account the interval between reproductive activities, it was set to approximately 2 years [30]. In addition, some females may lose their offspring during lactation or due to stillbirths or neonatal deaths and come into estrus sooner afterward, therefore reducing the inter-birth interval. It was assumed that 50% of adult females were breeding annually in the baseline scenario. In other scenarios, the upper and lower limits were 30% to 70%, respectively. An average of 70% of the adult males were capable of reproducing each year. An EV of 10% was used in the simulation. Density-dependent reproduction was not included in the baseline model.

2.4. Mortality Rates

The mortality rate of YFPs in the Yangtze River is possibly 20% for the 0–1 age group, 20% for the 1–2 age group, and 15% for the other age groups [22]. YFP 0–1, 1–2, 2–3, 3–4, 4–5, and >5 age groups mortality was 13.66, 12.37, 11.47, 10.92, 10.68, and 13.15, respectively, in the Tian-e-zhou semi-natural ex situ reserve [29]. However, the mortality rate of the YFPs at the age of 0–2 was estimated at 30.8% [31]. Mei et al. (2012) calculated that the mortality rate of YFPs at the age of 0–1 was 22.4–27.7% [32]. For this modeling, the YFP 0–1,

1–2, and other age groups' mortality was 25%, 20%, and 10%, respectively. Mortality rates for all age classes were considered to be equivalent between males and females. Standard deviations were set to 5% of the mean values for 0–2 age groups and 3% of the mean values for other age groups, which seemed to provide an appropriate interval.

2.5. Population Description

There was a historical large-scale movement of porpoises between Poyang Lake and the main stem of the Yangtze River [33,34]. However, following the construction of two bridges at the mouth of Poyang Lake in 2000 and 2008, this appeared to be no longer the case [35,36]. YFPs showed a seasonal movement pattern consistent with water level fluctuations. They entered the tributaries from the main lake body in low and medium water levels and returned to the main lake during high water level periods [37]. In the baseline model, only one population was considered, with no immigration or emigration of individuals. A total of 457 individuals of the initial YFP population size estimated by Huang et al. (2020) were used in the model [20]. We set 2017 as the starting year for evaluating population viability. The 2022 data can be used to verify which simulation scenario is closer to the real population dynamics. The environmental capacity of YFPs in the Yangtze River was 5000 [22]. Hao et al. (2006) and Li (2017) estimated the environmental capacity to be 50–100 YFP individuals in the Yangtze River Tian-e-zhou ancient channel, whose total area was about 20 km^2 [27,29]. The prediction results of the PVA model were credible when the maximum amount ever distributed in history was considered as the environmental capacity [7]. To be conservative, we set the environmental capacity to 1250 individuals in Poyang Lake.

The YFP population in Poyang Lake was seriously threatened by extreme weather, low water levels, water pollution, and sand-dredged shipping, which altered and destroyed the animal's habitat. It is reported that the extreme low temperatures in 2008 affected the reproductive rate and survival rate of YFPs in the Tian-e-Zhou nature reserve by 71.43% and 82.86%, respectively. The combined effect of multiple adverse factors can be expected to be greater; therefore, for this modeling, catastrophes were included with an annual probability of 1.75%, causing a 50% decrease in survival and reproduction [4]. A compilation of all parameters used in the baseline model is presented in Table 1. A sensitivity analysis was performed to evaluate the effect of reproductive parameters, mortality initial population size, catastrophes, and carrying capacity parameters on the stochastic growth rate (Stoch-r) of the YFP population in Poyang Lake.

Table 1. Summary of parameter input values used in the VORTEX baseline model for YFP populations in Poyang Lake.

Parameter	Baseline Value
Number of populations	1
Initial population size (N)	457
Carrying capacity (K)	1250
Inbreeding depression	3.14 LE
% of the effect of inbreeding due to recessive lethal alleles	50
Breeding system	Polygynous
Age of first reproduction by males/females	5/4 years
Maximum reproductive age	18 years
Annual % of adult females breeding (SD)	50% (10%)
Density-dependent reproduction?	No
Maximum litter size	1
Overall offspring sex ratio	50:50
Adult males in the breeding pool (%)	70

Table 1. *Cont.*

Parameter	Baseline Value
Mortality rates:	
% mortality from age 0–1 (SD)	25 (5)
% mortality from age 1–2 (SD)	20 (5)
% mortality from other ages (SD)	10 (3)
Catastrophe	1.75%; 50%, 50% 16.7%; 95%, 95%
Harvest	None
Supplementation	None

2.6. Sensitivity Analysis and Conservation Scenario

Sensitivity analysis was used to elucidate the effects of environmental parameters on the viability of the population over the next 100 years. Fourteen alternative scenarios were created (Table 2), and the levels in each scenario were selected appropriately. The deterministic growth rate (Det-r), stochastic annual population growth rate (Stoch-r), and YFP population size change (%) of each scenario were compared to the corresponding value of the basic scenario.

Table 2. The deterministic growth rate (Det-r), stochastic annual population growth rate (Stoch-r), and YFP population size change (%) in 100 years under different scenarios.

Projects	Det-r	Stoch-r	Decrease
Model 1 (baseline scenario)	−0.023	−0.0338	91.5%
Model 2 (Model 1 + maximum reproductive age 16 years)	−0.0305	−0.0436	96.6%
Model 3 (Model 1 + maximum reproductive age 20 years)	−0.0177	−0.0274	85.0%
Model 4 (Model 1 + breeding rate 30%)	−0.0758	−0.0846	100.0%
Model 5 (Model 1 + breeding rate 70%)	0.0149	0.0082	−70.2%
Model 6 (Model 1 + mortality rate of 0–1 age 30%, 1–2 age group 25%)	−0.0373	−0.0495	98.4%
Model 7 (Model 1 + mortality rate of 0–1 age 20%, 1–2 age group 15%)	−0.0092	−0.0184	64.2%
Model 8 (Model 1 + initial population size 329 individuals)	−0.023	−0.0352	92.40%
Model 9 (Model 1 + initial population size 634 individuals)	−0.023	−0.0325	90.92%
Model 10 (Model 1 +sex ratio at birth, male, 60%)	−0.0467	−0.0580	99.5%
Model 11 (Model 1 + sex ratio at birth, male, 40%)	−0.0028	−0.0111	36.8%
Model 12 (Model 1 + carrying capacity, 625 individuals)	−0.023	−0.034	91.8%
Model 13 (Model 1 + carrying capacity, 2500 individuals)	−0.023	−0.034	91.5%
Model 14 (Model 1 + effects of natural catastrophes, 0.25)	−0.0279	−0.0451	94.7%
Model 15 (Model 1 + effects of man-made catastrophes, 0.90)	−0.0323	−0.0451	97.1%

The baseline model was used to analyze different scenarios and conservation challenges faced by the YFP in Poyang Lake and surrounding habitat fragments. The scenarios modeled included (i) the impact of gene flow and (ii) the effect of the establishment of protection measures. To simulate the interchange of YFPs, annual capture and supplement of an equal number of 2–3 aged individuals with a sex ratio of 1:1 were set, and different numbers of interchange of YFPs (2, 4, 8, 16, and 32 individuals) were used to represent gene flow at different scales. Meanwhile, to reduce the direct impact of capture and replenishment, such interchange of YFPs was modeled to take place only when the population was larger than 50 individuals.

Based on the above simulation analysis, it was found that the most realistic choice for YFP protection in Poyang Lake was to adopt a comprehensive protection scenario; that is, the main sensitive and rigid factors were all improved marginally but effectively compared with the baseline scenario. The establishment of protection measures was tested, and the impact of increased gene flow (interchange of YFPs was set as 16 male individuals aged

2–3 years), increased breeding rate (increased to 55%), reduced mortality rate (the mortality of 0–1 age, 1–2 age, and other age groups females was reduced to 22.5%, 18%, and 9%), and reduction of the impact of natural catastrophes (reduced to 55%) on the viability of the YFP populations was analyzed.

3. Results

3.1. Population Trend of YFPs in Poyang Lake

The baseline model resulted in a deterministic growth rate of −0.0230 (λ = 0.9773) over 100 years, representing a potential annual decline of 2.3%. Generation time was estimated to be approximately 9 years for both sexes. The ratio of adult males to adult females was 0.878. Stoch-r, PE (probability of extinction), and N-extant (number of extant individuals) were −0.0338, 0.2414, and 50 YFPs, with 86% of genetic diversity remaining (Figure 2). However, in the complete absence of catastrophes and threats, Stoch-r, PE, and N-extant of this population were −0.0072, 0.0004, and 278 YFPs, respectively, with 96% of genetic diversity remaining.

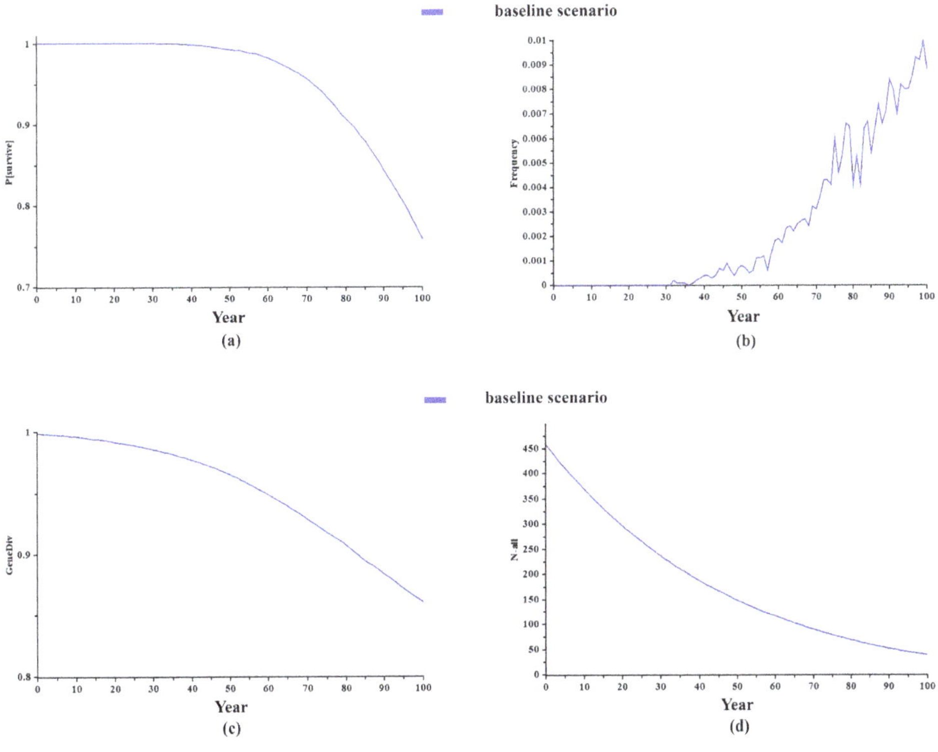

Figure 2. Population viability dynamics of YFPs in Poyang Lake ((**a**) Survival probability; (**b**) Extinction time-frequency; (**c**) Genetic diversity; (**d**) Population size).

3.2. Sensitivity Analysis and Conservation Scenario

Sensitivity analysis indicated that the breeding rate, the sex ratio at birth, and mortality rate were the parameters that had the strongest influence on the dynamics of YFP populations, as shown in Table 2. Further, the increase in mortality in females had an impact on population stability, and compared with young individuals, the increase in mortality in sexually mature females had a greater impact on population stability, as shown in Figure 3.

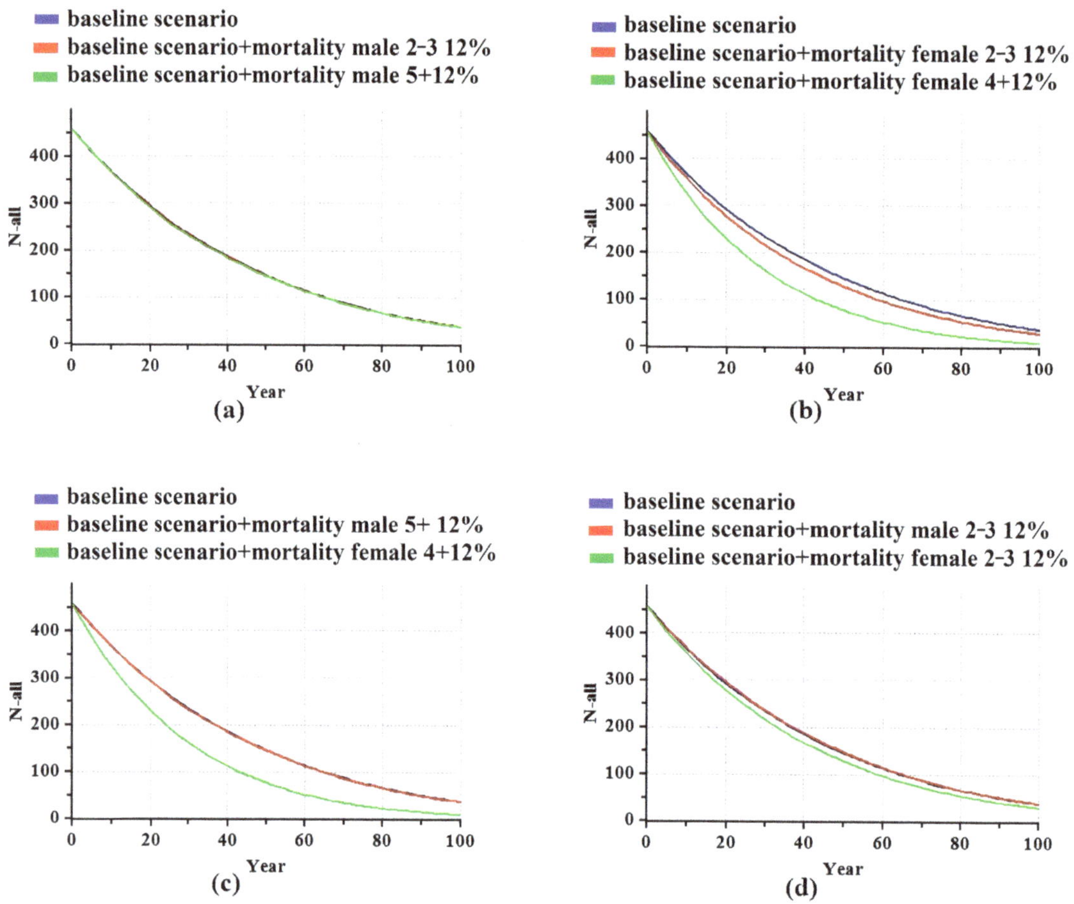

Figure 3. Population size dynamics of YFPs in Poyang Lake in different mortality rates scenarios ((**a**) Male groups of different ages; (**b**) Female groups of different ages; (**c**) Gender differences in the sexually mature group; (**d**) Gender differences in 2 to 3 age groups).

The results of gene flow simulation at different scales showed that the population viability was not affected when the size of individuals exchanged was less than 8 individuals. However, when more than 16 individuals were exchanged, the rate of population decline began to slow down, as shown in Table 3 and Figure 4. In a comprehensive protection scenario, the probability of extinction was 0.0028, the genetic diversity was maintained at approximately 0.996, and the population size increased relatively steadily by 13.3% in 100 years.

Table 3. The deterministic growth rate (Det-r), stochastic annual population growth rate (Stoch-r), and YFP population size change (%) in 100 years under different interchange scenarios.

Projects	Det-r	Stoch-r	Decrease
Model 1 (baseline scenario, no interchange)	−0.023	−0.0338	91.5%
Model 2 (Model 1 + 16 individuals interchange + age 2–3 years + sex ratio 1:1)	−0.023	−0.0125	65.0%

Table 3. *Cont.*

Projects	Det-r	Stoch-r	Decrease
Model 3 (Model 2 + all male)	−0.023	−0.0134	68.2%
Model 4 (Model 2 + all female)	−0.023	−0.0049	35.5%
Model 5 (Model 3 + age after 5 years)	−0.023	−0.0330	89.8%
Model 6 (Model 4 + age after 4 years)	−0.023	−0.0169	55.3%

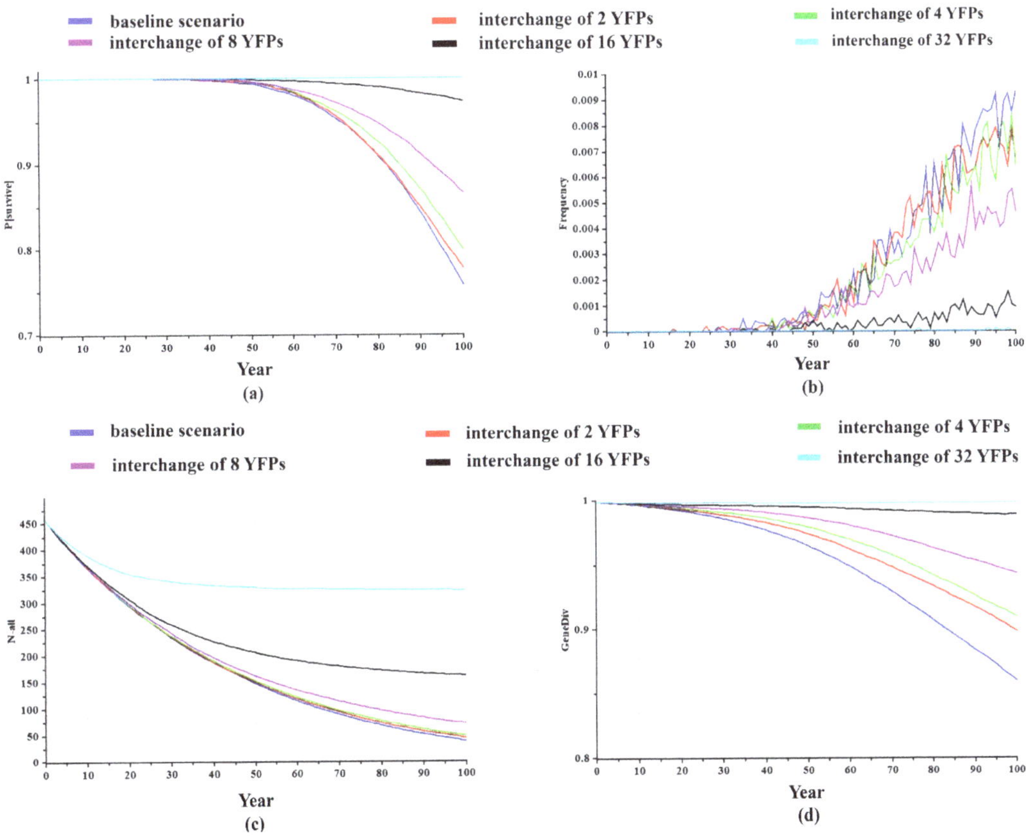

Figure 4. Population viability dynamics of YFPs in Poyang Lake in different interchanges of YFP scenarios ((**a**) Survival probability; (**b**) Extinction time-frequency; (**c**) Population size; (**d**) Genetic diversity).

4. Discussion

4.1. Vulnerability of YFPs in Poyang Lake

Direct and indirect evidence of significantly reduced movements between Poyang and Dongting Lakes and the Yangtze main stem indicated the fragmented distribution of porpoises between these three water bodies [33–36,38–40]. Although the Poyang Lake population remained relatively stable from 2006 to 2017, the fragmentation of porpoise populations in the Yangtze River main stem and significantly reduced migratory movements between Poyang Lake and the Yangtze main stem pose a continuous threat to the long-term viability of this species [20]. In the present paper, the baseline model resulted in a potential annual growth rate of the YFP population in Poyang Lake of −2.3%, which is higher than the value of −3.5% estimated using life table modeling [41]. Nevertheless, this growth rate provided evidence that the YFP population in Poyang Lake would still decline even in the absence of additional catastrophes and threats.

Franklin (1980) proposed the 50/500 rule, which is still useful in conservation biology [42] as long as it is used as a guiding principle to indicate when genetic concerns are likely to have an important role in the short- and long-term viability of populations [43]. The population size of the YFP population in Poyang Lake has been lower than 500 individuals in recent years [20,44], indicating some genetic vulnerability in the population in Poyang Lake. As evidenced by the VORTEX model simulations, the probability of extinction was 24.14%, and 14% of genetic diversity was lost at the end of 100 years. The loss of genetic diversity was, in part, due to the relatively small population size and due to the selection of a closed population. Recent genetic analyses suggested a highly asymmetric gene flow between YFP populations in Poyang Lake and the Yangtze River [40].

4.2. *Conservation Implications for YFPs in Poyang Lake*

The potential conservation scenario in this study examined a case in which gene flow, breeding rate, the mortality rate of young females, and the impacts of natural catastrophes were improved. Flow velocity and distance to fish spawning grounds were the primary variables influencing porpoise distribution during the dry season [15]. Unstable hydrological regimes may force the porpoises to live in habitats with lower water depths for suitable flow velocity conditions in the dry season, and habitats are increasingly infringed by grassland and mudflats [45]. These findings suggest management actions that would have the same effect, such as improving habitat quality in Poyang Lake through the establishment of feed supplements during the dry season, without making YFPs vulnerable to disaster mortality. The findings of this case study of YFPs in Poyang Lake highlight that many species-specific considerations need to be made when selecting gene flow as a conservation strategy. At the metapopulation level, habitat restoration and expansion are critical issues for facilitating dispersal to promote the interchange of YFPs between the Poyang Lake and Yangtze River mainstream YFP populations. Further, the establishment of a hydraulic project may not substantially benefit the porpoise, as it will block the only natural migration channel between Poyang Lake and the Yangtze River [46]. For neutralizing the genetic communication barriers, the interchange of YFPs is an alternative measure to maintain a viable YFPs population in Poyang Lake. Conservation actions for the YFP can be implemented through exchanges among the natural populations in Poyang Lake, Dongting Lake, and the mainstream of the Yangtze River. Additionally, coordinated conservation efforts can be established between the Poyang Lake YFP population and the four ex situ conservation areas, namely, Hubei Shishou Yangtze Tianezhou Dolphin National Nature Reserve, Anhui Tongling Freshwater Dolphin National Nature Reserve, Hubei Jianli Hewangmiao/Hunan Huarong Jichengwan Yangtze Finless Porpoise Provincial Nature Reserve, and Anhui Anqing Xijiang Yangtze Finless Porpoise Municipal Nature Reserve. The VORTEX model provided evidence that the Yangtze finless porpoise (YFP) population in Poyang Lake is vulnerable. The probability of its extinction is significantly influenced by extreme environmental fluctuations, which can sharply reduce the population size either over a short-term or long-term period. In November 2022, the Jiangxi Provincial Department of Agriculture and Rural Affairs orchestrated the relocation of 111 YFP individuals from the southern sand pit of Songmen Mountain. This group comprised 66 male porpoises, 45 female porpoises, 11 mother/calf pairs, 8 females confirmed pregnant via B-ultrasound, and 7 suspected to be pregnant. Given that the Poyang Lake YFP population was approximately 492 in 2022, the severe drought that year placed over one-fifth of the local YFP population in extreme peril. The drought exacerbated risks such as food scarcity and heightened stranding incidents. Consequently, the principle of "protecting water resources equates to safeguarding the YFP in Poyang Lake" has emerged as a national consensus. In the potential conservation scenario, the declining trend of the YFP population in Poyang Lake was changed to a stable and increasing trend. We suggest that the analysis and improvement of sensitivity parameters closely related to environmental conditions and the formulation of a complete set of protection schemes may be the most important insights based on the population viability analysis of the YFP population in Poyang Lake.

There is a strong coupling relationship between the YFP and the fish community in Poyang Lake. The fishing yield and the population of the YFP showed a relatively high tendency of coincidence; that is, both of them first decreased, then increased, and have been stable and increasing since 2012. The areas with a high probability of YFP occurrence overlapped with the areas with a high fish diversity index and catch per unit of effort. It is suggested that the YFP has a certain indicator function to fish diversity and fish density, and higher fish diversity and density can ensure the nutrient balance and improve the feeding efficiency of the YFP. It is important to note that the declining trend of fish diversity in Poyang Lake and its estuaries has not changed. The aggressive predatory fish were recovering rapidly, while the survival of small fish was under great pressure. Seasonal and regional food shortages might be one of the limiting factors to the recovery of the YFP population in Poyang Lake. However, the population of *Coilia nasus* in Poyang Lake was in an undeveloped state, and the population recovery trend was obvious. This was good for increasing the source of food for YFPs in Poyang Lake [47,48]. It should be noted in particular that the presence of ships was found to significantly alter the behavior patterns of YFPs. Specifically, the proportion of foraging behavior decreased from 48% to 33%, while the proportion of travel behavior increased from 32% to 43%. The distance between vessels and YFPs emerged as a critical factor influencing their behavioral responses. When vessels were positioned more than 300 m away, YFPs exhibited a neutral response in 95% of the observed cases, with no discernible escape behavior. Conversely, as the distance closed to within 300 m, the frequency of escape reactions among YFPs increased progressively, reaching approximately 30% when vessels were within 100 m. In contrast, vessel type and cluster size of YFPs had only minor and statistically non-significant effects on their behavioral responses [49]. These findings clearly demonstrate that the presence and proximity of vessels pose a substantial threat to the normal activities of YFPs. To safeguard the survival and well-being of this critically endangered species, it is imperative to implement strict regulations on navigation routes. In particular, maritime traffic should be restricted from traversing nearshore areas where YFPs are densely concentrated, thereby optimizing conservation efforts and enhancing the protection efficiency for YFPs.

Author Contributions: Conceptualization, B.W.; methodology, B.W. and Y.W.; software, B.W. and Y.W.; validation, B.W. and Z.Z.; formal analysis, B.W. and Y.W.; investigation, W.W. and B.W.; resources, Z.Z. and B.W.; data curation, B.W. and W.W.; writing-original draft preparation, B.W.; writing-review and editing, B.W. and Z.Z.; visualization, W.W. and B.W.; supervision, B.W. and Z.Z.; project administration, Z.Z. and B.W.; funding acquisition, Z.Z. All authors have read and agreed to the published version of the manuscript.

Funding: This research was funded by the Global Environment Facility (GEF) pilot project on wetland reserve system in Jiangxi province, grant number GCP/CPR/052/GEF.

Institutional Review Board Statement: This study is a model-based theoretical research, and ethical review and approval are not required.

Data Availability Statement: All data information of this study is included in the article.

Acknowledgments: We would like to express our gratitude to the relevant leaders and teachers from the Jiangxi Fisheries Research Institute, College of Life Sciences at Nanjing Normal University, Institute of Biological Resources of Jiangxi Academy of Sciences, and Jiangxi Provincial Aquatic Biology Protection and Rescue Center for their selfless assistance and meticulous guidance.

Conflicts of Interest: The authors declare no conflict of interest.

References

1. Shaffer, M.L. Minimum population sizes for species conservation. *BioScience* **1981**, *31*, 131–134. [CrossRef]
2. Cassel, A.; Windig, J.; Nylin, S.; Wiklund, C. Effects of population size and food stress on fitness-related characters in the scarce heath, a rare butterfly in Western Europe. *Conserv. Biol.* **2001**, *15*, 1667–1673. [CrossRef]
3. Keller, L.F.; Waller, D.M. Inbreeding effects in wild populations. *Trends Ecol. Evol.* **2002**, *17*, 230–241. [CrossRef]
4. Reed, D.H.; Frankham, R. Correlation between fitness and genetic diversity. *Conserv. Biol.* **2003**, *17*, 230–237. [CrossRef]
5. Vilas, C.; San Miguel, E.; Amaro, R.; Garcia, C. Relative contribution of inbreeding depression and eroded adaptive diversity to extinction risk in small populations of shore campion. *Conserv. Biol.* **2006**, *20*, 229–238. [CrossRef] [PubMed]
6. Brook, B.W.; Burgman, M.; Frankham, R. Differences and congruencies between PVA packages: The importance of sex ratio for predictions of extinction risk. *Conserv. Ecol.* **2000**, *4*, 920. [CrossRef]
7. Brook, B.W.; O'Grady, J.J.; Chapman, A.P.; Burgman, M.A.; Akçakaya, H.R.; Frankham, R. Predictive accuracy of population viability analysis in conservation biology. *Nature* **2000**, *404*, 385–387. [CrossRef]
8. Beissinger, S.R.; McCullough, D.R. *Population Viability Analysis*; The University of Chicago Press: Chicago, IL, USA, 2001.
9. Pilleri, G. Contribution to the knowledge of the cetaceans of Pakistan with particular reference to the genera Neomeris, Sousa, Delphinus and Tursiops and description of a new Chinese porpoise (*Neomeris asiaeorientalis*). *Investig. Cetacea* **1972**, *4*, 107–162.
10. Zhou, X.; Guang, X.; Sun, D.; Xu, S.; Li, M.; Seim, I.; Jie, W.; Yang, L.; Zhu, Q.; Xu, J.; et al. Population genomics of finless porpoises reveal an incipient cetacean species adapted to freshwater. *Nat. Commun.* **2018**, *9*, 1276. [CrossRef]
11. Gao, A.; Zhou, K. Geographical variation of external measurements and three subspecies of *Neophocaena phocaenoides* in Chinese waters. *Acta Theriol. Sin. China* **1995**, *15*, 81–92.
12. Yang, J.; Wang, K.; Mei, Z.; Xu, J.; Zheng, J.; Wan, X.; Hao, Y.; Wang, K.; Wang, D. Temporal variation in the diet of Yangtze finless porpoise calls for conservation of semi-migratory fish. *Freshw. Biol.* **2021**, *66*, 992–1001. [CrossRef]
13. Li, F.G.; Sun, Y.W.; Chen, K.; Yu, J.X.; Ruan, J.M.; Kui, L.L.; Liang, X.M.; Wang, Y.H.; Dai, Y.G. Analysis of Population Genetic Differences of Yangtze Finless Porpoise in Poyang Lake Basin. *Acta Agric. Univ. Jiangxiensis* **2024**, *46*, 1298–1307.
14. Mei, Z.G.; Hao, Y.J.; Zheng, J.S.; Wang, Z.T. Population status and conservation outlooks of Yangtze finless porpoise in the Lake Poyang. *Chin. J. Lake Sci.* **2021**, *33*, 1289–1298.
15. Li, Q.; Lai, G.; Liu, Y.; Devlin, A.T.; Zhan, S.; Wang, S. Identifying the seasonal characteristics of likely habitats for the Yangtze finless porpoise in Poyang Lake. *Aquat. Conserv. Mar. Freshw. Ecosyst.* **2022**, *32*, 523–536. [CrossRef]
16. Li, Q.; Li, W.; Lai, G.; Liu, Y.; Devlin, A.T.; Wang, W.; Zhan, S. Identifying High Stranding Risk Areas of the Yangtze Finless Porpoise via Remote Sensing and Hydrodynamic Modeling. *Remote Sens.* **2022**, *14*, 2455. [CrossRef]
17. Li, C.L.; Kuang, R.Y.; Wu, Q.W. Analysis of information extraction and monitoring of Poyang Lake wetland during low flow period in recent ten years. *J. Jiangxi Univ. Sci. Technol.* **2019**, *40*, 30–37.
18. Dai, Z.; Xia, L.; Kong, P.; Zhan, M. Remote Sensing Monitoring and Analysis of Poyang Lake Water Area Changes over the Past 20 Years. *Meteorol. Disaster Reduct. Res. China* **2021**, *44*, 127–132.
19. Lacy, R.C. Structure of the VORTEX simulation model for population viability analysis. *Ecol. Bull.* **2000**, *48*, 191–203.
20. Huang, J.; Mei, Z.; Chen, M.; Han, Y.; Zhang, X.; Moore, J.E.; Zhao, X.; Hao, Y.; Wang, K.; Wang, D. Population survey showing hope for population recovery of the critically endangered Yangtze finless porpoise. *Biol. Conserv.* **2020**, *241*, 108315. [CrossRef]
21. Chen, M.; Zheng, Y.; Hao, Y.; Mei, Z.; Wang, K.; Zhao, Q.; Zheng, J.; Wang, D. Parentage-based group composition and dispersal pattern studies of the Yangtze Finless Porpoise population in Poyang Lake. *Int. J. Mol. Sci.* **2016**, *17*, 1268. [CrossRef]
22. Zhang, X.F.; Wang, K.X. Viability analysis of finless porpoise populations in the Yangtze River. *Chin. Acta Ecol. Sin.* **1999**, *19*, 529–533.
23. Wu, B.; Wang, W.P.; He, G.; Wang, H.H. Analysis of population viability analysis of Yangtze finless porpoise in different simulated scenarios. *Prog. Fish. Sci.* **2021**, *42*, 28–35. (In Chinese)
24. Burgman, M.A.; Ferson, S.; Akçakaya, H.R. *Risk Assessment in Conservation Biology*; Chapman and Hall: New York, NY, USA, 1993.
25. Ginzburg, L.R.; Slobodkin, L.B.; Johnson, K. Quasi-extinction probabilities as a measure of impact on population growth. *Risk Anal.* **1982**, *2*, 171–181. [CrossRef]
26. Ralls, K.; Ballou, J.D.; Templeton, A. Estimates of lethal equivalents and the cost of inbreeding in mammals. *Conserv. Biol.* **1988**, *2*, 185–193. [CrossRef]
27. Hao, Y.J.; Wang, D.; Zhang, X.F. Review on breeding biology of Yangtze finless porpoise (*Neophocaena phocaenoides asiaeorientalis*). *Acta Theriol. Sin. China* **2006**, *26*, 191–200.

28. Wu, B.; Li, H.; He, G. Reproductive characteristics and conservation implications of the Yangtze finless porpoise in Poyang Lake. *Bull. Biol.* **2021**, *56*, 3–5. (In Chinese)
29. Li, Y.T. Study on the Habitat Selection, Environmental Capacity, and Population Viability of the Yangtze Finless Porpoises in Tian-E-Zhou Semi-Natural Ex-Situ Reserve. Ph.D. Thesis, Institute of Hydrobiology, the Chinese Academy of Sciences, Wuhan, China, 2017.
30. Zhang, X.F. Study on age identification, growth, and reproduction of finless porpoises. *Chin. Acta Hydrobiol. Sin.* **1992**, *16*, 289–297. [CrossRef]
31. Yang, G.; Zhou, K.Y.; Gao, A.L.; Chang, Q. Study on life table and population dynamics of finless porpoises. *Acta Zool. Sin.* **1998**, *18*, 1–7. (In Chinese)
32. Mei, Z.; Huang, S.L.; Hao, Y.; Turvey, S.T.; Gong, W.; Wang, D. Accelerating population decline of Yangtze finless porpoise (*Neophocaena asiaeorientalis asiaeorientalis*). *Biol. Conserv.* **2012**, *153*, 192–200. [CrossRef]
33. Xiao, W.; Zhang, X. Distribution and population size of Yangtze finless porpoise in Poyang Lake and its branches. *Acta Theriol. Sin.* **2002**, *22*, 7–14.
34. Wei, Z.; Wang, D.; Zhang, X.; Zhao, Q.Z.; Wang, K.X.; Kuang, X.A. Population size, behavior, movement pattern and protection of Yangtze finless porpoise at Balijiang section of the Yangtze River. *Resour. Environ. Yangtze Basin* **2002**, *11*, 427–432.
35. Dong, S. Studies on the Distribution and Movement of the Yangtze Finless Porpoises (*Neophocaena phocaenoides asiaeorientalis*) in Hukou Area Obtained by Acoustic Method. Master's Thesis, Institute of Hydrobiology, the Chinese Academy of Sciences, Wuhan, China, 2009.
36. Kimura, S.; Akamatsu, T.; Li, S.; Dong, L.; Wang, K.; Wang, D.; Arai, N. Seasonal changes in the local distribution of Yangtze finless porpoises related to fish presence. *Mar. Mammal Sci.* **2012**, *28*, 308–324. [CrossRef]
37. Liu, X.; Mei, Z.; Zhang, J.; Sun, J.; Zhang, N.; Guo, Y.; Wang, K.; Hao, Y.; Wang, D. Seasonal Yangtze finless porpoise (*Neophocaena asiaeorientalis asiaeorientalis*) movements in the Poyang Lake, China: Implications on flexible management for aquatic animals in fluctuating freshwater ecosystems. *Sci. Total. Environ.* **2022**, *807*, 150782. [CrossRef]
38. Zhang, X.Q. Population Ecology of Yangtze Finless Porpoise in Dongting Lake and Adjacent Waters. Ph.D. Thesis, Institute of Hydrobiology, the Chinese Academy of Sciences, Wuhan, China, 2011.
39. Wang, Z. Study on the Impacts of Sand Mining Activities in Dongting Lake on Yangtze Finless Porpoise. Master's Thesis, Institute of Hydrobiology, the Chinese Academy of Sciences, Wuhan, China, 2015.
40. Chen, M.; Fontaine, M.C.; Ben Chehida, Y.; Zheng, J.; Labbé, F.; Mei, Z.; Hao, Y.; Wang, K.; Wu, M.; Zhao, Q.; et al. Genetic footprint of population fragmentation and contemporary collapse in a freshwater cetacean. *Sci. Rep.* **2017**, *7*, 14449. [CrossRef] [PubMed]
41. Mei, Z. Study on Population Dynamics and Endangered Mechanism of Yangtze Finless Porpoise. Ph.D. Thesis, Institute of Hydrobiology, the Chinese Academy of Sciences, Wuhan, China, 2013.
42. Franklin, I.R. Evolutionary change in small populations. In *Conservation Biology—An Evolutionary-Ecological Perspective*; Soule, M.E., Wilcox, B.A., Eds.; Sinauer Associates: Sunderland, MA, USA, 1980; pp. 135–149.
43. Jamieson, I.G.; Allendorf, F.W. How does the 50/500 rule apply to MVPs? *Trends Ecol. Evol.* **2012**, *27*, 578–584. [CrossRef]
44. Mei, Z.; Zhang, X.; Huang, S.-L.; Zhao, X.; Hao, Y.; Zhang, L.; Qian, Z.; Zheng, J.; Wang, K.; Wang, D. The Yangtze finless porpoise: On an accelerating path to extinction? *Biol. Conserv.* **2014**, *172*, 117–123. [CrossRef]
45. Li, Q.; Deng, M.; Li, W.; Pan, Y.; Lai, G.; Liu, Y.; Devlin, A.T.; Wang, W.; Zhan, S. Habitat configuration of the Yangtze finless porpoise in Poyang Lake under a shifting hydrological regime. *Sci. Total. Environ.* **2022**, *838*, 155954. [CrossRef] [PubMed]
46. Li, Q.; Lai, G.; Liu, Y.; Devlin, A.T.; Zhan, S.; Wang, S. Assessing the impact of the proposed Poyang lake hydraulic project on the Yangtze finless porpoise and its calves. *Ecol. Indic.* **2021**, *129*, 107873. [CrossRef]
47. Wu, B.; Zhang, H.X.; Wu, Z.J.; Zhang, G.F.; Fu, P.F.; Wan, C.L.; Zhang, Y.P.; Fu, H.Y. Fishery resources survey in Poyang Lake and its estuary from 2020 to 2021. *J. Hydroecol. Chin.* **2025**, *46*, 99–110.
48. Wu, B. Study on Population Dynamics of Yangtze Finless Porpoise (YFP) and Fish Community in Its Habitat in Poyang Lake. Ph.D. Thesis, Nanjing Normal University, Nanjing, China, 2023.
49. Chen, B.Y.; Chen, L.W.; Zhang, J.; You, Q.Y.; Hao, X.Q.; Liu, S. Analysis of the behavioral impacts of vessels on Yangtze finless porpoises (*Neophocaena asiaeorientalis*) in the Nanjing section of the Yangtze River. *Acta Hydrobiol. Sin.* **2024**, *48*, 1672–1679. (In Chinese)

diversity

MDPI

Article

Spartina alterniflora-Derived Biochar Alters Biomass Allocation and Root Traits of Native *Scirpus mariqueter*

Yaoyao Tang [1], Jingwen Gao [2], Pengcheng Jiang [2], Junzhen Li [2], Ming Wu [2], Shengwu Jiao [2], Long Zhang [2], Niu Li [2,*] and Xuexin Shao [2,*]

[1] College of Forestry and Biotechnology, Zhejiang Agriculture and Forestry University, Hangzhou 311300, China; tangyaoyao@stu.zafu.edu.cn
[2] Wetland Ecosystem Research Station of Hangzhou Bay, Research Institute of Subtropical Forestry, Chinese Academy of Forestry, Hangzhou 311400, China; jwgao@caf.ac.cn (J.G.); jpcacn@stu.zafu.edu.cn (P.J.); 22461337055@stu.wzu.edu.cn (J.L.); mingwu@caf.ac.cn (M.W.); Jiao@caf.ac.cn (S.J.); Zhang@caf.ac.cn (L.Z.)
* Correspondence: niuli@caf.ac.cn (N.L.); shaoxuexin@126.com (X.S.); Tel./Fax: +86-0571-63342931 (X.S.)

Abstract: Coastal wetlands provide vital ecosystem services, yet large-scale removal of invasive *Spartina alterniflora* disrupts soil carbon pools and fragments habitats. Converting this biomass to biochar may enhance restoration outcomes, though ecological effects remain poorly understood. We evaluated how *Spartina alterniflora*-derived biochar (0%, 0.5%, 1%, and 3%) influences growth performance, clonal reproduction, root morphology, and rhizosphere properties of native *Scirpus mariqueter*. Moderate biochar addition (1%) significantly boosted plant performance, increasing total biomass by 64.5%, aboveground biomass by 36.7%, and belowground biomass by 115.0%, while root length increased by 135.8%. Biochar improved soil moisture and nutrient availability, including nitrate nitrogen (NO_3^--N), ammonium nitrogen (NH_4^+-N), and available phosphorus (AP), while stimulating nitrification and promoting clonal propagation. In contrast, high-dose biochar (3%) elevated soil salinity and electrical conductivity, leading to suppressed plant growth and reproductive allocation. Correlation analysis revealed strong positive associations between root volume and soil nutrient levels. Our findings demonstrate that moderate application of *Spartina alterniflora*-derived biochar enhances plant productivity and soil function, potentially improving carbon sequestration in restored coastal wetlands. This study provides insights into ecological recycling of invasive biomass and supports biochar as a viable tool for sustainable wetland restoration, though potential risks at high concentrations warrant further investigation.

Keywords: allochthonous inputs; carbon sequestration; coastal wetlands; invasive species; native halophyte; root morphology

1. Introduction

Coastal wetlands are among the most productive and ecologically valuable ecosystems on Earth, offering critical ecosystem services such as biodiversity conservation, climate regulation, and water purification. Owing to their high primary productivity and efficient carbon sequestration capabilities, coastal wetlands are widely recognized as significant natural carbon sinks [1,2]. However, these ecosystems are highly vulnerable to external disturbances. The aggressive invasion of *Spartina alterniflora* has led to a range of ecological consequences, including obstruction of tidal channels, displacement of native species, loss of biodiversity, and degradation of habitats for waterbirds [3,4]. Remote sensing data indicate that approximately 48.3% of China's coastal salt marshes have been invaded

https://doi.org/10.3390/d17050357

by this species [5]. *Spartina alterniflora* has become the dominant invasive species in the Hangzhou Bay New District, currently occupying approximately 4670 hectares—nearly one-third of the total invaded area in Zhejiang Province. Following its initial appearance in the mid-1990s, the species expanded rapidly, reaching over 5000 hectares by the early 2000s. Although its distribution temporarily declined due to land reclamation, recent monitoring indicates a marked resurgence [6]. These patterns highlight both the aggressive expansion capacity of *Spartina alterniflora* and the urgent need for sustainable, long-term management strategies in heavily invaded coastal regions.

In response, the Chinese government has launched large-scale eradication programs to control the spread of *Spartina alterniflora*. Integrated management strategies have demonstrated greater efficacy than single-method approaches in controlling *Spartina alterniflora*. Techniques combining mowing with subsequent flooding, soil tilling, or shading have proven particularly effective in limiting its spread across both salt marsh and mangrove ecosystems [7]. However, these interventions have also led to unintended ecological consequences, including habitat fragmentation, altered landscape structures, and significant reductions in above- and belowground carbon stocks [8]. These outcomes compromise the carbon sequestration capacity of coastal wetlands and pose challenges to national carbon neutrality targets. Among the various management strategies for *Spartina alterniflora*, converting its biomass into biochar offers a sustainable, dual-purpose approach—supporting both invasive species control and ecological restoration. Recent studies have highlighted its potential to improve soil properties, enhance plant growth, and increase carbon sequestration in coastal salt marshes [9,10]. Evaluating the restoration potential of *Spartina alterniflora*-derived biochar requires examining its effects on native keystone species that are ecologically sensitive to soil and hydrological changes.

Scirpus mariqueter is a perennial halophytic sedge endemic to the coastal wetlands of China, where it functions as a keystone and pioneer species during the early stages of intertidal zone colonization, particularly in the Yangtze River Estuary and Hangzhou Bay. Through its extensive aboveground biomass and belowground rhizome networks, *Scirpus mariqueter* delivers a range of vital ecosystem services, including shoreline stabilization, sediment accretion, wave attenuation, and carbon sequestration. Additionally, it provides essential foraging and nesting habitats for migratory waterbirds, thereby contributing to regional biodiversity and wetland resilience [11,12]. Despite its ecological importance, the distribution of *Scirpus mariqueter* has declined significantly due to escalating anthropogenic pressures, biological invasion by *Spartina alterniflora*, and rising sea levels [13,14]. Recent studies have highlighted the importance of understanding species-specific responses to environmental gradients in order to inform effective restoration strategies [15]. For *Scirpus mariqueter*, factors such as soil salinity, tidal inundation frequency, nutrient availability, and interspecific competition can significantly influence its growth performance, clonal expansion, and reproductive success [16,17]. In the context of climate change and coastal degradation, elucidating the physiological thresholds and adaptive capacity of *Scirpus mariqueter* is essential to support scalable nature-based solutions that enhance ecosystem functionality and contribute to long-term sustainability and blue carbon goals [18,19].

Invasive weeds, once viewed solely as ecological threats, are now recognized as promising feedstocks for biochar production. Biochar derived from such species has been shown to improve soil physicochemical properties, enhance nutrient retention, and promote plant growth, offering a sustainable pathway for biomass valorization and ecosystem restoration [20–24]. In particular, *Spartina alterniflora*-derived biochar transforms invasive biomass into a valuable resource for coastal rehabilitation [25–27]. However, most existing studies have focused on soil nutrient enhancement or individual plant traits, with limited insight into how varying application rates influence plant–soil interactions. The underlying mechanisms by which biochar affects rhizosphere nutrient dynamics and plant-mediated carbon sequestration remain poorly understood [28,29].

In this study, we conducted a controlled pot experiment to quantify the effects of four biochar application rates (0%, 0.5%, 1%, and 3%) on the performance of *Scirpus mariqueter*. These application levels were determined based on insights from previous studies and the projected requirements for enhancing soil carbon sequestration [22,30]. Specifically, we assessed plant biomass production, reproductive allocation, root morphological traits, and associated changes in soil physicochemical properties. The objectives of this study were twofold: (1) to identify the optimal biochar concentration for enhancing the growth and clonal propagation of *Scirpus mariqueter*, and (2) to elucidate the underlying mechanisms by biochar may enhance plant-driven carbon sequestration. Findings from this research offer a scientific foundation for effective coastal wetland restoration and provide novel insights into the ecological reutilization of invasive plant biomass for sustainable blue carbon management.

2. Materials and Methods

2.1. Study Site and Sediment Collection

Surface sediments were collected from a representative coastal wetland in Hangzhou Bay, eastern China (30°25′–30°50′ N, 121°00′–121°30′ E). The region experiences a subtropical monsoon climate, with a mean annual temperature of 16.2 °C and average annual precipitation of approximately 1100 mm. The tidal regime is semidiurnal. In March 2024, surface sediments (0–20 cm depth) were sampled for use in the pot experiment.

2.2. Biochar Preparation

In 2022, *Spartina alterniflora* was collected from the same wetland. After thorough rinsing, the biomass was air-dried and cut into segments shorter than 5 cm. It was then oven-dried at 60 °C to constant weight, ground using a pulverizer, and sieved through a 60-mesh screen to ensure uniform particle size. The processed material was pyrolyzed at 500 °C for 2 h under oxygen-limited conditions in a muffle furnace to produce biochar. Prior to use, the physicochemical properties of the biochar were characterized. pH and electrical conductivity (EC) were determined in a 1:5 (w/v) biochar-to-deionized water suspension after shaking for 1 h, using a pH meter (FiveEasy Plus FP20, Mettler-Toledo International Inc., Greifensee, Switzerland) and a conductivity meter (DDSJ-319L, Shanghai INESA Scientific Instruments Co., Ltd., Shanghai, China), respectively. Total carbon (TC) and total nitrogen (TN) were analyzed using an elemental analyzer (Vario EL III, Elementar Analysensysteme GmbH, Langenselbold, Germany), and total phosphorus (TP) was measured via colorimetric analysis following acid digestion. The resulting biochar demonstrated a pH of approximately 9.5 and an electrical conductivity (EC) of 2.07 mS/cm. The concentrations of total carbon (TC), total nitrogen (TN), and total phosphorus (TP) were determined to be 490 g/kg, 12.75 g/kg, and 1.68 g/kg, respectively.

2.3. Experimental Design

The experiment was initiated in March 2024 to evaluate the effects of different biochar application rates (0%, 0.5%, 1%, and 3%, w/w) on the growth of *Scirpus mariqueter* and associated soil properties. Biochar amendment treatments were designated as MBCX% (where X = 0.5, 1, or 3% w/w), representing a gradient of application rates. The unamended soil served as the control and was labeled as CK. A 2% application rate was not included in the design in order to establish a clear low–medium–high gradient and to avoid redundant intermediate levels, which have shown marginal differentiation in effect across soil and plant parameters in previous studies [30,31]. Similar dosing schemes (e.g., 0%, 1%, and

3%) have been widely adopted in recent biochar trials to ensure practical feasibility and statistical efficiency. Each treatment, including the control, was replicated five times ($n = 5$), resulting in a total of 20 independent experimental pots. Each pot (55 cm × 40 cm × 35 cm) was filled with 15 kg of homogenized sediment thoroughly mixed with the designated biochar concentration. Fifteen healthy and morphologically uniform *Scirpus mariqueter* seedlings (5–8 cm in height) were transplanted into each pot. For each pot, data were collected from 15 seedlings, and the mean value was used as one statistical replicate. Accordingly, the number of statistical replicates was consistent with the number of biological replicates ($n = 5$). Specifically, we assessed plant biomass production, reproductive allocation, root morphological traits, and associated changes in soil physicochemical properties. The pots were maintained in a greenhouse at the Wetland Ecosystem Research Station of Hangzhou Bay under natural light and ambient temperature. Soil moisture was kept at saturation through regular watering. Aside from the biochar treatments, all other management practices were standardized. To minimize edge effects, pots were randomly arranged and rotated 90° clockwise each week. No additional fertilizers or pesticides were applied throughout the experiment to avoid external influences on soil and plant responses.

2.4. Plant and Soil Measurements

After 240 days of growth (harvested on 30 November 2024), plants were harvested and separated into aboveground and belowground parts. Each component was oven-dried at 65 °C to constant weight to determine total, shoot, and root biomass. The number of ramets and clonal propagules (corms) per pot was also recorded. Root morphological traits—including total length, surface area, volume, and average diameter—were assessed using the WinRHIZO image analysis system (Regent Instruments Inc., Québec City, QC, Canada).

Soil samples were collected from each pot and analyzed for a range of physicochemical properties. Soil pH and EC were measured as described in Section 2.2. Gravimetric water content was determined by drying soil at 105 °C. Nitrate nitrogen (NO_3^--N) and ammonium nitrogen (NH_4^+-N) were extracted with 2 mol·L^{-1} KCl and quantified using a continuous flow analyzer (AA3, Norderstedt Instrumental Analysis GmbH, Germany). Available phosphorus (AP) was extracted using 0.5 mol·L^{-1} $NaHCO_3$ and determined by the molybdenum blue colorimetric method. The potential nitrification rate (PNR) was evaluated via the chlorate inhibition technique, wherein soil samples were incubated with ammonium chloride and sodium chlorate. Accumulated nitrite (NO_2^--N) was quantified spectrophotometrically to estimate PNR. All chemical analyses were conducted following standard protocols described in the *Handbook of Soil Agricultural Chemistry Analysis* [32]. The surface sediments in the study area are predominantly composed of silt-sized particles ranging from 3.9 to 62.5 μm, characterized by an approximate water-holding capacity of 27%. Physicochemical analyses revealed that these sediments exhibit a mean pH of 8.42 ± 0.15, salinity of 0.36 ± 0.05 g/kg, and electrical conductivity (EC) of 96.7 μS/cm. The sediments also contain an organic carbon content of 5.48 ± 0.29 g/kg, with ammonium nitrogen (NH_4^+-N) and nitrate nitrogen (NO_3^--N) concentrations of 21.63 ± 0.70 mg/kg and 12.72 ± 0.58 mg/kg, respectively.

2.5. Statistical Analysis

The raw data were checked for normality, and natural-log or squareroot transformation was performed when necessary before analysis. One-way analysis of variance (ANOVA) was conducted to assess the effects of different biochar treatments on plant and soil parameters. Significant differences between treatment means were identified using Tukey's HSD post hoc test, with a significance threshold of $p < 0.05$. Pearson correlation analysis was also performed to explore relationships between root morphological traits and soil nutrient availability. All statistical analyses were conducted in R version 4.2.2 [33].

3. Results

3.1. Effects of Spartina alterniflora-Derived Biochar on the Plant Traits of Scirpus mariqueter

As illustrated in Figure 1, the total and belowground biomass of *Scirpus mariqueter* exhibited significant responses ($p < 0.05$) to increasing concentrations of *Spartina alterniflora*-derived biochar, displaying a unimodal trend. A similar pattern was observed for aboveground biomass. Compared to CK, biochar additions at 0.5% and 1% increased aboveground biomass by 3.09% and 36.65%, and belowground biomass by 15.14% and 115.04%, respectively. The 1% biochar treatment significantly enhanced both total and belowground biomass ($p < 0.05$). In contrast, the 3% biochar application led to a marked decline in belowground biomass.

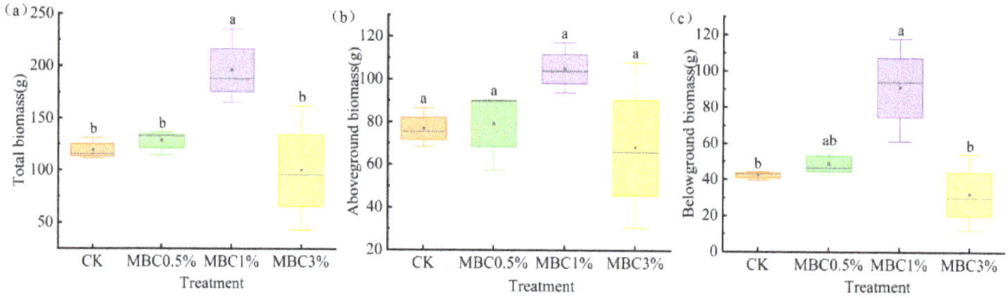

Figure 1. Effects of *Spartina alterniflora*-derived biochar on the biomass of *Scirpus mariqueter*: (**a**) total biomass, (**b**) aboveground biomass, and (**c**) belowground biomass. MBC0.5%, MBC1%, and MBC3% denote biochar application rates of 0.5%, 1%, and 3% (w/w), respectively, with control plots (CK) receiving no biochar amendment. Different lowercase letters indicate significant differences among treatments ($p < 0.05$).

In terms of vegetative growth, plant height increased significantly with rising concentrations of *Spartina alterniflora*-derived biochar (Figure 2a), indicating enhanced vegetative growth. In terms of reproductive traits, biochar addition had distinct effects on both asexual and sexual reproduction (Figure 2b–f). Among asexual traits, the number of ramets significantly decreased with increasing biochar concentration ($p < 0.05$; Figure 2b). The number of corms varied significantly among treatments ($p < 0.05$; Figure 2c), with increases of approximately 39.80% and 26.02% under the 0.5% and 3% treatments, respectively, compared to the CK. No significant difference was observed under the 1% treatment. Corm biomass did not differ significantly among treatments ($p > 0.05$; Figure 2d), although the 0.5% and 3% biochar additions resulted in increases of approximately 39.07% and 25.47% relative to the CK. Regarding sexual reproduction, both seed number and seed biomass varied significantly across biochar concentrations ($p < 0.05$; Figure 2e–f). Compared to CK, the 3% treatment significantly reduced both seed number and seed biomass ($p < 0.05$), while the 0.5% treatment had no significant effect. Notably, the 1% treatment significantly increased seed biomass ($p < 0.05$).

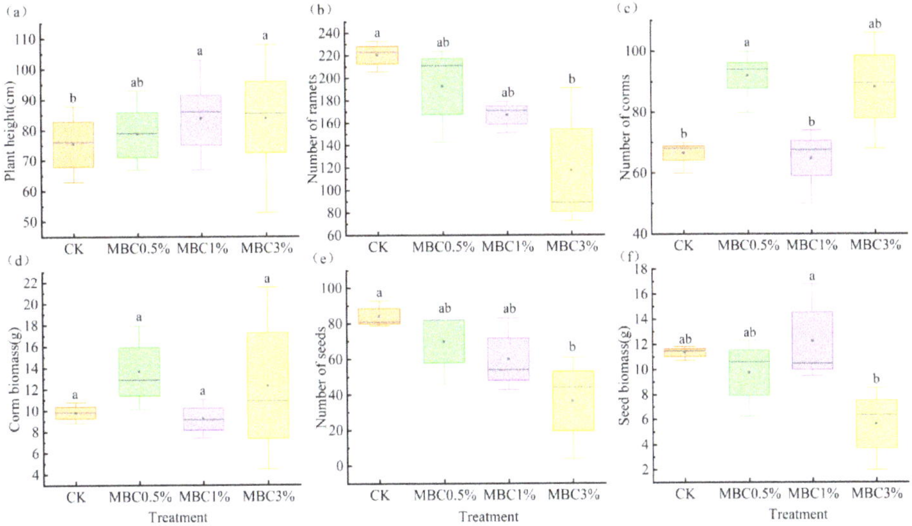

Figure 2. Effects of *Spartina alterniflora*-derived biochar on the vegetative and reproductive traits of *Scirpus mariqueter*: (**a**) plant height; (**b**) number of ramets; (**c**) number of corms; (**d**) corm biomass; (**e**) number of seeds; (**f**) seed biomass. Different lowercase letters indicate significant differences among treatments ($p < 0.05$).

The proportion of sexual reproductive biomass (seeds) relative to total plant biomass varied significantly among the *Spartina alterniflora*-derived biochar treatments ($p < 0.05$; Figure 3a), showing a clear decreasing trend with increasing biochar concentration. Compared to CK, this proportion declined by 21.54%, 41.18%, and 39.01% under the 0.5%, 1%, and 3% treatments, respectively, with the lowest value observed at the 1% level ($p < 0.05$). The proportion of asexual reproductive biomass (corms) also differed significantly among treatments ($p < 0.05$; Figure 3b). The 3% treatment exhibited the highest proportion, significantly exceeding those of the CK and 1% treatments, while the 1% treatment showed the lowest proportion among all groups ($p < 0.05$). In addition, the ratio of sexual to asexual reproductive biomass (seed/corm ratio) decreased significantly with increasing biochar concentration ($p < 0.05$; Figure 3c). Both the 0.5% and 3% treatments resulted in lower ratios compared to CK, with the 3% treatment displaying the lowest value, significantly lower than all other treatment groups.

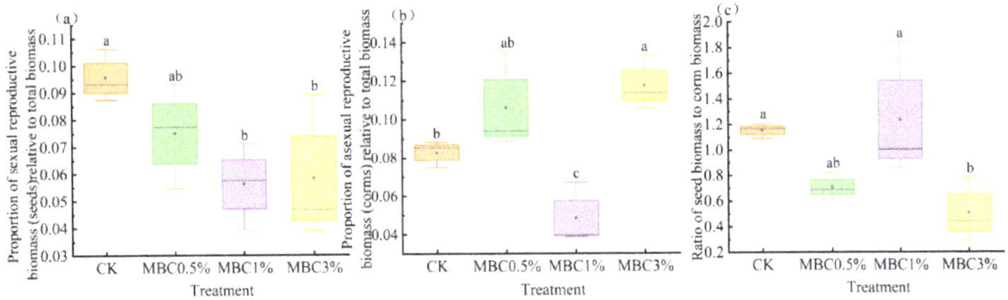

Figure 3. Effects of *Spartina alterniflora*-derived biochar on the reproductive allocation of *Scirpus mariqueter*: (**a**) proportion of seed biomass relative to total biomass; (**b**) proportion of corm biomass relative to total biomass; (**c**) ratio of seed biomass to corm biomass. Different lowercase letters indicate significant differences among treatments ($p < 0.05$).

3.2. Effects of *Spartina alterniflora*-Derived Biochar on Soil Physicochemical Properties

The application of *Spartina alterniflora*-derived biochar significantly influenced soil physicochemical properties in *Scirpus mariqueter* habitats ($p < 0.05$; Table 1). Soil moisture content increased progressively with higher biochar concentrations, reaching a maximum under the 3% treatment (12.54% higher than CK). Soil pH significantly increased under the 0.5% and 1% treatments but declined at the 3% level, while both EC and salinity exhibited nonlinear responses—initially decreasing and then increasing—with the highest values observed under the 3% treatment, where EC was 28.04% greater than CK. AP followed a similar pattern, increasing by 20.15% and 41.36% under the 1% and 3% treatments, respectively. Biochar addition also significantly enhanced soil nitrogen availability: NH_4^+-N peaked at the 0.5% level with a 26.84% increase, whereas NO_3^--N increased by 14.35% and 18.57% under the 1% and 3% treatments, respectively. Moreover, TC rose markedly with increasing biochar application, with the 0.5%, 1%, and 3% treatments showing increases of 10.94%, 14.24%, and 42.18%, respectively, compared to CK.

Table 1. Effects of *Spartina alterniflora*-derived biochar on soil physicochemical properties in *Scirpus mariqueter* habitats.

Treatment	CK	MBC0.5%	MBC1%	MBC3%
pH	7.67 ± 0.08 a	7.82 ± 0.03 a	7.75 ± 0.04 a	7.49 ± 0.01 b
Moisture (%)	46.48 ± 0.89 b	48.81 ± 0.87 ab	51.96 ± 0.98 a	52.31 ± 2.09 a
EC (μS/cm)	136.57 ± 9.03 b	121.20 ± 2.60 b	133.03 ± 7.79 b	174.87 ± 3.06 a
Salinity (g/kg)	0.27 ± 0.02 b	0.24 ± 0.01 b	0.27 ± 0.02 b	0.35 ± 0.01 a
AP (mg/kg)	84.40 ± 13.49 ab	83.88 ± 2.74 b	101.40 ± 3.02 ab	119.30 ± 1.95 a
NH_4^+-N (mg/kg)	26.16 ± 0.84 b	33.18 ± 3.13 a	30.39 ± 0.81 ab	30.86 ± 0.73 ab
NO_3^--N (mg/kg)	15.02 ± 0.57 c	15.69 ± 0.66 bc	17.17 ± 0.17 ab	17.81 ± 0.36 a
TC (g/kg)	53.89 ± 0.87 c	59.78 ± 1.60 b	61.56 ± 1.07 b	76.62 ± 2.17 a

Note: Abbreviations: EC, electrical conductivity; AP, available phosphorus; NH_4^+-N, ammonium nitrogen; NO_3^--N, nitrate nitrogen; TC, total carbon. Different lowercase letters indicate significant differences among treatments ($p < 0.05$).

3.3. Effects of *Spartina alterniflora*-Derived Biochar on Root Morphology of *Scirpus mariqueter*

The application of *Spartina alterniflora*-derived biochar significantly enhanced root development in *Scirpus mariqueter* ($p < 0.05$; Table 2). With increasing biochar concentration, root length, surface area, and volume displayed a unimodal (hump-shaped) response, peaking under the 1% treatment and subsequently declining. Compared to CK, the 1% biochar addition resulted in substantial increases in root length (135.84%), surface area (128.39%), and volume (82.26%) ($p < 0.05$). Root average diameter also varied significantly among treatments ($p < 0.05$), with the highest value observed under the 3% treatment—23.68% greater than CK—while the value under the 1% treatment was slightly lower than that of CK.

Table 2. Root morphological traits of *Scirpus mariqueter* in response to *Spartina alterniflora*-derived biochar application.

Treatments	Root Length (cm)	Root Surface Area (cm^2)	Root Volume (cm^3)	Root Avgrage Diameter (mm)
CK	101.02 ± 20.47 b	18.25 ± 3.14 b	0.26 ± 0.04 b	0.58 ± 0.02 b
MBC0.5%	108.13 ± 21.42 b	21.62 ± 3.20 b	0.35 ± 0.03 ab	0.65 ± 0.04 ab
MBC1%	238.25 ± 34.52 a	41.67 ± 3.68 a	0.48 ± 0.09 a	0.56 ± 0.03 b
MBC3%	133.40 ± 26.18 b	29.84 ± 5.09 ab	0.42 ± 0.06 ab	0.72 ± 0.02 a

Note: Values are presented as mean ± standard error. Different lowercase letters indicate significant differences among treatments ($p < 0.05$).

3.4. Correlation Analysis Among Plant Traits, Root Morphology, and Soil Properties

Correlation analysis revealed that most plant biomass traits were not significantly related to soil nutrients ($p > 0.05$), except for seed biomass, which was negatively correlated with TC ($p < 0.05$). Biomass traits showed positive correlations with root length and surface area ($p < 0.05$). Plant height was positively correlated with soil NO_3^--N, AP, EC, salinity, and TC ($r = 0.74$, $p < 0.01$), but negatively with seed biomass, ramet number, and pH ($p < 0.05$). Sexual traits were positively associated with ramet number, while seed number was negatively correlated with soil moisture ($p < 0.05$). Asexual traits showed weak correlations, though ramet number was positively correlated with pH and negatively with TC ($p < 0.05$).

4. Discussion

This study demonstrates that biochar derived from the invasive species *Spartina alterniflora* has significant, concentration-dependent effects on the growth, reproductive strategies, and root architecture of *Scirpus mariqueter*, while simultaneously altering the physicochemical properties of coastal wetland soils. These findings provide new insights into the use of biochar as a nature-based restoration tool, highlighting its potential to enhance native plant productivity and improve soil functioning in degraded tidal wetlands.

Among the tested treatments, moderate biochar application (1%) most effectively promoted plant growth, with total, shoot, and root biomass increasing by 64.53%, 36.65%, and 115.04%, respectively, relative to CK (Figure 1). This growth enhancement is likely driven by improved soil structure, increased nutrient retention, and enhanced rhizosphere conditions—consistent with previous studies indicating that biochar can enhance nutrient bioavailability and soil water-holding capacity [34,35]. Previous studies have shown that the abundant microporous structure of biochar enhances its capacity to adsorb and retain nutrient ions in the soil, thereby reducing nutrient leaching and improving plant nutrient use efficiency [36,37]. Moreover, biochar can indirectly promote nutrient uptake by modulating soil pH and electrical conductivity, which subsequently influences the composition and functional dynamics of rhizosphere microbial communities [38]. In contrast, the 3% biochar treatment negatively impacted plant biomass, particularly belowground, possibly due to elevated soil salinity and EC (Table 1), which can induce osmotic stress, restrict nutrient uptake, and inhibit root elongation [39,40]. Root morphological responses support this interpretation: the 1% treatment significantly improved root length, surface area, and volume, suggesting improved soil–root interactions, whereas the 3% treatment increased root diameter but reduced absorptive capacity, likely reflecting a stress-induced thickening response [41,42].

Biochar application also influenced reproductive allocation, shifting the balance between sexual and asexual strategies in *Scirpus mariqueter* (Figures 2 and 3). While low biochar levels promoted the development of clonal organs (corms), higher concentrations reduced seed production and sexual reproductive biomass, indicating a trade-off potentially driven by nutrient enrichment and salinity stress. This pattern aligns with previous findings that nutrient-rich conditions tend to favor vegetative propagation, whereas increased salinity and EC can reduce investment in sexual reproduction [2,43]. Further analysis suggests that biochar may influence plant reproductive strategies through multiple interacting mechanisms. First, due to its high specific surface area and porous structure, biochar can adsorb nutrients and improve soil physical properties, thereby enhancing soil water and nutrient retention capacity [44,45]. This, in turn, facilitates more efficient acquisition of belowground resources and promotes the development of roots and corms. Second, biochar has been shown to regulate soil pH and salinity levels [46,47], which may help alleviate the inhibitory effects of salt stress on sexual reproduction. However, at higher application rates, biochar may negatively affect plant growth and seed formation by disrupting rhizosphere microbial communities or inducing ion toxicity [48]. Additionally, biochar may influence reproductive allocation by modulating plant hormone levels, such as auxin and abscisic acid [49]. Correlation analysis further supports this mechanism: root volume was positively associated with NH_4^+-N and NO_3^--N, indicating a key role of nitrogen in promoting root development and clonal expansion (Figure 4). However, the weak or inconsistent correlations between asexual reproductive traits and soil variables suggest that other ecological factors—such as light availability, hydrological conditions, and biotic interactions—may also regulate reproductive plasticity, highlighting the need for further field-based research [50].

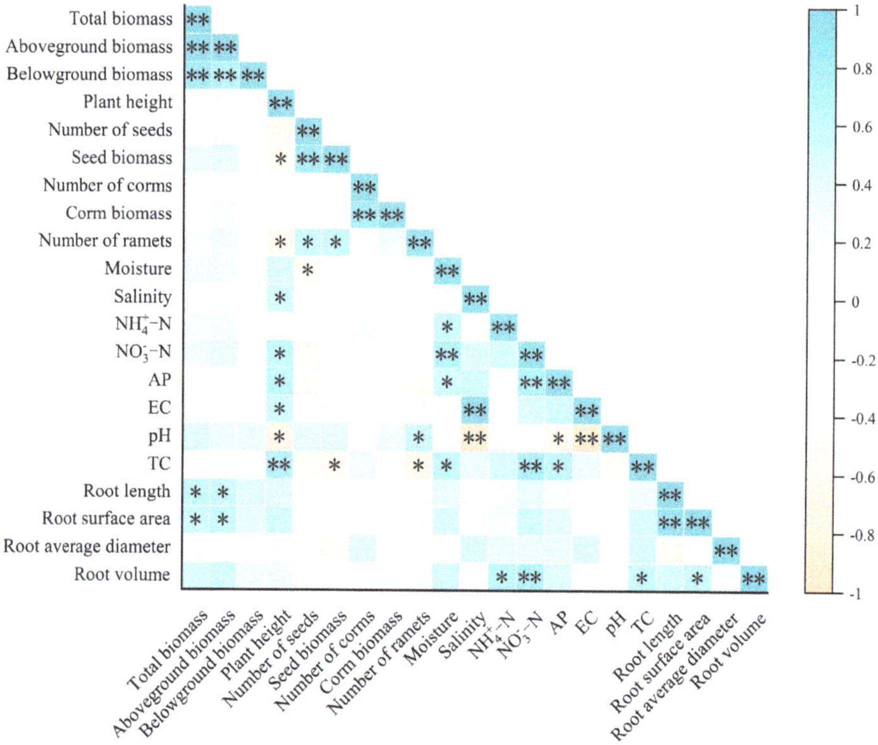

Figure 4. Spearman's correlation heatmap between plant traits of *Scirpus mariqueter* and soil nutrient variables. Colors indicate the strength of correlation coefficients. Asterisks indicate levels of statistical significance (* $p < 0.05$; ** $p < 0.01$).

The observed improvements in soil properties underscore biochar's multifaceted role in enhancing ecosystem functions (Table 1). In particular, increased soil moisture, available nutrients (AP, TC, NH_4^+-N, NO_3^--N), and potential nitrification under the 1% treatment suggest enhanced microbial activity and nutrient cycling, which are critical for sustaining plant productivity and facilitating soil carbon input [51,52]. In addition, soluble organic compounds and leachates released from biochar can serve as readily available carbon sources for soil microorganisms, thereby stimulating microbial metabolic activity and accelerating the mineralization of nitrogen and phosphorus, as well as the decomposition of organic matter [53,54]. These processes collectively contribute to improved nutrient availability and uptake by plants. Furthermore, biochar may indirectly influence the composition and functional gene expression of soil microbial communities. For example, it has been shown to increase the relative abundance of ammonia-oxidizing bacteria (AOB) and denitrifying bacteria, thereby enhancing soil nitrification potential and nitrogen use efficiency [49,55]. In coastal wetlands—key blue carbon ecosystems—belowground biomass is closely linked to carbon sequestration potential [4,8]. Thus, moderate biochar addition could simultaneously support vegetation recovery and long-term carbon storage.

However, the accumulation of salts and reduction in pH observed under the 3% treatment highlight potential ecological risks associated with high biochar dosages, such as shifts in microbial community composition and inhibited plant performance [1,43]. Therefore, while this study supports the use of *Spartina alterniflora*-derived biochar as a viable strategy for wetland restoration, its application should be carefully tailored to site-specific conditions. Consideration of appropriate dosage thresholds, hydrological context, and species-specific tolerance is essential to maximize ecological benefits while minimizing unintended consequences.

Invasive weeds have recently garnered increasing attention as promising feedstocks for biochar production due to their high biomass availability, elevated carbon content, and potential to improve soil fertility and structure. A growing body of research has demonstrated that, when applied at appropriate rates, weed-derived biochars can significantly enhance soil physicochemical properties and promote plant growth [24,56]. Collectively, these findings highlight the potential of weed-derived biochar as a sustainable tool for soil remediation and nutrient management.

While earlier studies have primarily emphasized the general benefits of invasive plant-derived biochar for soil improvement, the novelty of the present study lies in its focus on the biochar's role in enhancing both the growth and reproductive allocation of *Scirpus mariqueter*, a native coastal wetland species. By elucidating the relationships between soil nutrient dynamics and shifts in plant reproductive strategies across a gradient of biochar application rates, this study contributes new insights into the ecological functions of biochar in coastal wetland restoration. Furthermore, the findings underscore the "dual value" of invasive weeds: mitigating ecological threats through biomass valorization, while simultaneously supporting native species recovery and advancing blue carbon management strategies.

However, a critical limitation of this study is the lack of investigation into the dynamics of labile carbon fractions leached from invasive plant-derived biochar and their potential priming effects on the mineralization of native soil organic matter (SOM). Such processes may increase dissolved organic carbon levels in adjacent water bodies, with potential downstream impacts on aquatic ecosystem structure and function [57]. Future research should therefore aim to quantify carbon leaching and associated priming effects following biochar application in coastal wetlands. A mechanistic understanding of these pathways will be essential for assessing ecological risks and developing optimized biochar application strategies that maximize ecological benefits while minimizing unintended environmental consequences.

5. Conclusions

This study provides empirical evidence that biochar derived from the invasive plant *Spartina alterniflora* can significantly enhance the growth performance, reproductive traits, and root morphology of the native halophyte *Scirpus mariqueter*, while concurrently improving soil moisture, nutrient availability, and nitrification activity in coastal wetland sediments. Notably, a moderate biochar application rate (1%) was most effective in maximizing plant biomass and reproductive efficiency, indicating an optimal threshold for balancing vegetation growth and soil improvement. In contrast, excessive application (3%) increased soil salinity and electrical conductivity, leading to suppressed plant development and altered reproductive allocation, thereby emphasizing the critical importance of dosage control in ecological applications. These findings highlight the potential of *Spartina alterniflora*-derived biochar as a sustainable and circular resource for coastal wetland restoration, offering dual benefits in terms of invasive biomass utilization and enhanced blue carbon sequestration. Nevertheless, the negative impacts observed at higher concentrations underscore the need for careful ecological risk assessment. Future research should incorporate long-term field trials and integrate soil microbial community analyses to better elucidate the mechanisms and ecological thresholds governing biochar performance in tidal wetland ecosystems.

Author Contributions: Project administration, N.L.; conceptualization, N.L. and Y.T.; investigation, J.G., P.J. and J.L.; methodology, Y.T.; data curation, Y.T.; formal analysis, Y.T. and M.W.; visualization, Y.T.; writing—original draft, Y.T.; writing—review and editing, S.J., L.Z. and X.S., N.L. and X.S. contributed equally to this work. All authors have read and agreed to the published version of the manuscript.

Funding: This study was supported by the Pioneer and Leading Goose R&D Program of Zhejiang Province (Grant No. 2024C02002); the Zhejiang Province Commonwealth Projects (Grant No. LQ23C030003); the Cooperation Program between Zhejiang Province and the Chinese Academy of Forestry (Grant No. 2023SY11); the Special Fund for Basic Scientific Research Operations of Central Public Welfare Research Institutes (Grant No. CAFYBB2024MA029); the Zhejiang Provincial Science and Technology Program (Grant No. 2023C03120); the Pioneer and Leading Goose R&D Program of Zhejiang Province (Grant Nos. 2025C02050 and 2025C02230); and the National Key Research and Development Program of China (Grant No. 2023YFE0101700).

Institutional Review Board Statement: Not applicable.

Data Availability Statement: The datasets used and analyzed during the current study are available from the corresponding author.

Acknowledgments: We express our sincere gratitude to Xiaohong Zhu from Wetland Ecosystem Research Station of Hangzhou Bay for his invaluable assistance and insightful suggestions during the study.

Conflicts of Interest: The authors declare no conflicts of interest.

References

1. Barbier, E.B.; Hacker, S.D.; Kennedy, C.; Koch, E.W.; Stier, A.C.; Silliman, B.R. The value of estuarine and coastal ecosystem services. *Ecol. Monogr.* **2011**, *81*, 169–193. [CrossRef]
2. Murtaza, G.; Deng, G.; Usman, M.; Jamil, A.; Qasim, M.; Iqbal, J.; Ercisli, S.; Akram, M.I.; Rizwan, M.; Elshikh, M.S. Impact of Acacia-derived biochar to mitigate salinity stress in Zea mays L. by morpho-physiological and biochemical indices. *Sci. Rep.* **2024**, *14*, 31883. [CrossRef] [PubMed]
3. Chmura, G.L.; Anisfeld, S.C.; Cahoon, D.R.; Lynch, J.C. Global carbon sequestration in tidal, saline wetland soils. *Glob. Biogeochem. Cycles* **2003**, *17*, 1111. [CrossRef]
4. Mcleod, E.; Chmura, G.L.; Bouillon, S.; Salm, R.; Björk, M.; Duarte, C.M.; Lovelock, C.E.; Schlesinger, W.H.; Silliman, B.R. A blueprint for blue carbon: Toward an improved understanding of the role of vegetated coastal habitats in sequestering CO_2. *Front. Ecol. Environ.* **2011**, *9*, 552–560. [CrossRef]
5. Hu, Y.; Tian, B.; Yuan, L.; Li, X.; Huang, Y.; Shi, R.; Jiang, X.; Wang, L.; Su, C. Mapping coastal salt marshes in China using time series of Sentinel-1 SAR. *ISPRS J. Photogramm. Remote Sens.* **2021**, *173*, 122–134. [CrossRef]
6. Nan, L.; Longwei, L.; Yinlong, Z.; Ming, W. Monitoring of the invasion of Spartina alterniflora from 1985 to 2015 in Zhejiang Province, China. *BMC Ecol.* **2020**, *20*, 7. [CrossRef]
7. Wang, S.; Martin, P.A.; Hao, Y.; Sutherland, W.J.; Shackelford, G.E.; Wu, J.; Ruiting, J.; Zhou, W.; Li, B. A global synthesis of the effectiveness and ecological impacts of management interventions for Spartina species. *Front. Environ. Sci. Eng.* **2023**, *17*, 141. [CrossRef]
8. Duarte, C.M.; Losada, I.J.; Hendriks, I.E.; Mazarrasa, I.; Marbà, N. The role of coastal plant communities for climate change mitigation and adaptation. *Nat. Clim. Change* **2013**, *3*, 961–968. [CrossRef]

9. Feng, L.J.; Wei, L.Q.; Qin, G.J.; Ge, F.J.; Ya, Z.X.; Jing, H.Y.; Hai, Y.F. Biochar-compost addition benefits Phragmites australis growth and soil property in coastal wetlands. *Sci. Total Environ.* **2021**, *769*, 145166. [CrossRef]
10. Huang, Y.; Xiao, X. Role of biochar in raising blue carbon stock capacity of salt marshes. *Pedosphere* **2024**, *34*, 19–22. [CrossRef]
11. Li, C.; Chen, J.; Liao, X.; Ramus, A.P.; Angelini, C.; Liu, L.; Silliman, B.R.; Bertness, M.D.; He, Q. Shorebirds-driven trophic cascade helps restore coastal wetland multifunctionality. *Nat. Commun.* **2023**, *14*, 8076. [CrossRef]
12. Yuan, L.; Liu, D.; Tian, B.; Yuan, X.; Bo, S.; Ma, Q.; Wu, W.; Zhao, Z.; Zhang, L.; Keesing, J.K. A solution for restoration of critical wetlands and waterbird habitats in coastal deltaic systems. *J. Environ. Manag.* **2022**, *302*, 113996. [CrossRef] [PubMed]
13. Xu, M.; Gu, J.; Yun, P.; Gan, J.; Fan, W.; Chen, Z.; Shao, J.; Zhang, G.; Zeng, J. Spatiotemporal dynamics and influencing factors of native and invasive saltmarshes in a rapidly silting bay during 1985–2023. *Front. Mar. Sci.* **2025**, *11*, 1459935. [CrossRef]
14. Zheng, X.; Javed, Z.; Liu, B.; Zhong, S.; Cheng, Z.; Rehman, A.; Du, D.; Li, J. Impact ofSpartina alternifloraInvasion in Coastal Wetlands of China: Boon or Bane? *Biology* **2023**, *12*, 1057. [CrossRef] [PubMed]
15. Li, B.; Liao, C.; Zhang, X.; Chen, H.; Wang, Q.; Chen, Z.; Gan, X.; Wu, J.; Zhao, B.; Ma, Z. Spartina alterniflora invasions in the Yangtze River estuary, China: An overview of current status and ecosystem effects. *Ecol. Eng.* **2009**, *35*, 511–520. [CrossRef]
16. Li, L.; Li, D.; Kong, R.; Ren, Z.; Liu, L.; Zhu, Y.; Sun, Y.; Peng, N.; He, J.; Ji, Y. Strategies in growth and reproduction of the native endangered plant species Scripus mariqueter and the driving factors in a coastal salt marsh wetland, eastern China. *Acta Oecologica* **2024**, *122*, 103979. [CrossRef]
17. Temmerman, S.; Meire, P.; Bouma, T.J.; Herman, P.M.; Ysebaert, T.; De Vriend, H.J. Ecosystem-based coastal defence in the face of global change. *Nature* **2013**, *504*, 79–83. [CrossRef]
18. Jiang, C.; Li, X.Z.; Xue, L.M.; Yan, Z.Z.; Liang, X.; Chen, X.C. Pioneer salt marsh species Scirpus mariqueter disperses quicker in summer with seed contribution from current and last year. *Estuar. Coast. Shelf Sci.* **2022**, *264*, 107682. [CrossRef]
19. Qun, Z.; Bo, L. Field practice of Scirpus mariqueter restoration in the bird habitats of Chongming Dongtan Wetland, China. *Chin. J. Appl. Ecol./Yingyong Shengtai Xuebao* **2023**, *34*, 2663–2671. [CrossRef]
20. Antonangelo, J.A.; Sun, X.; Eufrade-Junior, H.D. Biochar impact on soil health and tree-based crops: A review. *Biochar* **2025**, *7*, 51. [CrossRef]
21. Cohen-Shacham, E.; Walters, G.; Janzen, C.; Maginnis, S. Nature-based solutions to address global societal challenges. *IUCN Gland Switz.* **2016**, *97*, 2036. [CrossRef]
22. Han, L.; Lu, C.; Chen, L.; Wang, F.; Gao, K.; Yu, Y.; Xu, C. Carbon sequestration potential of biochar in soil from the perspective of organic carbon structural modification. *Appl. Soil Ecol.* **2024**, *198*, 105389. [CrossRef]
23. Ghosh, D.; Maiti, S.K. Effect of invasive weed biochar amendment on soil enzymatic activity and respiration of coal mine spoil: A laboratory experiment study. *Biochar* **2021**, *3*, 519–533. [CrossRef]
24. Gupta, I.; Singh, R.; Kaushik, A.; Kohli, R.K.; Singh, H.P.; Batish, D.R. Impact of Broussonetia papyrifera Biochar on the Biological Attributes of Cajanus cajan and Soil Enzymatic Activities. *J. Soil Sci. Plant Nutr.* **2024**, *24*, 1990–2007. [CrossRef]
25. Cai, J.F.; Jiang, F.; Liu, X.S.; Sun, K.; Wang, W.; Zhang, M.X.; Li, H.L.; Xu, H.F.; Kong, W.J.; Yu, F.H. Biochar-amended coastal wetland soil enhances growth of Suaeda salsa and alters rhizosphere soil nutrients and microbial communities. *Sci. Total Environ.* **2021**, *788*, 147707. [CrossRef]
26. Wang, P.; Wang, S.; Chen, F.; Zhang, T.; Kong, W. Preparation of two types plant biochars and application in soil quality improvement. *Sci. Total Environ.* **2024**, *906*, 167334. [CrossRef]
27. Zhang, M.; Xu, S.; Gao, Z.; Jia, H.; Du, D.; Eltawab, R.; Abdelfattah, A. Innovative solutions for coastal wetlands pollution: Application of Spartina alterniflora Loisel.-biochar for cadmium removal and carbon sequestration. *J. Environ. Manag.* **2025**, *373*, 123444. [CrossRef]
28. Ghulam, M.; Zeeshan, A.; Eldin, S.M.; Basharat, A.; Sami, B.; Muhammad, U.; Rashid, I.; Dhurba, N.; Abd, U.; Ahmad, K.; et al. Biochar-Soil-Plant interactions: A cross talk for sustainable agriculture under changing climate. *Front. Environ. Sci.* **2023**, *11*, 1059449. [CrossRef]
29. Zhao, Z.; Cheng, L.; He, C.; Wang, F.; Liu, J.; Li, Y.; Chen, X.; Liu, X.; Lv, G.; Wang, D. Spartina alterniflora Invaded Coastal Wetlands by Raising Soil Sulfur Contents: A Meta-Analysis. *Water* **2022**, *14*, 1633. [CrossRef]
30. Hussein, F.; Hasan, F.A. Biochar Enhanced Chemical and Biological Properties of Contaminated Soils with Lead. *IOP Conf. Ser. Earth Environ. Sci.* **2023**, *1259*, 012024. [CrossRef]
31. Chen, X.; Liu, L.; Yang, Q.; Xu, H.; Shen, G.; Chen, Q. Optimizing Biochar Application Rates to Improve Soil Properties and Crop Growth in Saline–Alkali Soil. *Sustainability* **2024**, *16*, 2523. [CrossRef]
32. Bao, S.D. *Soil Agricultural Chemistry Analysis*, 3rd ed.; China Agriculture Press: Beijing, China, 2000.
33. R Core Team. *R: A Language and Environment for Statistical Computing*; R Foundation for Statistical Computing: Vienna, Austria, 2022.
34. Lehmann, J.; Rillig, M.C.; Thies, J.; Masiello, C.A.; Hockaday, W.C.; Crowley, D. Biochar effects on soil biota—A review. *Soil Biol. Biochem.* **2011**, *43*, 1812–1836. [CrossRef]

35. Piccolo, E.L.; Becagli, M.; Lauria, G.; Cantini, V.; Ceccanti, C.; Cardelli, R.; Massai, R.; Remorini, D.; Guidi, L.; Landi, M. Biochar as a soil amendment in the tree establishment phase: What are the consequences for tree physiology, soil quality and carbon sequestration? *Sci. Total Environ.* **2022**, *844*, 157175. [CrossRef]
36. Jikai, L.; Yina, L.; Junlin, H.; Bingyan, H.; Bing, W.; Kenji, O.; Jian, Z.; Hongyu, S. The effect of biochar on the migration theory of nutrient ions. *Sci. Total Environ.* **2022**, *845*, 157262. [CrossRef]
37. Rasse, D.P.; Weldon, S.; Joner, E.J.; Joseph, S.; Kammann, C.I.; Liu, X.; O'Toole, A.; Pan, G.; Kocatürk Schumacher, N.P. Enhancing plant N uptake with biochar-based fertilizers: Limitation of sorption and prospects. *Plant Soil* **2022**, *475*, 213–236. [CrossRef]
38. Wang, X.; Riaz, M.; Babar, S.; Eldesouki, Z.; Liu, B.; Xia, H.; Li, Y.; Wang, J.; Xia, X.; Jiang, C. Alterations in the composition and metabolite profiles of the saline-alkali soil microbial community through biochar application. *J. Environ. Manag.* **2024**, *352*, 120033. [CrossRef]
39. Ahmed, M.; Hasanuzzaman, M.; Raza, M.A.; Malik, A.; Ahmad, S. Plant nutrients for crop growth, development and stress tolerance. *Sustain. Agric. Era Clim. Change* **2020**, 43–92. [CrossRef]
40. Parihar, P.; Singh, S.; Singh, R.; Singh, V.P.; Prasad, S.M. Effect of salinity stress on plants and its tolerance strategies: A review. *Environ. Sci. Pollut. Res.* **2015**, *22*, 4056–4075. [CrossRef]
41. Laghari, M.; Mirjat, M.S.; Hu, Z.; Fazal, S.; Xiao, B.; Hu, M.; Chen, Z.; Guo, D. Effects of biochar application rate on sandy desert soil properties and sorghum growth. *Catena* **2015**, *135*, 313–320. [CrossRef]
42. Subedi, R.; Bertora, C.; Zavattaro, L.; Grignani, C. Crop response to soils amended with biochar: Expected benefits and unintended risks. *Ital. J. Agron.* **2017**, *12*, 794. [CrossRef]
43. Zhu, Z.; Zhang, Y.; Tao, W.; Zhang, X.; Xu, Z.; Xu, C. The Biological Effects of Biochar on Soil's Physical and Chemical Characteristics: A Review. *Sustainability* **2025**, *17*, 2214. [CrossRef]
44. Edeh, I.G.; Mašek, O.; Buss, W. A meta-analysis on biochar's effects on soil water properties—New insights and future research challenges. *Sci. Total Environ.* **2020**, *714*, 136857. [CrossRef]
45. Woo, K.M.; Mesenbet, Y.; Hyun, K.Y.; Jin, O.S.; Cheol, L.J.; Kwon, E.E.; Soo, L.S. Enhancement of soil physical properties and soil water retention with biochar-based soil amendments. *Sci. Total Environ.* **2022**, *836*, 155746. [CrossRef]
46. Jin, F.; Piao, J.; Miao, S.; Che, W.; Li, X.; Li, X.; Shiraiwa, T.; Tanaka, T.; Taniyoshi, K.; Hua, S.; et al. Long-term effects of biochar one-off application on soil physicochemical properties, salt concentration, nutrient availability, enzyme activity, and rice yield of highly saline-alkali paddy soils: Based on a 6-year field experiment. *Biochar* **2024**, *6*, 40. [CrossRef]
47. Liu, Q.; Meki, K.; Zheng, H.; Yuan, Y.; Shao, M.; Luo, X.; Li, X.; Jiang, Z.; Li, F.; Xing, B. Biochar application in remediating salt-affected soil to achieve carbon neutrality and abate climate change. *Biochar* **2023**, *5*, 45. [CrossRef]
48. Das, S.K.; Ghosh, G.K.; Avasthe, R. Ecotoxicological responses of weed biochar on seed germination and seedling growth in acidic soil. *Environ. Technol. Innov.* **2020**, *20*, 101074. [CrossRef]
49. Wiszniewska, A.; Dziurka, K.; Dziurka, M.; Rodrigues, A.F.; Latawiec, A.E. Biochars as culture medium additives influence organogenic potential of plant explants through changes in endogenous phytohormone and carbohydrate contents in Daphne species. *Plant Cell Tissue Organ Cult.* **2022**, *152*, 45–66. [CrossRef]
50. Wang, Y.; Zhang, M. Modeling hydrodynamic and hydrological processes in tidal wetlands. *Wetlands* **2022**, *42*, 1. [CrossRef]
51. Mohan, D.; Abhishek, K.; Patel, M.; Pittman, C.U., Jr. Harnessing biochar in contaminated soil for heavy metal immobilization, soil health enhancement, and carbon sequestration. *Ind. Eng. Chem. Res.* **2024**, *63*, 10380–10396. [CrossRef]
52. Woolf, D.; Amonette, J.E.; Street-Perrott, F.A.; Lehmann, J.; Joseph, S. Sustainable biochar to mitigate global climate change. *Nat. Commun.* **2010**, *1*, 56. [CrossRef]
53. Hu, J.; Tang, H.; Wang, Y.Z.; Yang, C.; Gao, M.-T.; Tsang, Y.F.; Li, J. Effect of dissolved solids released from biochar on soil microbial metabolism. *Environ. Sci. Process. Impacts* **2022**, *24*, 598–608. [CrossRef]
54. Sadia, A.; Shabana, B.; Mehedi, H.M.; Partha, B.; Ishtiaq, A.M.; Muhammad, B.; Hitesh, C.; Nobendu, M.; Swastika, M. A review on influence of biochar amendment on soil processes and environmental remediation. *Biotechnol. Genet. Eng. Rev.* **2023**, *40*, 31–35. [CrossRef]
55. Ding, J.; Yu, S. Impact of Biochar on Nitrogen-Cycling Functional Genes: A Comparative Study in Mollisol and Alkaline Soils. *Life* **2024**, *14*, 1631. [CrossRef]
56. Kassa, Y.; Amare, A.; Nega, T.; Alem, T.; Gedefaw, M.; Chala, B.; Freyer, B.; Waldmann, B.; Fentie, T.; Mulu, T.; et al. Water hyacinth conversion to biochar for soil nutrient enhancement in improving agricultural product. *Sci. Rep.* **2025**, *15*, 1820. [CrossRef]
57. Yang, F.; Wang, C.; Sun, H. A comprehensive review of biochar-derived dissolved matters in biochar application: Production, characteristics, and potential environmental effects and mechanisms. *J. Environ. Chem. Eng.* **2021**, *9*, 105258. [CrossRef]

diversity

MDPI

Article

Effects of Phenotypic Plasticity and Genetic Variation on Plant Growth and Litter Decomposition in a Widespread Wetland Grass

Wei Wei [1,†], Qishen Wen [1,†], Hong Zhu [1], Huijia Song [2,*], Xiya Zhang [1], Wenyi Sheng [1], Liujuan Xie [3], Xiao Guo [4], Yaolin Guo [5], Siyuan Ye [3], Yuzhi Wang [1], Lele Liu [1] and Weihua Guo [1,*]

1 Qingdao Key Laboratory of Ecological Protection and Restoration, Ministry of Natural Resources Key Laboratory of Ecological Prewarning, Protection and Restoration of Bohai Sea, School of Life Sciences, Shandong University, 72 Binhai Road, Qingdao 266237, China; weiv@mail.sdu.edu.cn (W.W.); wqshakurei@gmail.com (Q.W.); 202332502@mail.sdu.edu.cn (H.Z.); 17853728970@163.com (X.Z.); shwy@sdu.edu.cn (W.S.); wangyuzhi@sdu.edu.cn (Y.W.); liulele@sdu.edu.cn (L.L.)

2 Natural History Museum of China, 126 Tianqiao South Street, Beijing 100050, China

3 Key Laboratory of Coastal Wetland Biogeosciences, Qingdao Institute of Marine Geology, China Geological Survey, Qingdao 266000, China; xieliujuan1102@163.com (L.X.); siyuanye@hotmail.com (S.Y.)

4 College of Landscape Architecture and Forestry, Qingdao Agricultural University, 700 Changcheng Road, Qingdao 266109, China; guoxiao018@163.com

5 School of Renewable Natural Resources, Louisiana State University Agricultural Center, Baton Rouge, LA 70803, USA; yaolinguo22@gmail.com

* Correspondence: songhuijia@nnhm.org.cn (H.S.); guowh@email.sdu.edu.cn (W.G.)

† These authors contributed equally to this work.

Abstract: Wetlands are crucial ecosystems that provide a wide range of ecological services, such as water purification, flood control, and carbon sequestration, where the diversity of wetland plants is fundamental to maintaining these functions. *Phragmites australis* is a globally widespread wetland grass with a high genetic diversity, exhibiting strong intraspecific variation across environmental gradients. While both phenotypic plasticity and genetic variation are recognized drivers of plant adaptation, their relative importance in mediating the growth and decomposition traits of wetland plants remains debated. Here, we surveyed the growth and litter traits of two lineages (haplotypes) of *P. australis* in two common gardens in eastern China. The leaf litter of *P. australis* was also collected from the field in two provinces where the two common gardens are located. Microcosm experiments were carried out to explore the litter decomposition ability. We found that the common garden (growth environment) significantly affected the growth performance (shoot diameter and height), leaf litter stoichiometric traits (contents of N and P), and the leaf decomposition over 180 days. The N content in the leaf litter from the higher-latitude province of Liaoning was greater than that from the lower-latitude province of Shandong, regardless of whether it was collected from the field or the common garden. The litter N and P contents were the key factors affecting the decomposition during the 180-day experiment. However, we did not find the effects of lineage (genetic variation) on the variation of these traits. The climatic factors of the genotype origin were significantly correlated with the growth traits but not the litter traits or decomposition rates. The findings indicate that *P. australis* in eastern temperate China primarily adapts through phenotypic plasticity rather than genetic variation, which is crucial for wetland resilience in a changing climate. This study underscores the pivotal role of environmental factors and phenotypic plasticity in *P. australis* growth and decomposition, suggesting that conservation efforts should prioritize the local environment over genetic variation for effective wetland management.

 https://doi.org/10.3390/d17040282

Keywords: *Phragmites australis*; functional traits; litter decomposition; stoichiometric characteristics; intraspecific variation; adaptive strategies

1. Introduction

Wetlands provide numerous ecosystem services, ranging from nutrient retention and floodwater storage to climate regulation and biodiversity conservation [1]. Wetlands play a very essential role in biogeochemical cycling. Their high productivity, combined with hypoxic and moist environments, enables wetlands to accumulate large amounts of soil organic carbon [2,3]. The carbon sink capacity of wetlands is variable and sensitive, depending on their age, type, and environmental conditions such as climate and location [4,5]. Wetlands in high-latitude regions typically have strong carbon storage capacity due to their low-temperature environments. However, rising temperatures and altered precipitation patterns caused by climate change may weaken their carbon sink functions. In contrast, wetlands in low-latitude regions are more likely to be carbon sources due to the high temperatures and complex precipitation patterns. Climate change further exacerbates their potential for carbon emissions, reducing their carbon sequestration capacity [4]. Plants are the fundamental components of wetlands, especially large wetland plants. They are not only the primary producers in the wetland ecosystem but also the key contributors of these essential functions [6–8]. Many of these dominant plants are widely distributed. Simultaneously, many wetland plants establish mono-dominant communities, significantly influencing the community structure and ecosystem function.

The common reed, *Phragmites australis* (Cav.) Trin. Ex Steud., is a critical species in wetland ecosystems, often forming extensive mono-dominant stands that dominate large areas [9,10]. The pronounced phenotypic plasticity and genetic variability exhibited by *P. australis* endow it with the capacity to adapt to both short-term and long-term environmental changes, thus rendering it a suitable model species for investigating the mechanisms underlying the plant responses to environmental gradients [9,11]. These traits not only enable it to thrive in diverse conditions but also contribute to its significant role in the wetland carbon cycle. *Phragmites australis* exhibits strong carbon sequestration abilities [7] and actively participates in the carbon cycle of wetlands [12]. The intraspecific variation within *P. australis* significantly influences its ecological strategies and resilience levels, thereby affecting wetlands' overall functions [13,14]. There are significant differences in the carbon sequestration capacity of wetlands dominated by *P. australis* across different environments [15]. Latitude can alter the functional traits of *P. australis* [13,16] and the rate of litter decomposition [17], thereby affecting its carbon sequestration capacity.

Intraspecific variation arises from two primary sources: genetic adaptation and phenotypic plasticity [13]. Genetic adaptation might result in long-term differentiation among populations, while plasticity enables rapid trait adjustment within a population under varying conditions. Genetic adaptation involves the rapid selection of adaptive genotypes, often driven by high levels of genetic diversity [18–20]. Meanwhile, phenotypic plasticity, which refers to the ability of a single genome to produce multiple phenotypes, enables plants to maintain fitness components without solely relying on evolutionary adaptation in certain scenarios [21,22]. This capacity for phenotypic variation is particularly relevant when examining the intraspecific diversity of plants such as *P. australis*. In temperate China, *P. australis* comprises two relatively independent lineages, haplotype O and haplotype P, which differ in ploidy levels and morphological traits [23]. These genetic differences influence their growth rates and litter quality, which in turn affect their decomposition rates and carbon cycling. This unique system provides an ideal framework for investigating the roles of genetic and environmental factors in intraspecific variation, as well as their influence on plant growth and function [13].

Plant growth is influenced by varying environmental conditions, and plants ensure their survival and reproduction success by adapting their leaf traits, growth rates, and biomass allocation in response to environmental changes [24]. Some variation is driven by genetic adaptation, with intraspecific differences often correlating with latitude-related factors such as temperature and water availability [12,18]. Additionally, phenotypic plasticity, particularly in leaf traits and growth rates, plays a significant role in plant growth [11,22]. These mechanisms not only provide a buffer against rapid environmental changes but also facilitate rapid adaptation, thereby promoting the development of ecological strategies [25]. Populations have evolved various strategies to cope with environmental factors, such as the leaf economic spectrum and CSR strategies [26,27]. The potential for evolutionary responses is likely to play a critical role in shaping population dynamics and adaptive strategies. As a dominant primary producer, the growth of *P. australis* is essential for carbon deposition and the overall health of wetland ecosystems.

Litter decomposition provides the energy for microbial growth, and alters the microbial activity, microbial abundances, and carbon cycling [28]. The carbon pool in wetlands comprises two main components: the vegetation biomass carbon pool and the sediment organic carbon pool. The plant litter input is an important driver of soil and/or sediment organic carbon turnover [29]. Studies have shown that higher species and genetic diversity can negatively impact plant litter decomposition [30]. Changes in plant diversity, including genetic diversity, can alter the litter decomposition rates and, consequently, affect the carbon sink of wetland ecosystems [31,32]. Previous studies have shown that *P. australis* influences decomposition rates compared to other wetland species. For example, its decomposition rate is faster than that of *Triarrhena lutarioriparia* but slower than that of *Carex cinerascens* [33]. Additionally, the decomposition rate of *P. australis* is positively correlated with the initial nitrogen content and negatively correlated with the C/N ratio [34]. These findings highlight the importance of *P. australis* in nutrient cycling within wetland ecosystems. However, there is a lack of integrated studies on both plant growth and decomposition under common garden conditions across different lineages. Our study aims to fill this gap by examining these processes in two distinct lineages of *Phragmites australis* under controlled conditions [35,36].

In the present study, we therefore sampled *P. australis* from the Liaohe Wetland and the Yellow River Delta, and planted in common gardens located in Panjin and Qingdao. We hypothesized that different lineages of *P. australis* would exhibit significant variations in litter decomposition and plant growth, driven by both phenotypic plasticity and genetic adaptation. Our study aims to test these hypotheses by comparing these traits across lineages under the common garden conditions. We anticipate that our findings will elucidate the mechanisms underlying the observed variations, thereby enhancing our understanding of the carbon and nutrient cycling in wetland ecosystems. Specifically, by identifying the relative contributions of phenotypic plasticity and genetic adaptation, our results will provide actionable insights for the effective management of wetland resources. Additionally, the knowledge gained will support evidence-based conservation strategies and contribute to the sustainable development of wetland ecosystems.

2. Materials and Methods

2.1. Common Garden Experiment

Common gardens were established at Shandong University in Qingdao (36.36° N, 120.69° E) and at the Wetland Scientific Research Institute in Panjin (121.3° E, 41.22° N) (Figure 1A) [11]. The common garden in Qingdao has an average annual precipitation of

662.1 mm and an average annual temperature of 12.7 °C, while the common garden in Panjin has an average annual precipitation of 650 mm and an average annual temperature of 10.5 °C. To capture a representative genetic variation found in temperate China wetlands, we selected nine genotypes of *P. australis*, encompassing tetraploid haplotype O (YRD3, YRD4, YRD5, YRD9) and octoploid haplotype P (CN2014, CN2019, CN2021, CN2024, CN2026). In February 2023, the rhizomes from these nine genotypes were transplanted into pots in two common gardens. Each pot (0.4 m in diameter and 0.5 m in height) was filled with a substrate composed of 50% peat soil and 50% quartz sand (particle size: 0~2 mm). Pots were spatially arranged with 1 m intervals and buried to ensure uniformity in the soil surface levels. Manual weeding and pest control were carried out, and the plants were irrigated with municipal water to maintain the optimal moisture. Drainage holes in each pot facilitated the removal of excess water.

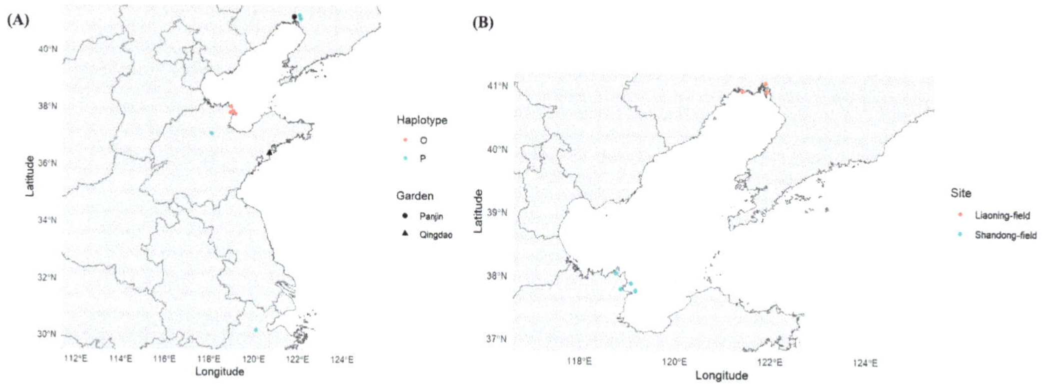

Figure 1. Location of *Phragmites australis* sampling sites. Black dot represents Panjin common garden in Liaoning, and black triangle dot represents Qingdao common garden in Shandong. Colored dots indicate sampling points, with red dots representing haplotype O and green dots representing haplotype P (**A**). Red dots indicate the location of field sites in Liaoning, and the green dots indicate the location of field sites in Shandong (**B**).

In October 2023, the functional traits of *P. australis* within the common gardens were determined. To focus on the most competitive individuals, we selected the five tallest plants from each pot. We measured the plant height from the soil surface to the apex using a tape measure and assessed the base diameter with a vernier caliper. The total number of individual plants and inflorescences were counted, and the aboveground biomass was harvested by cutting the stems at the soil level.

2.2. Litter Collection

To explore the variations in the stoichiometric characteristics of *P. australis* litter, litter samples in both common gardens and the field were collected. Leaf litter was collected in the Panjin common garden on 14 October 2023 and the Qingdao common garden on 20 November 2023. To focus on the typical litterfall dynamics of *Phragmites australis*, we collected only the leaf blades that had fallen but were still attached to the plant, excluding leaf sheaths and other parts. Wilted litter leaves were collected in kraft paper bags, ensuring that they did not make contact with the ground. Stem litter was collected at the Panjin common garden on 14 October and the Qingdao common garden on 20 November 2023. The five tallest dry plants were selected from each common garden and cut off from the base, with any attached leaves removed.

We also conducted litter collection in the field: the Yellow River Delta Wetland in Dongying City, Shandong Province (118.7~119.2° E, 37.7~38.1° N) and the Liaohe Delta Wetland in Panjin City, Liaoning Province (121.3~121.9° E, 40.9~41.2° N) (Figure 1B). These areas are characterized by a temperate monsoon climate that features distinct seasonal variability and considerable precipitation, providing diverse wetland ecosystems conducive to the growth of *P. australis* [14]. The Yellow River Delta Wetland is known for its complex interactions between natural conditions (e.g., vegetation, soil properties, and hydrology) and anthropogenic activities (e.g., agriculture, road construction, and oil exploration). In contrast, the Liaohe Delta Wetland is characterized by its unique ecological features, such as different soil nutrient compositions and vegetation types. These two wetlands were selected to provide a comparative study of *P. australis* under contrasting environmental pressures and latitudes. Leaf litter was collected in Liaohe Delta on 14 October 2023 and Yellow River Delta area on 21 November 2023. Five samples were collected from each site, and the withered reed leaves of the same year were selected. The collected litter was labeled by lineage, thoroughly washed, and dried in an oven at 65 °C for 72 h before proceeding with the analysis of the litter stoichiometric characteristics and conducting decomposition experiments.

The nitrogen content was analyzed using the Kjeldahl method [37]. Dried litter samples were ground and subjected to sulfuric acid digestion, with the nitrogen concentration quantified using a fully automatic Kjeldahl nitrogen analyzer (K9860, Hanon, Jinan, China). The phosphorus concentration was determined by using a spectrophotometer (UV-9000s, Metash, Shanghai, China) according to the molybdenum blue colorimetric method. The carbon content was assessed via the potassium dichromate titration method, where samples were digested using potassium dichromate and sulfuric acid prior to titration with a standard ferrous sulfate solution.

2.3. Decomposition Experiment

For the decomposition assessment, 18 samples from nine genotypes (two lineages: CN and YRD) were selected from the litter of the common gardens. A decomposition experiment was carried out in the greenhouse at Shandong University on 8 May 2024, utilizing three biological replicates for each treatment. Approximately 0.5 g of dry, shredded litter was placed into nylon bags with a 0.5 mm mesh size. Decomposition occurred within plastic containers (height: 15.4 cm, upper diameter: 11.8 cm, lower diameter: 8.0 cm) containing 50 g of soil. The litter bags were buried in the soil, and each container was covered with a lid featuring ventilation holes. To accurately simulate the coastal wetland environment, the soil moisture level was maintained above the soil surface by 1 cm, and the water lost through evaporation was replenished every 2 days. To ensure experimental consistency, the spatial arrangement of microcosm systems was randomized within the greenhouse. Furthermore, the microcosm systems were subjected to weekly random relocation to mitigate positional bias. Field microcosm experiments included five litter samples from both Shandong and Liaoning, resulting in a total of 10 samples and three biological replicates, leading to 30 microcosm systems. After 60 days on 8 July 2024 and 180 days on 8 November 2024, litter bags were retrieved, washed, dried, and weighed to determine the residual mass.

The initial mass of the litter placed in the nylon bags was recorded using an electronic balance. Following the decomposition phase, the litter was washed and dried, and final mass measurements were taken. The litter mass loss and decomposition percentage were calculated based on the initial and final mass.

2.4. Statistical Analysis

Data were analyzed using R software (v4.2.2). We used independent samples *t*-tests to detect the differences in *P. australis* traits and litter characteristics across varying environmental conditions. Two-way ANOVA and principal component analysis were employed to discern the factors influencing the variations in *P. australis* traits. We used the Shapiro–Wilk test to assess the normality of the data. For the differences in growth traits and litter chemical traits between two independent groups (e.g., between the Qingdao and Panjin common gardens), Student's *t*-test was used. For multiple group comparisons (e.g., the differences in litter decomposition rates across different sites), Duncan's multiple comparison test was employed to analyze the differences among the groups. Significant results were denoted by different lowercase letters ($p < 0.05$). Pearson correlation analysis was conducted to examine the relationships between the growth traits of *P. australis* and associated litter stoichiometric characteristics and decomposition rates.

3. Results

Two-way ANOVA showed no significant differences in the stoichiometric characteristics of the litter and plant growth traits among different *P. australis* lineages. However, significant differences were observed in the plant diameter ($p = 0.046$), height ($p = 0.004$), leaf nitrogen (LN) ($p < 0.001$), leaf phosphorus (LP) ($p < 0.001$), and carbon-to-nitrogen ratio (C/N) ($p < 0.001$), as well as the mass loss ratios (MLRs) of the leaf litter from the 180-day harvest (LeafMLR180) ($p = 0.026$) among *P. australis* in different common gardens (Table 1).

Table 1. *F* values of two-way ANOVA of effects of garden, haplotype, and their interactions on growth traits, litter stoichiometric characteristics, and mass loss ratios (MLRs) of *Phragmites australis*. (* $p < 0.05$; ** $p < 0.01$; *** $p < 0.001$).

Parameters	*F* (Garden)	*F* (Haplotype)	*F* (Garden × Haplotype)
Diameter	4.867 *	0.445	0.080
Biomass	4.581	0.008	0.440
Height	12.281 **	1.933	1.098
Shoot Density	0.089	0.291	0.659
Leaf Litter N	44.779 ***	0.0000	0.986
Leaf Litter P	46.481 ***	1.174	0.058
Leaf Litter C	1.410	0.908	1.306
Leaf Litter C/N	24.210 ***	0.911	1.674
Leaf Litter N/P	3.769	0.677	0.594
Leaf MLR 60 days	1.004	0.005	0.382
Leaf MLR 180 days	6.302 *	2.836	0.027

For the growth traits, the average base diameter of *P. australis* in the Qingdao common garden was significantly higher than that in the Panjin common garden (Figure 2A). The average height (H) of *P. australis* in the two common gardens showed significant differences, with the *P. australis* in the Qingdao common garden being 27% taller than that in the Panjin common garden (Figure 2B). Compared with the Panjin common garden, the *P. australis* in the Qingdao common garden had a higher biomass and lower density, but there were no significant differences statistically. Significant differences in LN and LP were noted between the two common gardens, and LN was significantly different between the two wild populations. Notably, LP in the Qingdao common garden was significantly higher than that of the wild *P. australis* in Shandong. In contrast, no significant differences were observed among LP and LN between the Panjin common garden and the wild populations in Liaoning (Figure 2C,D). For the litter stoichiometric characteristics, there was no significant difference between the two common gardens and two wild populations.

In the decomposition study of *P. australis* leaf litter, after 60 days of incubation, the mass loss rates were approximately 25% for lineage O and 25% for lineage P in the Qingdao common garden, and about 27% for lineage O and 26% for lineage P in the Panjin common garden (Figure 3A). After 180 days of incubation, the mass loss rates were about 55% for lineage O and 58% for lineage P in the Qingdao common garden, and around 50% for lineage O and 53% for lineage P in the Panjin common garden (Figure 3B). The results indicate that the mass loss rate of leaf litter significantly increased with an extended incubation time, and the mass loss rate of the lineage P was slightly higher than that of the lineage O.

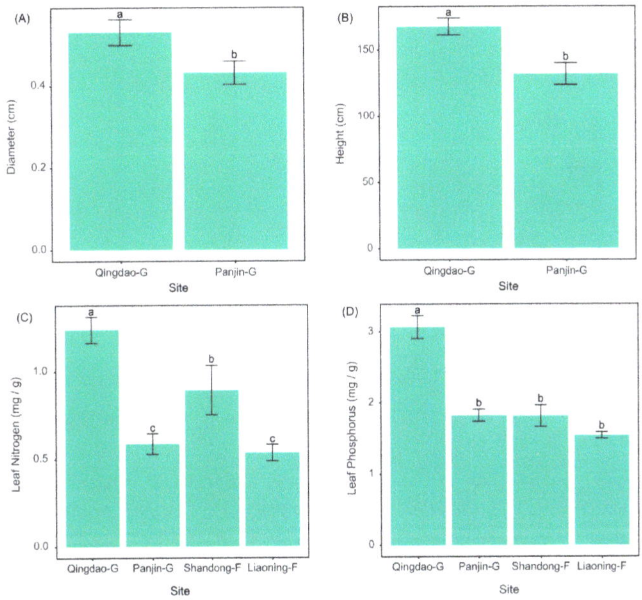

Figure 2. Diameter (**A**), height (**B**), leaf nitrogen (**C**), leaf phosphorus (**D**) of *Phragmites australis* in two common gardens (G) of Qingdao and Panjin and in the field (F) of Shandong and Liaoning (values are means ± standard error). Different letters denote significant differences among groups: (**A**,**B**) as determined by one-way ANOVA followed by Student's *t*-test ($p < 0.05$); (**C**,**D**) as determined by Duncan's multiple range test ($\alpha = 0.05$).

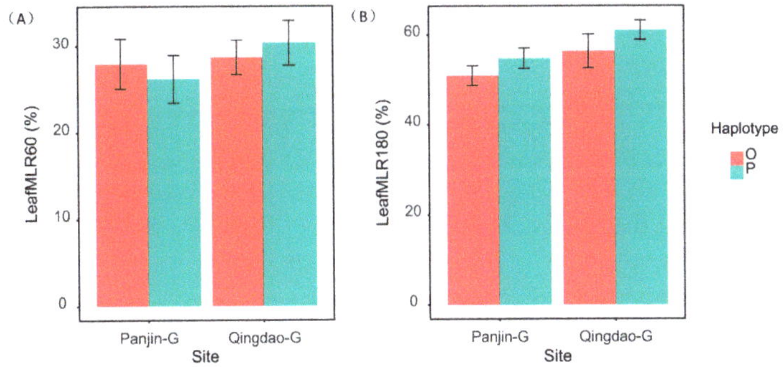

Figure 3. The mass loss ratios of leaf litter from the 60-day harvest (LeafMLR60, **A**) and the mass loss ratios of leaf litter from the 180-day harvest (LeafMLR180, **B**) of *Phragmites australis* litter in Qingdao and Panjin common garden (G) (values are mean ± standard error).

Pearson correlation analysis revealed a strong correlation between LP and LN, a significant positive correlation between LP and height, and a strong negative correlation with C/N. Additionally, LP showed positive correlations with Leaf Resource (LeafMLR 60 and LeafMLR 180). LN exhibited a significant positive correlation with height and diameter, a strong negative correlation with C/N, and a significant positive correlation with LeafMLR 180. Diameter showed positive correlations with biomass and height, and a significant negative correlation with density. Finally, biomass was significantly positively correlated with height (Figure 4).

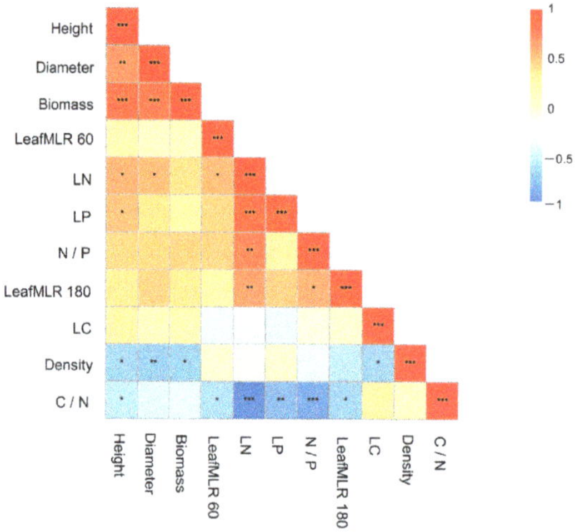

Figure 4. Pearson correlation analysis between growth traits and stoichiometric characteristics. Average diameter, biomass, density, height; carbon to nitrogen ratio (C/N), percentage of carbon content (LC), nitrogen content (LN), phosphorus content (LP), nitrogen to phosphorus ratio (N/P), the mass loss ratios of leaf litter from the 60-day harvest (LeafMLR60), the mass loss ratios of leaf litter from the 180-day harvest (LeafMLR180) (*, $p < 0.05$; **, $p < 0.01$; ***, $p < 0.001$).

Pearson correlation analysis revealed correlations between several bioclimatic variables and the growth traits of *P. australis*. Specifically, a significant positive correlation was found between the mean diurnal temperature range and the diameter of *P. australis* ($p < 0.05$). The mean diurnal temperature range and the mean temperature of the wettest quarter were positively correlated with *P. australis* biomass and negatively correlated with *P. australis* density, although these correlations were not statistically significant (Figure S1).

4. Discussion

Phenotypic plasticity is a key mechanism through which *P. australis* adapts to the environmental conditions. In previous studies, it was found that salinization shifts the phenotypic variability and intraspecific variation in the foundation plant species in estuarine wetlands [38]. In our study, the location of the common garden environment significantly influenced the growth performance (shoot diameter and height), leaf litter stoichiometric traits (contents of N and P), and the leaf decomposition in 180 days. *Phragmites australis* of different lineages were found to have a variation in growth traits in the Panjin and Qingdao gardens. *Phragmites australis* adapts to climate factors through phenotypic plasticity and then has possible effects on population functioning [39–41]. *Phragmites australis* showed an increase in plant height and basal diameter with increasing latitude in the common gardens, as found in other studies [11,42]. Plants often face trade-offs between the energy allocation and evolutionary potential. The role of phenotypic plasticity is constrained by resources, genetic factors, and other limitations, ultimately restricting its effectiveness [25]. Correlation analysis also confirmed the relationship between growth traits and temperature. This adaptability allows organisms to improve their survival and reproductive success under varying environmental conditions. Adaptive plasticity is helpful to the rapid and low-cost adaptive adjustment of plants, thereby enhancing their capacity to sustain fitness across a larger region or a spectrum of conditions.

Genetic variation, including epigenetic variation, serves as a key mechanism of adaptation for many species. Previous studies on *P. australis* showed significant differences between haplotype O and haplotype P in plant functional traits (e.g., specific leaf area and leaf length) [23,40,43], which may affect the decomposition rate of leaf litter. The correlation between species diversity and litter decomposition has been found in other studies, and higher species and genetic diversity had negative effects on plant litter decomposition. The litter mass loss of four-species mixtures was significantly lower compared to that of one-species and two-species mixtures ($p < 0.05$) [30]. Intraspecific variation plays a critical role in maintaining ecosystem function and resilience. This diversity enables populations to adapt to changing environmental conditions, thereby contributing to the health and sustainability of ecosystems. However, the differences in plant growth and litter decomposition between haplotype O and haplotype P were not observed in our study. Plant growth and little decomposition are affected by many factors, and the samples in our study came from a similar environment or the same ecotype. Natural selection, gene flow, or clone reproduction may lead to a similar genetic background, which in turn affects the phenotype. It is worth noting that conducting the decomposition experiment in natural field settings, rather than in microcosms, might have revealed additional differences that were not detected in our current experimental design. This is because microbial communities in natural soils play a crucial role in the decomposition processes and are influenced by the litter characteristics. Other factors, such as population size, environmental conditions, and the short duration of the research, may have also contributed to an incomplete assessment of the influence of genetic variation. Additionally, local adaptation to regional factors, such as salinity, may have diminished the phenotypic differences among different lineages in this species [14].

Plants employ a variety of growth strategies to optimize their survival and reproduction in diverse environments. These strategies often involve trade-offs between resource allocation to different structures and functions. Some plants exhibit rapid growth and early reproduction (r-strategists) in unstable or disturbed habitats [26,27]. In our study, the diameter of *P. australis* is negatively correlated with density and also shows a correlation with latitude, indicating that intraspecific variation exists in the adaptation strategies of *P. australis*. The adaptation strategy of *P. australis* population is rapid growth and dispersal in cold climates, consistent with r strategists. The litter leaf N content in Qingdao is higher than that in Panjin, a result consistent across both the common garden and field experiments. It has also been found in other studies that the leaf N and P content is correlated with the annual average temperature and precipitation, which may be due to the fact that plants at higher latitudes allocate more energy to their leaves, forming a larger leaf area and increasing the chlorophyll content [44,45]. This is further supported by the significant positive correlation between the leaf N content and leaf area. Evidence suggests that leaves are a stronger sink for N than fine roots, and the N might transfer from high-N to low-N components [46,47]. *Phragmites australis* employs different adaptation strategies to balance the relationship between economic traits. The leaf economics spectrum is also present within the *P. australis* population. Additionally, the correlation of the leaf N and P contents with plant height and basal stem also confirms that the adjustment of leaf traits and growth rate is the main survival strategy for plants to adapt to environmental changes [48].

The dynamics of nitrogen and phosphorus nutrients are the main factors influencing the carbon balance and litter decomposition rate in ecosystems. Nitrogen and phosphorus, as key nutrients, affect the carbon cycle. Nitrogen affects photosynthesis and biomass production, thus affecting carbon fixation. Phosphorus is also involved in energy transfer and nucleic acid synthesis, which are very important for the activities of plants and microorganisms [6,49]. The transfer of N is likely associated with the transfer of C. As a primary pathway for soil carbon input, the litter induces an excitation effect, which is linked to the soil nitrogen availability [15,29]. The decomposition rate of leaf litter is positively correlated with the N and P content of leaf litter and negatively correlated with C/N, indicating that a lower C/N of leaf litter facilitates the rapid decomposition of leaf litter. The dynamic change of the nutrient content of carbon, nitrogen, and phosphorus in litter shows a trend of promoting decomposition. Nitrogen and phosphorus in litter limit the decomposition rate of litter and affect the dynamic balance of the carbon cycle in the environment [45,50]. The experimental results also confirmed the coupling effect of nitrogen and phosphorus. Additionally, based on the decomposition ability of leaves, the decomposition ability in gardens is significantly higher than that in the wild. Nitrogen and phosphorus in soil are limiting nutrients for plant growth, and *P. australis* in the wild may be constrained by the nitrogen and phosphorus availability. Litter decomposition is an important process in the cycling of biogeochemical elements and energy flow in wetland ecosystems, serving as a primary mechanism for the carbon pool balance. In future research, we should comprehensively consider the interaction among plant organs, litter, and soil stoichiometric characteristics, as well as the influence of environmental variables (such as climate, soil types and interference factors) on the overall nutrient cycle, and strengthen the fitting of large-scale data models. Thus, the influence of the environment on the stoichiometric characteristics and nutrient migration can be understood more deeply.

5. Conclusions

This study demonstrates that environmental factors, rather than genetic variation (lineage), predominantly shape the growth traits, litter stoichiometry, and decomposition dynamics of *P. australis* in eastern temperate China. The growth performance and litter nutrient content (N and P) were significantly influenced by the common garden environment, with the leaf litter from higher-latitude regions exhibiting a consistently higher N content, irrespective of its origin. The litter decomposition rates were strongly driven by the N and P availability, underscoring the role of the nutrient stoichiometry in ecosystem processes. Notably, climatic conditions from the genotype's origin were linked to growth traits, but not to litter traits or decomposition, indicating that *P. australis* may locally adapt genetically for diverse environments without influencing the associated ecological processes of litter decomposition. These findings emphasize the importance of environmental context and plant plasticity in sustaining wetland ecosystem functions. For conservation and management, prioritizing the habitat conditions over genetic diversity may enhance the resilience of *P. australis* populations and wetland ecosystems under global climate change.

Supplementary Materials: The following supporting information can be downloaded at: https://www.mdpi.com/article/10.3390/d17040282/s1. Figure S1: Pearson correlation analysis between climate environmental factors and the traits and stoichiometric characteristics of *Phragmites australis*. Diameter, biomass, density, height; carbon-to-nitrogen ratio (C/N), percentage of carbon content in litter (LC), nitrogen content in litter (LN), phosphorus content in litter (LP), nitrogen-to-phosphorus ratio (N/P), the mass loss ratios of leaf litter from the 60-day harvest (LeafMLR60), the mass loss ratios of leaf litter from the 180-day harvest (LeafMLR180). bio1, Annual mean temperature; bio2, Mean diurnal temperature range; bio3, Isothermality; bio4, Temperature seasonality coefficient; bio5, Maximum temperature of warmest month; bio6, Minimum temperature of coldest month; bio7,

Temperature annual range; bio8, Mean temperature of wettest quarter; bio9, Mean temperature of driest quarter; bio10, Mean temperature of warmest quarter; bio11, Mean temperature of coldest quarter; bio12, Annual precipitation; bio13, Precipitation of wettest month; bio14, Precipitation of driest month; bio15, Precipitation seasonality; bio 16, Precipitation of wettest quarter; bio 17, Precipitation of driest quarter; bio 18, Precipitation of warmest quarter; bio 19, Precipitation of coldest quarter (*, $p < 0.05$).

Author Contributions: Conceptualization, H.S., W.G. and L.L.; methodology, W.W., Q.W., H.S. and L.L.; software, W.W., Q.W., L.L. and W.S.; validation, H.Z., W.S. and X.Z.; formal analysis, W.W. and Q.W.; investigation, Q.W., H.Z., X.Z. and L.X.; resources, X.G., Y.G., S.Y. and W.G.; data curation, W.W., Q.W. and L.L.; writing—original draft preparation, W.W., Q.W. and H.S.; writing—review and editing, W.W., H.S., W.S., L.X., Y.W. and L.L.; visualization, W.W.; supervision, X.G., Y.G., S.Y., L.L., Y.W. and W.G.; project administration, H.S., L.L. and W.G.; funding acquisition, X.G. and W.G. All authors have read and agreed to the published version of the manuscript.

Funding: This research was funded by the National Natural Science Foundation of China (grant number: 32301317; 32470388; 32271588; U22A20558), and Shandong Provincial Natural Science Foundation (ZR2024QC197).

Institutional Review Board Statement: Not applicable.

Data Availability Statement: Data will be made available upon request.

Conflicts of Interest: The authors declare no conflicts of interest.

References

1. McLaughlin, D.L.; Cohen, M.J. Realizing Ecosystem Services: Wetland Hydrologic Function along a Gradient of Ecosystem Condition. *Ecol. Appl.* **2013**, *23*, 1619–1631. [CrossRef] [PubMed]
2. Means, M.M.; Ahn, C.; Korol, A.R.; Williams, L.D. Carbon Storage Potential by Four Macrophytes as Affected by Planting Diversity in a Created Wetland. *J. Environ. Manag.* **2016**, *165*, 133–139. [CrossRef]
3. Ma, S.; Creed, I.F.; Badiou, P. New Perspectives on Temperate Inland Wetlands as Natural Climate Solutions under Different CO_2-Equivalent Metrics. *NPJ Clim. Atmos. Sci.* **2024**, *7*, 222. [CrossRef]
4. Kayranli, B.; Scholz, M.; Mustafa, A.; Hedmark, Å. Carbon Storage and Fluxes within Freshwater Wetlands: A Critical Review. *Wetlands* **2010**, *30*, 111–124. [CrossRef]
5. Lu, Q.; Gao, Z.; Zhao, Z.; Ning, J.; Bi, X. Dynamics of Wetlands and Their Effects on Carbon Emissions in China Coastal Region—Case Study in Bohai Economic Rim. *Ocean Coast. Manag.* **2014**, *87*, 61–67.
6. Were, D.; Kansiime, F.; Fetahi, T.; Hein, T. Soil Organic Carbon Storage in a Tropical Freshwater Wetland: The Influence of Vegetation Type. *Afr. J. Aquat. Sci.* **2021**, *46*, 161–172. [CrossRef]
7. Wang, Q.; Xie, H.; Ngo, H.H.; Guo, W.; Zhang, J.; Liu, C.; Liang, S.; Hu, Z.; Yang, Z.; Zhao, C. Microbial Abundance and Community in Subsurface Flow Constructed Wetland Microcosms: Role of Plant Presence. *Environ. Sci. Pollut. Res.* **2016**, *23*, 4036–4045. [CrossRef] [PubMed]
8. Zhang, Y.; Ding, W.; Luo, J.; Donnison, A. Changes in Soil Organic Carbon Dynamics in an Eastern Chinese Coastal Wetland Following Invasion by a C4 Plant *Spartina alterniflora*. *Soil Biol. Biochem.* **2010**, *42*, 1712–1720. [CrossRef]
9. Eller, F.; Skálová, H.; Caplan, J.S.; Bhattarai, G.P.; Burger, M.K.; Cronin, J.T.; Guo, W.-Y.; Guo, X.; Hazelton, E.L.G.; Kettenring, K.M.; et al. Cosmopolitan Species as Models for Ecophysiological Responses to Global Change: The Common Reed *Phragmites australis*. *Front. Plant Sci.* **2017**, *8*, 1833. [CrossRef]
10. Packer, J.G.; Meyerson, L.A.; Skálová, H.; Pyšek, P.; Kueffer, C. Biological Flora of the British Isles: *Phragmites australis*. *J. Ecol.* **2017**, *105*, 1123–1162. [CrossRef]
11. Song, H.; Jespersen, E.; Guo, X.; Du, N.; Xie, L.; Pei, L.; Ye, S.; Wang, R.; Brix, H.; Eller, F.; et al. Differences in Relative Air Humidity Affect Responses to Soil Salinity in Freshwater and Salt Marsh Populations of the Dominant Grass Species *Phragmites australis*. *Hydrobiologia* **2021**, *848*, 3353–3369. [CrossRef]
12. Maron, J.L.; Elmendorf, S.C.; Vilà, M. Contrasting Plant Physiological Adaptation to Climate in the Native and Introduced Range of Hypericum Perforatum. *Evol. Int. J. Org. Evol.* **2007**, *61*, 1912–1924. [CrossRef]
13. Ren, L.; Guo, X.; Sorrell, B.K.; Eller, F.; Brix, H. Responses to Cold Temperature Determine Clinal Patterns of Photosynthetic Acclimation of a Cosmopolitan Grass Genus and Challenge the Concept of Quantifying Phenotypic Plasticity. *Funct. Ecol.* **2025**, *39*, 583–595. [CrossRef]

14. Sheng, W.; Liu, L.; Wu, Y.; Yin, M.; Yu, Q.; Guo, X.; Song, H.; Guo, W. Exploring Salt Tolerance and Indicator Traits across Four Temperate Lineages of the Common Wetland Plant, Phragmites australis. *Sci. Total Environ.* **2024**, *912*, 169100. [CrossRef] [PubMed]
15. Silan, G.; Buosi, A.; Bertolini, C.; Sfriso, A. Dynamics and Drivers of Carbon Sequestration and Storage Capacity in *Phragmites australis*-Dominated Wetlands. *Estuar. Coast. Shelf Sci.* **2024**, *298*, 108640. [CrossRef]
16. Ren, L.; Guo, X.; Liu, S.; Yu, T.; Guo, W.; Wang, R.; Ye, S.; Lambertini, C.; Brix, H.; Eller, F. Intraspecific Variation in *Phragmites australis*: Clinal Adaption of Functional Traits and Phenotypic Plasticity Vary with Latitude of Origin. *J. Ecol.* **2020**, *108*, 2531–2543. [CrossRef]
17. Zhang, Y.; Pennings, S.C.; Liu, Z.; Li, B.; Wu, J. Consistent Pattern of Higher Lability of Leaves from High Latitudes for Both Native *Phragmites australis* and Exotic *Spartina alterniflora*. *Funct. Ecol.* **2021**, *35*, 2084–2093. [CrossRef]
18. Ravenscroft, C.H.; Fridley, J.D.; Grime, J.P. Intraspecific Functional Differentiation Suggests Local Adaptation to Long-term Climate Change in a Calcareous Grassland. *J. Ecol.* **2014**, *102*, 65–73. [CrossRef]
19. Kahl, S.M.; Lenhard, M.; Joshi, J. Compensatory Mechanisms to Climate Change in the Widely Distributed Species *Silene vulgaris*. *J. Ecol.* **2019**, *107*, 1918–1930. [CrossRef]
20. Sambatti, J.B.M.; Rice, K.J. Local adaptation, patterns of selection, and gene flow in the Californian serpentine sunflower (*Helianthus exilis*). *Evolution* **2006**, *60*, 696–710.
21. Kawecki, T.J.; Ebert, D. Conceptual Issues in Local Adaptation. *Ecol. Lett.* **2004**, *7*, 1225–1241. [CrossRef]
22. Cuma Mushagalusa, F.; Bauman, D.; Mujinya Bazirake, B.; Mleci, Y.; Kalenga, M.; Shutcha, M.N.; Meerts, P. Phenotypic Plasticity, Not Ecotype Differentiation, Explains the Broad Ecological Niche of a Tree Species in African Dry Woodlands. *Environ. Exp. Bot.* **2020**, *178*, 104186. [CrossRef]
23. Liu, L.; Yin, M.; Guo, X.; Wang, J.; Cai, Y.; Wang, C.; Yu, X.; Du, N.; Brix, H.; Eller, F.; et al. Cryptic Lineages and Potential Introgression in a Mixed-ploidy Species (*Phragmites australis*) across Temperate China. *J. Syst. Evol.* **2022**, *60*, 398–410. [CrossRef]
24. Pastore, M.A. Bringing the Underground to the Surface: Climate Change Stressors Negatively Affect Plant Growth, with Contrasting above and Belowground Physiological Responses. *Plant Cell Environ.* **2022**, *45*, 2267–2270. [CrossRef]
25. Nicotra, A.B.; Atkin, O.K.; Bonser, S.P.; Davidson, A.M.; Finnegan, E.J.; Mathesius, U.; Poot, P.; Purugganan, M.D.; Richards, C.L.; Valladares, F.; et al. Plant Phenotypic Plasticity in a Changing Climate. *Trends Plant Sci.* **2010**, *15*, 684–692. [CrossRef]
26. Pierce, S.; Brusa, G.; Vagge, I.; Cerabolini, B.E.L. Allocating CSR Plant Functional Types: The Use of Leaf Economics and Size Traits to Classify Woody and Herbaceous Vascular Plants. *Funct. Ecol.* **2013**, *27*, 1002–1010. [CrossRef]
27. Zhang, Y.; Meiners, S.J.; Meng, Y.; Yao, Q.; Guo, K.; Guo, W.; Li, S. Temporal Dynamics of Grime's CSR Strategies in Plant Communities during 60 Years of Succession. *Ecol. Lett.* **2024**, *27*, e14446. [CrossRef]
28. Pei, Z.; Leppert, K.N.; Eichenberg, D.; Bruelheide, H.; Niklaus, P.A.; Buscot, F.; Gutknecht, J.L.M. Leaf Litter Diversity Alters Microbial Activity, Microbial Abundances, and Nutrient Cycling in a Subtropical Forest Ecosystem. *Biogeochemistry* **2017**, *134*, 163–181. [CrossRef]
29. Ding, Y.; Wang, D.; Zhao, G.; Chen, S.; Sun, T.; Sun, H.; Wu, C.; Li, Y.; Yu, Z.; Li, Y.; et al. The Contribution of Wetland Plant Litter to Soil Carbon Pool: Decomposition Rates and Priming Effects. *Environ. Res.* **2023**, *224*, 115575. [CrossRef]
30. Yin, M.; Liu, L.; Wu, Y.; Sheng, W.; Ma, X.; Du, N.; Zhu, P.; Wang, C.; Cui, Z.; Brix, H.; et al. Effects of Litter Species and Genetic Diversity on Plant Litter Decomposition in Coastal Wetland. *Ecol. Indic.* **2022**, *144*, 109439. [CrossRef]
31. Mori, A.S.; Cornelissen, J.H.C.; Fujii, S.; Okada, K.; Isbell, F. A Meta-Analysis on Decomposition Quantifies Afterlife Effects of Plant Diversity as a Global Change Driver. *Nat. Commun.* **2020**, *11*, 4547. [CrossRef] [PubMed]
32. Yan, J.; Zhang, Y.; Crawford, K.M.; Chen, X.; Yu, S.; Wu, J. Plant Genotypic Diversity Effects on Soil Nematodes Vary with Trophic Level. *New Phytol.* **2021**, *229*, 575–584. [CrossRef] [PubMed]
33. Xiao, L.; Ma, Y.; Gan, Z.; Cai, R.; Li, Z.; Ge, G.; Wu, L. Influence of soil fauna on the litter decomposition of Lake Poyang Wetland in winter. *J. Lake Sci.* **2020**, *32*, 395–405.
34. Zhang, L.; Zhang, Y.; Zou, J.; Siemann, E. Decomposition of *Phragmites australis* litter retarded by invasive Solidago canadensis in mixtures: An antagonistic non-additive effect. *Sci Rep.* **2014**, *4*, 5488. [CrossRef] [PubMed]
35. Zhang, G.; Cai, Y.; Luo, H.; Hong, S.; Song, C.; Yan, J.; Fang, H. Effects of Abiotic Factors on Plant Community Composition and Functional Traits in the Liaohe Estuary. *J. Mar. Environ. Eng.* **2024**, *11*, 141–156. [CrossRef]
36. Liu, L.; Du, N.; Eller, F.; Ye, S.; Li, X.; Wei, J.; Guo, Y.; Brix, H.; Guo, W. Ecological Mechanisms of Carbon Sequestration in Vegetated Coastal Wetland Ecosystem: Exploring the Roles of Biodiversity and Environmental Changes. *J. Mar. Environ. Eng.* **2025**, *12*, 35–47.
37. Brabson, J.A. The Kjeldahl Method for Organic Nitrogen. *J. Assoc. Off. Anal. Chem.* **1966**, *49*, 481. [CrossRef]
38. Wu, Y.; Liu, L.; Yin, M.; Guo, W. Phylogenetic relationship and soil salinity shape intraspecific trait variability of *Phragmites australis* in the Yellow River Delta. *Front. Mar. Sci.* **2022**, *9*, 980695. [CrossRef]
39. Clevering, O.A.; Brix, H.; Lukavská, J. Geographic Variation in Growth Responses in *Phragmites australis*. *Aquat. Bot.* **2001**, *69*, 89–108. [CrossRef]

40. Mozdzer, T.J.; Caplan, J.S.; Hager, R.N.; Proffitt, C.E.; Meyerson, L.A. Contrasting Trait Responses to Latitudinal Climate Variation in Two Lineages of an Invasive Grass. *Biol. Invasions* **2016**, *18*, 2649–2660. [CrossRef]
41. Oostra, V.; Saastamoinen, M.; Zwaan, B.J.; Wheat, C.W. Strong Phenotypic Plasticity Limits Potential for Evolutionary Responses to Climate Change. *Nat. Commun.* **2018**, *9*, 1005. [CrossRef] [PubMed]
42. Lissner, J.; Schierup, H.-H.; Comín, F.A.; Astorga, V. Effect of Climate on the Salt Tolerance of Two *Phragmites australis* Populations. *Aquat. Bot.* **1999**, *64*, 317–333. [CrossRef]
43. Liu, L.; Wang, J.; Ma, X.; Li, M.; Guo, X.; Yin, M.; Cai, Y.; Yu, X.; Du, N.; Wang, R.; et al. Impacts of the Yellow River and Qingtongxia Dams on Genetic Diversity of *Phragmites australis* in Ningxia Plain, China. *Aquat. Bot.* **2021**, *169*, 103341. [CrossRef]
44. Kattge, J.; Knorr, W.; Raddatz, T.; Wirth, C. Quantifying Photosynthetic Capacity and Its Relationship to Leaf Nitrogen Content for Global-scale Terrestrial Biosphere Models. *Glob. Change Biol.* **2009**, *15*, 976–991. [CrossRef]
45. He, Y.; Xu, X.; Kueffer, C.; Zhang, X.; Shi, P. Leaf Litter of a Dominant Cushion Plant Shifts Nitrogen Mineralization to Immobilization at High but Not Low Temperature in an Alpine Meadow. *Plant Soil* **2014**, *383*, 415–426. [CrossRef]
46. Pregitzer, K.S.; Zak, D.R.; Talhelm, A.F.; Burton, A.J.; Eikenberry, J.R. Nitrogen Turnover in the Leaf Litter and Fine Roots of Sugar Maple. *Ecology* **2010**, *91*, 3456–3462. [CrossRef]
47. Berglund, S.L.; Ågren, G.I.; Ekblad, A. Carbon and Nitrogen Transfer in Leaf Litter Mixtures. *Soil Biol. Biochem.* **2013**, *57*, 341–348. [CrossRef]
48. Wright, I.J.; Reich, P.B.; Westoby, M. Strategy Shifts in Leaf Physiology, Structure and Nutrient Content between Species of High- and Low-rainfall and High- and Low-nutrient Habitats. *Funct. Ecol.* **2001**, *15*, 423–434. [CrossRef]
49. Sisto, M.L.D.; Macdougall, A.H. Effect of Terrestrial Nutrient Limitation on the Estimation of the Remaining Carbon budget. *Biogeosciences* **2024**, *21*, 21. [CrossRef]
50. Uddin, M.N.; Robinson, R.W. Responses of Plant Species Diversity and Soil Physical-Chemical-Microbial Properties to *Phragmites australis* Invasion along a Density Gradient. *Sci. Rep.* **2017**, *7*, 11007. [CrossRef]

diversity

MDPI

Systematic Review

Biodiversity Monitoring in Constructed Wetlands: A Systematic Review of Assessment Methods and Ecosystem Functions

Marvin John Uy, Miguel Enrico Robles, Yugyeong Oh, Md Tashdedul Haque, Cloie Chie Mueca and Lee-Hyung Kim *

Department of Civil and Environmental Engineering, Kongju National University, Cheonan 1223-24, Chungnamdo, Republic of Korea; marvinjohnfuy@smail.kongju.ac.kr (M.J.U.); roblesmiguel@smail.kongju.ac.kr (M.E.R.); yugyeong@smail.kongju.ac.kr (Y.O.); haquemdtashdedul@smail.kongju.ac.kr (M.T.H.); chiemueca@smail.kongju.ac.kr (C.C.M.)

* Correspondence: leehyung@kongju.ac.kr

Abstract: Constructed wetlands (CWs) are widely implemented as nature-based solutions for delivering essential ecosystem services such as water purification, carbon sequestration, and habitat provision. However, biodiversity monitoring within CWs remains limited and unevenly integrated into performance evaluations. This scoping review analyzed 76 peer-reviewed studies to assess current methods for biodiversity monitoring, explore linkages to ecosystem functions, and examine the diversity indices most frequently applied. Results revealed a predominant focus on microbial communities, primarily assessed through high-throughput sequencing and general ecological indices such as the Shannon–Wiener Diversity Index and Chao1 Richness Estimator, with limited taxonomic depth or functional specificity. Plant and animal biodiversity were addressed less frequently and were rarely linked to treatment outcomes or ecosystem services beyond regulation. Vertical subsurface flow systems were the most studied configuration, particularly in lab-scale studies, while free water surface systems exhibited greater microbial phylum richness. These findings highlight a critical need for CW-specific biodiversity monitoring frameworks that integrate microbial, plant, and faunal assessments using functionally relevant phylogenetic indices such as Rao's Quadratic Entropy and Faith's Phylogenetic Diversity. Emphasis on standardization, trait-based analyses, and mechanistic approaches is essential for enhancing ecological interpretation and ensuring biodiversity is recognized as a central component of CW design, performance, and resilience.

Keywords: assessment techniques; ecological indicators; ecosystem services; functional traits; microbial communities; phylogenetic diversity

1. Introduction

Biodiversity underpins the stability, resilience, and functionality of Earth's ecosystems. It plays a vital role in ecological processes such as resource capture, biomass production, nutrient recycling, and the maintenance of ecosystem services that support human well-being [1,2]. The concept, derived from "biological diversity", was introduced in 1985 and gained global recognition following the 1992 UN Earth Summit, which led to the establishment of international frameworks such as Agenda 21 and the Convention on Biological Diversity (CBD) [3]. In response to accelerating biodiversity loss, global frameworks including the Intergovernmental Science-Policy Platform on Biodiversity and Ecosystem Services (IPBES) have emphasized the importance of standardized biodiversity monitoring to support conservation and ecosystem management [4]. This international momentum was reinforced at the 16th Conference of the Parties (COP16), which promoted the Kumming-Montreal Global Biodiversity Framework and its target to protect 30% of ecosystems by 2030. These outcomes highlight the need to apply biodiversity indicators not only in natural but also in human-managed systems such as constructed wetlands (CWs). As engineered nature-based solutions (NBS), CWs simultaneously support conservation and ecosystem services such as water purification, carbon storage, and habitat provisioning [5], making them key candidates for advancing global biodiversity goals through integrative monitoring.

CWs have been widely adopted over the past four decades to manage wastewater, mitigate flooding, and restore ecological balance. Typically consisting of shallow vegetated basins populated with emergent macrophytes, CWs provide multiple ecological and biogeochemical services. Despite these multifunctional benefits, assessments of CW performance have traditionally focused on pollutant removal efficiencies [6,7]. As a result, biodiversity, a critical driver of ecosystem function and long-term system resilience, has remained underrepresented in the planning and management of CWs.

Higher levels of biodiversity are known to enhance carbon sequestration, support nutrient retention, improve water quality, and promote system stability [8]. However, the mechanisms linking biodiversity to these ecosystem services in CWs are not yet well understood. This gap is partly due to the use of biodiversity metrics that differ from those used to assess functional performance. Many commonly applied indices, such as the Shannon–Wiener Diversity Index (Shannon), Simpson's Index (Simpson), and Pielou's Evenness Index (Pielou's Evenness), were originally developed for general ecological assessments and are sensitive to sample size [9,10]. Consequently, they may require standardization or adjustment to adequately capture the engineered complexity of CWs [11].

Inconsistencies in data reporting across spatial and temporal scales have been caused by the current diversity of assessment methodologies. This fragmentation has hindered the development of cohesive monitoring frameworks that can effectively link biodiversity patterns with functional outcomes. Without standardized, functionally relevant, and sample-size independent metrics, it remains challenging to assess biodiversity's role in ecosystem services or compare findings across CW studies [12,13]. The urgent need for such standardization is compounded by the lack of indicators specifically developed for CWs and the continued reliance on legacy ecological approaches that do not reflect the unique design and performance goals of these engineered systems [14].

Given the increasing recognition of biodiversity's role in sustaining key ecosystem functions, there is a critical need for a standardized biodiversity monitoring framework specific to CWs. Thus, a scoping review was conducted to systematically map existing research, assess current methodologies, and identify gaps in biodiversity monitoring within CWs. This review addresses this need by pursuing three main objectives. First, to identify and describe the methods currently used to assess biodiversity in CWs, including field-based surveys, molecular tools, and bioindicator-based approaches. Second, to explore how different aspects of biodiversity specifically, microbial, plant, and faunal aspects, contribute to essential ecosystem functions such as carbon sequestration, water purification, and nutrient cycling. Third, to evaluate the biodiversity indices applied in CW research, with particular focus on their relevance, frequency of use, and applicability in capturing biodiversity's role in supporting ecosystem services. By fulfilling these objectives, this study provides a basis for developing harmonized biodiversity monitoring strategies aimed at improving CW design, performance, and ecological resilience.

2. Materials and Methods

2.1. Review Protocol

This scoping review followed the Preferred Reporting Items for Systematic Reviews and Meta-Analyses extension for Scoping Reviews (PRISMA-ScR) guidelines to ensure transparency and methodological rigor [15]. Additionally, the review protocol was pre-developed and refined iteratively to maintain consistency during the study selection and data extraction.

2.2. Eligibility Criteria

Studies were included if they met at least one of the following eligibility criteria: they employed assessment methods to measure biodiversity in CWs; examined the relationship between biodiversity and ecosystem functions such as nutrient removal, water purification, or carbon sequestration; or applied biodiversity indices to evaluate ecological patterns within CW systems.

The eligibility criteria were designed to capture a broad range of studies that reflect the multifaceted nature of biodiversity monitoring in CWs. This included research focusing on assessment methodologies, the functional roles of biodiversity in supporting ecosystem services, and the application of ecological indices to quantify biodiversity. These criteria were chosen to ensure comprehensive coverage of microbial, plant, and faunal components, and to highlight the diverse methodological approaches used to evaluate biodiversity within CW systems. A summary of the final inclusion and exclusion criteria is provided in Table 1.

Table 1. Inclusion and exclusion criteria used to select studies for the scoping review on biodiversity monitoring in CWs.

Inclusion Criteria	Exclusion Criteria
Studies conducted in CWs or similar engineered/treatment wetlands.	Studies did not assess biodiversity and focused on wetland hydrology or treatment performance.
Studies assessed biodiversity using quantitative indices such as Shannon, Simpson, Abundance-based Coverage Estimator (ACE), and Chao1	Studies focused solely on physical or chemical treatment processes without biodiversity considerations.
Studies linked biodiversity to ecosystem functions.	Studies were opinion pieces, non-peer-reviewed, or lacked a clear methodology.
Studies were peer-reviewed articles.	

2.3. Information Sources and Search Strategy

A systematic literature search was conducted in Scopus and Web of Science (WoS) between January and March 2025. The search strategy was developed iteratively over multiple days, during which various keyword combinations and Boolean operators (AND/OR) were tested and refined. It was ultimately structured into three thematic queries, designed to capture studies focused on biodiversity assessment methods, ecosystem functions, and the application of biodiversity indices within CWs. The search was limited to English-language publications that were either open-access or institutionally accessible. No publication year filter was applied to capture all relevant studies regardless of publication date, consistent with the scoping review objective of mapping the breadth and evolution of existing research. A summary of the refined search queries is presented in Table 2.

Table 2. Search queries and themes used for the systematic literature review.

Database	Query Type	Optimized Search Query
Scopus, WOS	Biodiversity Assessment Methods in CWs	"Biodiversity assessment", "species diversity", "functional diversity", "bioindicators", "community structure", "ecological indicators", "constructed wetland", "engineered wetland", "assessment", "survey", "monitoring", "evaluation"
Scopus, WOS	Biodiversity and Ecosystem Functions in CWs	"Biodiversity monitoring", "species diversity", "bioindicators", "community structure", "functional diversity", "plant diversity", "macroinvertebrate diversity", "microbial diversity", "vegetation diversity", "faunal diversity", "ecological diversity", "trophic diversity", "constructed wetland", "engineered wetland", "treatment wetland", "surface flow wetland", "subsurface flow wetland", "hybrid constructed wetland", "free water surface wetland", "wetland treatment system", "nature-based treatment system", "wetland restoration", "carbon sequestration", "water purification", "pollution removal", "nutrient removal", "nutrient retention", "self-purification", "biogeochemical cycling", "ecosystem services", "sediment retention", "ecological function", "primary productivity"
Scopus, WOS	Specific biodiversity indices in CWs	"Biodiversity index", "species richness", "diversity index", "ACE", "Shannon index", "Simpson index", "Phylogenetic diversity", "constructed wetland", "engineered wetland", "treatment wetland", "survey", "monitoring", "evaluation", "analysis"

2.4. Selection of Sources of Evidence

A total of 182 records were initially identified across Scopus (n = 102) and WOS (n = 80), and 40 duplicate entries were removed, resulting in 142 records for screening, as shown in Figure 1. Manual screening of titles and abstracts was then conducted on these 142 records to assess relevance, leading to the exclusion of 38 records that were irrelevant, inaccessible, or outside the scope of the review. The remaining 104 full-text articles were assessed using the Rayyan platform, an online tool for managing systematic reviews [16]. Titles and abstracts were screened in the first phase to exclude irrelevant records. Full-text articles were evaluated against the eligibility criteria in the second phase. Studies were considered eligible if they addressed biodiversity assessment methods, examined biodiversity–function relationships, or applied biodiversity indices within the CW context. Studies were excluded if they lacked biodiversity data, did not assess ecosystem functions, or focused solely on chemical or hydrological treatment without biodiversity analysis.

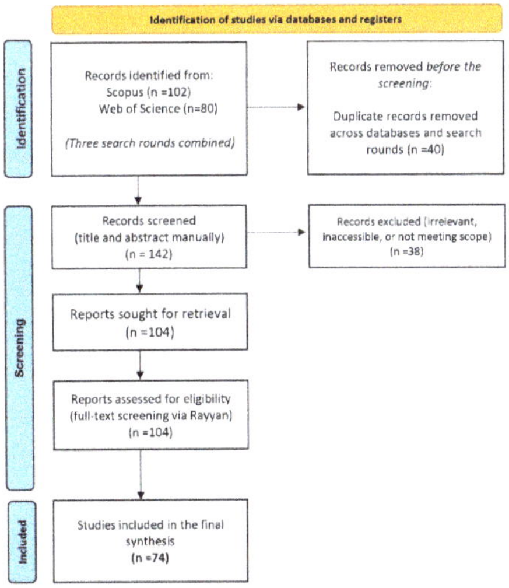

Figure 1. PRISMA 2020 flow diagram of the study selection process for the scoping review on biodiversity in CWs.

Although the review focused on CWs, a few studies on natural or restored wetlands were included for their methodological relevance in informing biodiversity monitoring in engineered systems. In total, 76 studies met the inclusion criteria and were selected for synthesis. These studies encompassed a wide range of wetland types, assessment techniques, and functional linkages.

2.5. Data Extraction Processing and Visualization

All eligible studies were manually reviewed, and relevant data were extracted manually into structured Microsoft Excel spreadsheets. Data were categorized according to wetland type, biodiversity focus, assessment methods, and ecosystem functions. Information related to microbial sequencing techniques, taxonomic resolution, and diversity indices was also extracted and classified. Diversity indices were grouped into alpha diversity metrics and less commonly applied phylogenetic or trait-based indices. Visualizations such as the Sankey diagram, heatmap, and bar charts were generated using OriginPro to illustrate co-occurrences and trends across CW types, scales, and microbial methods. No bibliometric mapping was used in this study, as the focus was on content-based synthesis of biodiversity monitoring methods and functional linkages, rather than citation network analysis.

3. Results

3.1. Study Landscape and Institutional Support

Over the past two decades, research on biodiversity in CWs has expanded across countries and institutions. As shown in Figure 2, publication activity steadily increased after 2013, with notable peaks in 2020 and 2023. China has emerged as the most prolific contributor, with 48 publications since 2012. Other key contributors include the United States (14 publications) and Spain (4 publications). This growth in China, supported by national legislation such as the Provisions on Administration of Wetland Protection [17], laid the groundwork for wetland governance and ecological research priorities. This positions China to influence global practices in CW biodiversity assessment, particularly in advancing microbial profiling techniques and promoting biodiversity-integrated, multifunctional CW system design.

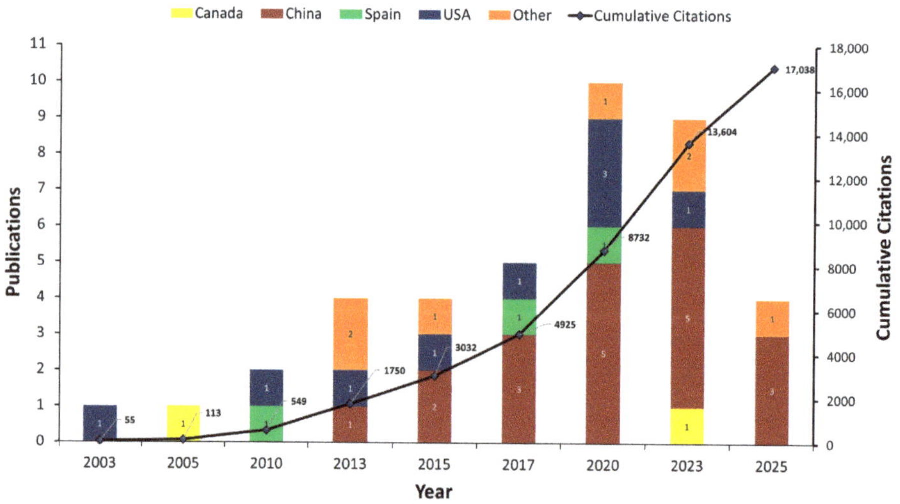

Figure 2. Annual publication count and cumulative citations of studies on biodiversity in CWs (2003–2025), categorized by country of origin.

A total of 204 funding acknowledgments were identified across the reviewed studies. As summarized in Table 3, the majority were attributed to national government programs (105), followed by regional and local authorities (47), universities (18), and international organizations (9). The National Natural Science Foundation of China (NSFC), cited in 28 studies, exemplifies a centralized and strategic approach to research funding. It supports international collaborations aligned with the Ramsar Convention's goals on wetland conservation, and provides targeted funding for less developed regions, thereby promoting widespread participation in biodiversity monitoring across CWs [18].

Table 3. Number of funding acknowledgments by type of institution across all reviewed studies.

Type of Institution/ Funding Body	Number of Entries
National Government/Program/Initiative	105
Regional or Local Authority/Program	47
University/Higher Education Institution	18
Others *	43

* Categories with fewer than 10 entries.

3.2. Overview of Constructed Wetland Types and Research Scope

A total of 76 studies were included in this review, encompassing various CW types, operational scales, and geographic settings. As summarized in Table 4, a total of 42 studies explicitly reported the flow configuration. Vertical Subsurface Flow (VSSF) wetlands were the most studied, appearing in 16 studies, 15 of which were lab-scale, likely due to their structural simplicity, compact design, and compatibility with controlled experiments [19]. Lab-scale VSSF wetlands have enabled detailed microbial biodiversity analysis, particularly through high-throughput sequencing (HTS). Enhanced aeration and subsurface flow may also create unique microhabitats that influence microbial community structure and functional diversity.

Table 4. Number and percentage of CW types by operational scale among studies.

Wetland Type	Full Scale (n, %)	Lab-Scale/Mesocosm (n, %)	Total (n, %)
FWS	7 (16.7%)	7 (16.7%)	14 (33.33%)
HSSF	4 (9.5%)	5 (11.9%)	9 (21.4%)
VSSF	1 (2.4%)	15 (35.7%)	16 (38.1%)
Hybrid	2 (4.8%)	1 (2.4%)	3 (7.1%)
Total	14 (33.33%)	28 (66.7%)	42 (100%)

Abbreviations: FWS = Free Water Surface; HSSF = Horizontal Subsurface Flow; VSSF = Vertical Subsurface Flow.

FWS systems appeared in 14 studies, evenly divided between lab- and full-scale systems. These wetlands offer strong biodiversity support and deliver ecosystem functions such as wildlife habitat provisioning and aesthetic value [20]. However, their open-water design also makes them more sensitive to seasonal temperature fluctuations. In contrast, Horizontal Subsurface Flow (HSSF) wetlands, reported in nine studies, are typically implemented in rural or expansive areas due to their greater land requirements and more complex hydraulic infrastructure [21]. HSSF systems offer advantages in phosphorus and pathogen removal, making them suitable for pollutant-focused treatment goals [20,22]. Hybrid systems, such as Integrated Constructed Wetlands (ICW), were the least represented, appearing in only three studies. Additionally, 34 studies did not specify a flow configuration, limiting comparative evaluation of design-related biodiversity and functional outcomes. System size was inconsistently reported across studies and thus not analyzed as a factor, though it may significantly influence biodiversity–function relationships.

Among the 76 studies, 60 addressed all three components of the review: biodiversity, ecosystem function, and assessment method. In contrast, only 13 studies explored ecosystem functions and monitoring methods without explicitly incorporating biodiversity data, and 1 study focused solely on biodiversity assessment, indicating a gap, as biodiversity is rarely studied as an independent focus in CW research, often treated as a supporting or indirect variable rather than a primary outcome [6] (see Figure 3).

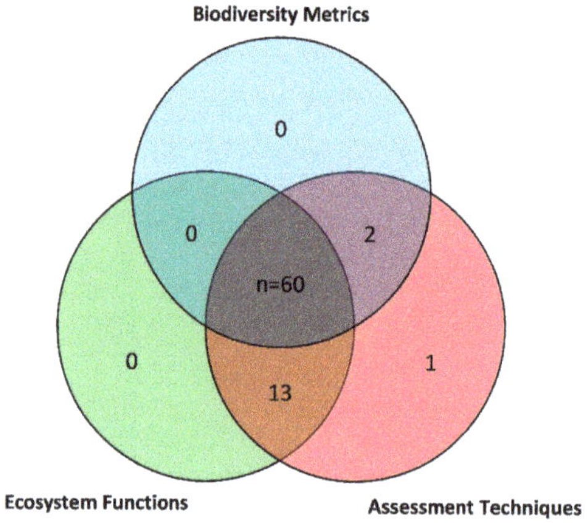

Figure 3. Venn diagram of 76 studies showing overlap among biodiversity metrics, ecosystem functions, and monitoring methods in CWs.

3.3. Microbial Sequencing Strategies and Diversity Metrics Across Constructed Wetlands

A Sankey diagram illustrating the connections between CW flow types, study scales, microbial analysis methods, and the biodiversity indices applied is presented in Figure 4. The Sankey diagram is a synthesis-based visualization constructed from manually extracted categorical data across the reviewed studies. It illustrates the co-occurrence frequency of flow types, study scales, microbial analysis methods, and diversity indices, rather than representing continuous or quantitative flow values. In the diagram, each colored block represents a category and the lines connecting them represent the number of studies that link one category to another. Thicker lines indicate stronger connections, reflecting higher frequencies of co-occurrence across the reviewed studies. This visualization highlights key trends in how microbial communities have been studied in CWs.

Microbial communities in CWs have been analyzed using various molecular techniques, including HTS, Sanger sequencing, qPCR, and metagenomics. HTS has become the predominant method due to its scalability and sensitivity to rare or low-abundance taxa. When paired with bioinformatics tools, HTS supports ecological interpretation across complex microbial communities [23]. Commonly applied diversity indices for microbial community analysis in CWs include Shannon, Simpson, Chao1 Richness Estimator (Chao1), and Abundance-based Coverage Estimator (ACE). These indices were most often applied in lab-scale studies where controlled conditions allow for detailed microbial analysis. In contrast, indices such as Faith's Phylogenetic Diversity (Faith's PD), Pielou's Evenness Index (Pielou's Evenness), and Good's Coverage Estimator (Good's Coverage) were used in fewer than 5% of reported cases.

The predominance of HTS reflects a broader shift in CW research toward high-resolution, data-rich techniques in CW microbial research. These methods enable the profiling of both culturable and unculturable taxa [24]. Given the central role of microbes in nutrient removal, organic degradation, and biogeochemical cycling [25], microbial profiling is key to optimizing CW performance. However, HTS using 16S rRNA amplicon sequencing is often limited to the genus-level identification, reducing the resolution of functional interpretations, especially when microbial traits vary within genera. As demonstrated by Lemos et al. [26], datasets with low sequence coverage yield insufficient taxonomic resolution and underestimate diversity, limiting the detection of rare or functionally critically taxa. Commonly used indices like Shannon and Chao1 emphasize richness and diversity but lack ecological depth. In contrast, less commonly used metrics like Faith's PD and Pielou's Evenness offer greater ecological relevance by incorporating evolutionary relationships and community balance [27]. Comparative studies suggest that these alternative indices may better detect ecological shifts or treatment impacts [28]. The absence of standardized protocols and baseline references hinders cross-study synthesis, highlighting the need for harmonized microbial monitoring frameworks.

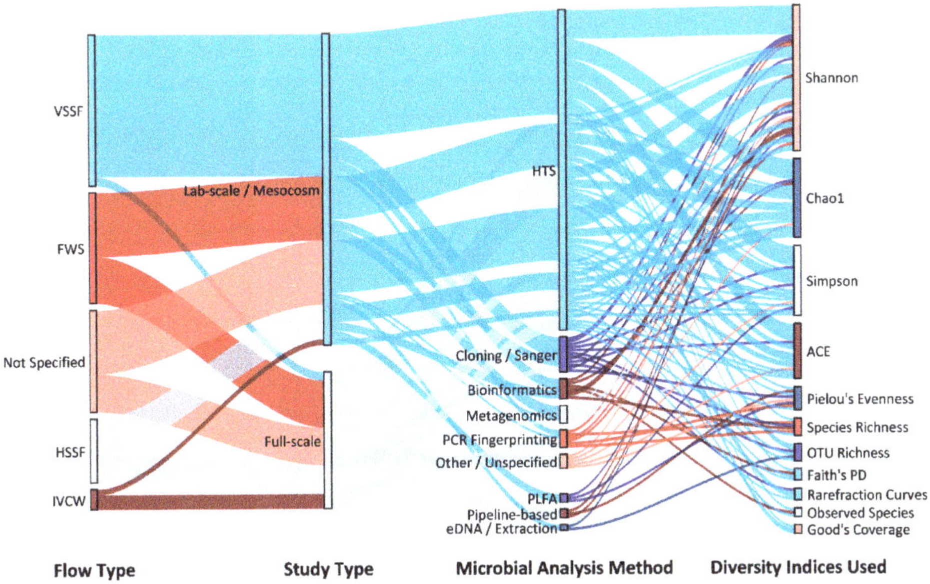

Figure 4. Sankey diagram illustrating the relationships between CW flow types, study scales, microbial analysis methods, and diversity indices used across reviewed studies.

Recent studies further highlight the limitations of standard alpha diversity metrics in linking microbial diversity to functional performance in CWs. For instance, Cakin et al. [29] demonstrated that on HSSF CWs treating distillery wastewater, Shannon and richness indices captured general trends but failed to reflect the functional roles of key taxa such as *Acidobacteriia*, *Planctomycetacia*, and *Rozellomycota* in driving COD, TON, and NH_4^+ removal. Similarly, Ibekwe et al. [30], noted strong correlations between microbial community composition and treatment efficiency, yet relied primarily on richness-based indices that lacked functional specificity.

This over-reliance on generalized indices may obscure functionally important dynamics within microbial communities, especially in systems where treatment performance depends on narrow or specialized microbial guilds. Despite their value, richness and evenness metrics often fall short in detecting ecological shifts that influence CW function, as noted in comparative assessments by Jost [27] and Feranchuk et al. [28]. By contrast, phylogenetic and trait-based indices, such as Faith's PD and Rao's Quadratic Entropy, have shown greater sensitivity to functional shifts in microbial communities and may be more appropriate for engineered ecosystems with complex pollutant loads.

To address this gap, future CW research should expand beyond taxonomic profiling and integrate multi-dimensional biodiversity indices that reflect both evolutionary history and ecological function. Developing CW-specific index frameworks and linking them directly to pollutant removal performance could significantly enhance the ecological relevance and management utility of microbial diversity assessments. Broader adoption of functional diversity indices, phylogenetic metrics, and trait-based indices tailored to CW processes is recommended. These metrics offer greater sensitivity to ecological shifts and may better capture the functional contributions of microbial communities to ecosystem services such as nutrient cycling, organic matter degradation, and carbon sequestration. Establishing standardized protocols that include these indices would improve comparability across studies and support the development of more predictive and performance-linked biodiversity monitoring frameworks. To advance the ecological relevance of microbial monitoring, future research should standardize sequencing protocols, adopt functionally relevant indices and expand the use of metagenomic tools capable of resolving microbial traits. Cross-validation between molecular findings and functional performance metrics will be key in optimizing CW design and management.

3.4. *Microbial Phyla and Taxonomic Resolution in Constructed Wetlands*

Lab-scale studies commonly featured more detailed microbial analyses and biodiversity metrics, whereas full-scale studies were less likely to report species-level resolution. This gap highlights the operational difficulty of biodiversity monitoring in real-world systems, particularly at microbial and invertebrate scales. Lab-scale systems have enabled finer assessments of microbial stratification and enzymatic activity under controlled conditions, whereas full-scale applications often face constraints related to sampling complexity, taxonomic resolution, and resource limitations [25]. Greater precision in reporting CW flow types and operational scales is needed to improve cross-study comparisons. Standardizing typologies would enhance meta-analyses and support global knowledge exchange on biodiversity–function relationships.

Microbial taxonomic resolution varied across CW types. This analysis focused on phylum-, genus-, and species-level classifications, which were the most consistently reported across the reviewed studies. Intermediate ranks such as class, order, and family were excluded due to inconsistent reporting and limited functional interpretation. Studies that focused on VSSF showed the lowest species-level identification (5.4%) and the highest genus-level classification (74.5%), while HSSF systems demonstrated a more balanced distribution across taxonomic ranks. FWS systems reported the highest proportion of phylum-level identifications (55.5%) and the greatest overall phylum richness, with 26 distinct groups. These patterns are illustrated in Figure 5, which compares taxonomic resolution across CW configurations.

The predominance of phylum and genus-level identifications reflects both methodological and technical constraints in microbial monitoring. Most studies employed 16S rRNA gene amplicon sequencing, which is effective for broad community profiling but typically limited to genus-level resolution. Even with more targeted approaches like clone libraries and phylogenetic reconstruction, as noted by Arroyo et al. [31], many sequences could only be assigned to the family or order level, with some remaining unclassified due to reference database gaps. These challenges were particularly evident in high-stress environments like swine wastewater systems, where microbial diversity was low and communities were dominated by beta-proteobacteria. Silveira et al. [32] similarly observed that microbial richness and structure varied spatially within horizontal flow CWs, yet even HTS failed to resolve finer community differences, highlighting the need for improved sampling strategies and deeper sequencing. These taxonomic limitations are not due to

HTS itself, but rather the reliance on marker genes with restricted resolution. Although advanced methods like shotgun metagenomics offer better taxonomic and functional detail, their high cost and technical complexity have limited routine use in CW studies.

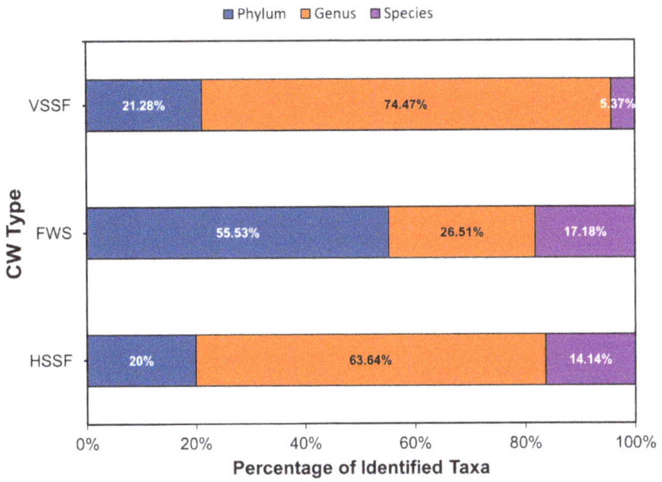

Figure 5. Taxonomic resolution of microbial identifications in constructed wetlands.

Following the resolution analysis, microbial composition at the phylum level displayed clear differences across wetland types. As shown in Figure 6, Proteobacteria were dominant in all systems, comprising 32.1% in HSSF, 32.4% in FWS, and 60.5% in VSSF wetlands. This group plays vital roles in treatment processes such as ammonification, nitrification, denitrification, and degradation of organic pollutants via biosorption and catalytic transformation [33]. Other phyla varied more widely, Euryarchaeota and Bacteroidetes were more prevalent in HSSF and VSSF systems, while Firmicutes and Actinobacteria appeared consistently across all types but at a lower relative abundance.

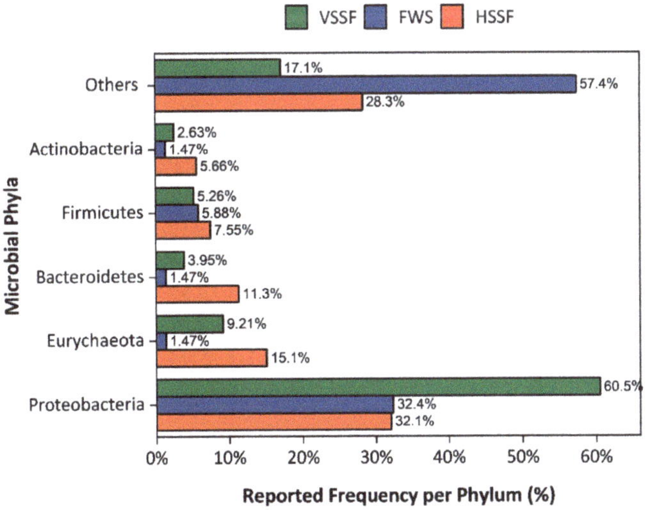

Figure 6. Relative frequency of reported microbial phyla across constructed wetland (CW) types.

In terms of richness, FWS systems exhibited the greatest phylum diversity (26), followed by VSSF (24) and HSSF (22). This richness in FWS wetlands likely results from their open structure and dense vegetation, which create heterogeneous microhabitats and oxygen gradients that favor diverse microbial communities [30]. In contrast, the more compact and oxygen-limited environments of HSSF and VSSF tended to support fewer but more specialized phyla. Some phyla were enriched or uniquely observed within specific systems, Ascomycota and Rozellomycota in HSSF, Caldiserica, Chloroflexi, and Chlamydiae in VSSF, and rare groups such as Nitrospirae, Synergistetes, and Verrucomicrobia in FWS. However, because most identifications were at the phylum level, finer-scale diversity may be underrepresented, reinforcing the need for higher-resolution methods and more robust taxonomic frameworks [24]. Moving forward, microbial community studies in CWs should incorporate deeper sequencing platforms such as shotgun metagenomics or full-length 16S approaches.

3.5. Plant and Animal Biodiversity in Constructed Wetlands

3.5.1. Types of Macrophytes and Floating Plants Used

A total of 13 macrophyte genera were reported in lab-scale studies, with *Phragmites, Typha, Canna, Acorus, Iris,* and *Thalia* being the most frequently used. In contrast, full-scale systems reported a broader diversity of 33 genera. The most cited were *Phragmites, Typha, Canna, Carex, Cyperus, Iris,* and *Scirpus,* along with other genera such as *Eleocharis, Juncus, Mentha, Sparganium,* and several others cited only once. This difference in diversity reflects the broader ecological and functional goals of full-scale systems compared to the more standardized species selection in lab-scale experiments. Mixed culture systems have been shown to consistently achieve higher nitrogen and phosphorus removal efficiencies. These improvements are linked to increased oxygen and organic carbon release in the rhizosphere, healthier root structures, and enhanced microbial communities that support nutrient cycling. Diverse plant configurations play a functional role in improving microbial nitrification and denitrification processes in constructed wetlands [34]. Figure 7 illustrates notable differences in macrophyte species usage between lab-scale/microcosm and full-scale CWs.

This disparity in macrophyte diversity reflects both methodological constraints in lab-scale experiments and broader ecological and functional goals in full-scale systems. Lab-scale studies often rely on a core group of robust, fast-growing species known for pollutant uptake efficiency and root oxygen release, such as *Phragmites, Typha,* and *Canna* [35]. These selections enable replicability and control in small-scale systems. In contrast, full-scale CWs integrate a much wider range of macrophytes to address landscape heterogeneity, site-specific hydrology, habitat provisioning, and long-term stability.

While dominant species often drive core treatment functions in CWs, multispecies assemblages have been shown to enhance system resilience and deliver additional co-benefits. These include improved nutrient removal efficiency, greater stability under fluctuating environmental conditions, and increased plant productivity. For instance, Brisson et al. [36] conducted a meta-analysis of 28 wetland studies and found that increased plant species richness modestly but significantly improved nutrient and organic pollutant removal, particularly for total nitrogen (TN) and chemical oxygen demand (COD). Similarly, Luo et al. [37] demonstrated that both species richness and plant growth forms significantly improved nitrogen and phosphorus removal in horizontal subsurface flow mesocosms, with mixed growth types outperforming single-form plantings. These findings suggest that lab-scale CWs would benefit from incorporating greater macrophyte diversity to more accurately reflect the complexity and multifunctionality of full-scale systems.

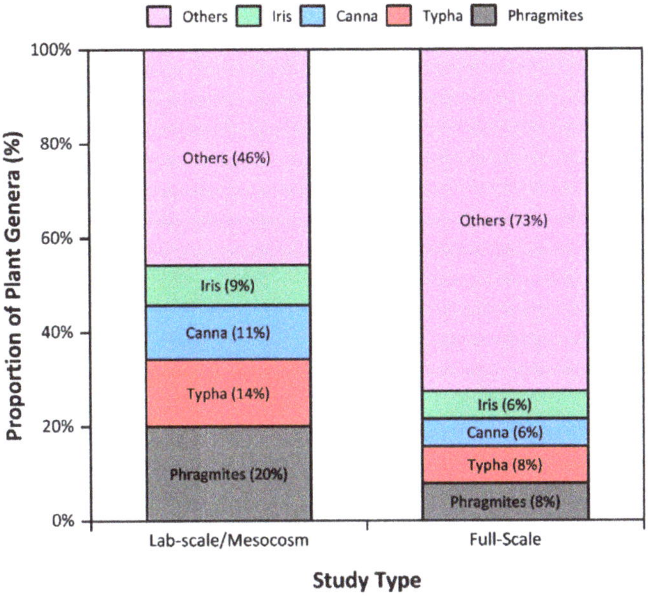

Figure 7. Comparison of macrophyte genera used in lab-scale/microcosm and full-scale constructed wetlands.

3.5.2. Animal Biodiversity and Functional Roles

CWs support a wide range of faunal biodiversity across multiple taxonomic groups (Table 5). Birds such as *Anas platyrhynchos*, *Fulica atra*, and *Ardea alba* were among the most frequently reported avifauna. These species were typically monitored using fixed-radius point counts, with observers recording all birds seen or heard within a 200 m radius during early morning surveys under favorable weather conditions [38]. Insect pollinators, including wild Hymenoptera and Diptera, were sampled using vane traps, pan traps, and targeted netting along 25 m floral transects, followed by species identification from frozen specimens in the lab [39]. Other faunal groups, such as fish, amphibians, mollusks, and macroinvertebrates, were generally documented through field observations or incidental reports, though their sampling was less consistently described.

To quantify biodiversity, studies commonly employed metrics such as species richness, the Shannon–Wiener Diversity Index, and the Simpson Index, particularly for bird and insect assemblages [38,39]. Some analyses also included Hill's diversity numbers (N1 and N2) to assess evenness and dominance in pollinator communities. While these indices offer valuable insights into faunal composition, most studies focus on single taxonomic groups in isolation [35], often without linking biodiversity data to wetland treatment performance, hydrological dynamics, or vegetation structure.

Faunal communities play a direct role in supporting ecosystem functioning CWs. Birds facilitate seed dispersal and nutrient redistribution, macroinvertebrates contribute to organic matter decomposition and nutrient cycling, and pollinators enhance plant reproductive success, promoting long-term vegetation diversity and system resilience. This was also emphasized by Zhu et al. [35], who found that diverse macrophyte compositions enhanced pollinator visitation and functional support, illustrating how biotic interactions can influence vegetation dynamics and treatment resilience in CW systems. Despite these contributions, faunal indicators remain underutilized in wetland monitoring and design frameworks. Very few studies explicitly connect animal diversity with treatment outcomes or use it as a feedback mechanism for adaptive management. A more integrated approach to faunal monitoring would support holistic evaluation of CW performance and strengthen their role as multifunctional NBS. Biodiversity monitoring in CWs should be expanded to systematically include faunal groups such as macroinvertebrates and birds, linking their ecological roles directly to treatment outcomes and landscape resilience. In addition, lab-scale experiments would benefit from the inclusion of greater plant species richness and functional traits to better reflect real-world complexity and ecological function.

Table 5. Representative faunal taxa reported in constructed wetlands, categorized by major taxonomic groups.

Taxonomic Group	Observed Taxa
Birds (Avifauna)	*Anas platyrhynchos, Fulica atra, Ardea alba, Charadrius dubius, Vanellus vanellus*, etc.
Reptiles and Amphibians	*Trachemys scripta, Chelydra serpentina, Emydoidea blandingii, Sternotherus odoratus*
Fish	*Anguilla reinhardtii, Macrobrachium novaehollandiae, Misgurnus anguillicaudatus, Terapon jarbua*
Crustaceans	*Gammarus pulex, Neosarmatium meinerti, Uca* spp.
Mollusks	*Potamopyrgus antipodarum, Radix balthica, Terebralia palustris, Saccostrea cucullata*
Insects and Arthropods	*Cloeon dipterum, Enallagma cyathigerum, Chaoborus crystallinus, Orthetrum* spp., Coleoptera, Plecoptera
Zooplankton and Protozoa	*Brachionus calyciflorus, Daphnia* spp., *Cyclops* spp., *Lecane* spp., *Polyarthra vulgaris*
Pollinators	*Apis mellifera, Bombus* spp., *Lasioglossum* spp., *Eristalis* spp., *Toxomerus* spp.
Annelids and Worms	*Eisenia foetida, Monopylephorus* sp., *Tubulanus, Micrura*
Other Invertebrates	*Capitella* sp., *Harpacticoid copepoda, Manayunkia aestuarina, Nereidae, Turbellaria*

3.6. Linking Biodiversity Types to Ecosystem Service Functions in Constructed Wetlands

Figure 8 presents a heatmap linking biodiversity types with key ecosystem service categories. Each cell value represents the raw number of studies that reported a link between a specific biodiversity type and an ecosystem service. The color gradient shown in the legend on the right side of the figure is not a correlation scale, but a visual representation of the number of supporting studies, ranging from 0 (dark blue) to 37 (deep red). Warmer colors indicate stronger associations, while cooler tones reflect weaker or absent linkages. This visual scale allows for rapid identification of well-studied connections and underexplored areas within the biodiversity–ecosystem service framework. To present these linkages more clearly, the ecosystem services were categorized following the Millennium Ecosystem Assessment framework into four types: regulating, supporting, cultural, and provisioning [40]. The heatmap was developed by systematically categorizing and counting studies that reported explicit links between biodiversity types and ecosystem services.

As reflected in the heatmap, biodiversity monitoring in CWs was most frequently associated with regulating services such as water purification and nutrient cycling. Microbial and plant diversity were strongly linked with pollutant attenuation, carbon sequestration, and hydrological regulation. For instance, Brisson et al. [36] demonstrated that plant species richness positively influenced nitrogen and COD removal, suggesting that vegetation diversity enhances the regulating functions of CWs. Similarly, studies on microbial communities have shown how root-associated bacteria contribute to pollutant breakdown and carbon transformations, further supporting the microbial role in biogeochemical regulation [35,41].

Supporting services, including habitat provision and ecological connectivity, were often assessed through faunal and plant biodiversity metrics. Macroinvertebrate assemblages were shown by Silveira et al. [32] to reflect both wetland ecological integrity and biodiversity value, even in artificial pond systems. In addition, Biervliet et al. [42] found that integrated CWs not only improved nutrient removal but also supported high bird species richness, highlighting the contribution of animal diversity to both supporting and cultural services. Cultural values were occasionally addressed, particularly in relation to avian diversity and landscape aesthetics, though often as secondary observations rather than core study objectives. Moreover, Almeida et al. [43] emphasized the multifunctional role of CWs by comparing bird and fish community composition across natural and CWs, showing that well-designed systems can deliver biodiversity benefits comparable to natural analogues. Their findings suggest that CWs can potentially provide regulating, supporting, and cultural services when designed with multiple taxonomic groups in mind. Notably, provisioning services such as biomass production or water supply were rarely addressed, revealing a consistent research gap. This heatmap highlights the current focus of biodiversity–function research and underscores the need for more integrated assessments that capture the full spectrum of ecosystem services delivered by CWs.

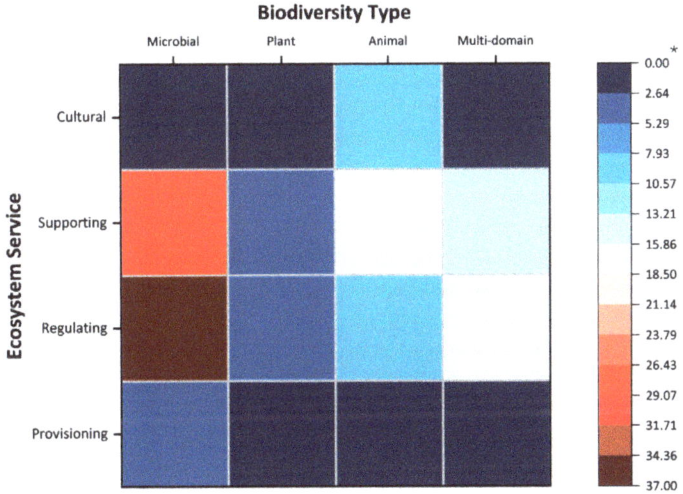

Figure 8. Heatmap showing the relationship between biodiversity types and ecosystem service categories. * Note: color scale represents number of studies (frequency), not correlation coefficients.

While this review identified numerous studies linking biodiversity with ecosystem functions such as nutrient removal and carbon sequestration, most findings remain correlational rather than mechanistic. This reflects a broader limitation in current CW research, where functional pathways connecting specific taxa or community traits to ecosystem outcomes are rarely explored in depth. Future studies should aim to incorporate functional trait analysis, experimental manipulations, or metagenomic profiling to better establish causal relationships between biodiversity and ecosystem service delivery. Future CW research should integrate ecosystem service frameworks more explicitly, using multi-taxa biodiversity data to assess regulating, supporting, provisioning, and cultural service outcomes. Functional trait analysis and process-based modeling should be prioritized to uncover the mechanistic links between biodiversity and service delivery across different CW types and contexts.

4. Conclusions

This scoping review examined 76 studies to assess current practices in biodiversity monitoring within CWs, with a focus on methods, functional linkages, and the application of biodiversity indices. The review highlighted the growing yet uneven integration of biodiversity across microbial, plant, and faunal domains. While microbial diversity was the most frequently assessed, evaluations often lacked functional depth and relied on generalized indices not tailored to engineered ecosystems. Plant and faunal biodiversity remained underrepresented, particularly in relation to supporting and cultural ecosystem services. Most studies continued to treat biodiversity as a secondary outcome, prioritizing treatment performance. Overreliance on indices such as Shannon and Chao1 limited cross-study comparability and reduced ecological interpretability. Consequently, biodiversity's role in sustaining ecosystem services was often acknowledged but rarely demonstrated with functional specificity. These findings underscore the need for more targeted, CW-specific biodiversity evaluation strategies.

Future research should prioritize integrative monitoring frameworks that combine microbial, plant, and animal assessments aligned with ecosystem service delivery. Functional and phylogenetic indices could improve sensitivity to ecological change, while standardized and cross-scalar protocols would support data integration. Greater adoption of trait-based approaches and experimental designs can help move beyond correlations to identify causal biodiversity–function relationships. Addressing these gaps would reinforce biodiversity as a core metric in CW performance assessments and help position CWs more effectively as multifunctional, climate-resilient NBS. In doing so, biodiversity monitoring can shift from a supporting role to a central pillar in sustainable wetland management.

Author Contributions: Conceptualization, L.-H.K.; Supervision, L.-H.K.; Writing—original draft, M.J.U.; Writing—review and editing, L.-H.K. and M.E.R.; Data curation, M.T.H., Y.O. and C.C.M. All authors have read and agreed to the published version of the manuscript.

Funding: The research received no external funding.

Data Availability Statement: No new data were created or analyzed in this study. Data sharing is not applicable to this article.

Acknowledgments: This work was supported by the research grant of Kongju National University in 2024.

Conflicts of Interest: The authors declare no conflicts of interest. The funders had no role in the design of the study; in the collection, analyses, or interpretation of data; in the writing of the manuscript; or in the decision to publish the results.

Abbreviations

The following abbreviations are used in this manuscript:

ACE	Abundance-based Coverage Estimator
CBD	Convention on Biological Diversity
COP	Conference of the Parties
CWs	Constructed Wetlands
FWS	Free Water Surface Wetland
HSSF	Horizontal Subsurface Flow Wetland
HTS	High-Throughput Sequencing
ICW	Integrated Constructed Wetland
IPBES	Intergovernmental Science-Policy Platform on Biodiversity and Ecosystem Services
PD	Phylogenetic Diversity
PRISMA-ScR	Preferred Reporting Items for Systematic Reviews and Meta-Analyses Extension for Scoping Reviews
qPCR	Quantitative Polymerase Chain Reaction
VSSF	Vertical Subsurface Flow Wetland

References

1. Loreau, M.; de Mazancourt, C. Biodiversity and Ecosystem Stability: A Synthesis of Underlying Mechanisms. *Ecol. Lett.* **2013**, *16*, 106–115. [CrossRef] [PubMed]
2. Vačkář, D.; Ten Brink, B.; Loh, J.; Baillie, J.E.M.; Reyers, B. Review of Multispecies Indices for Monitoring Human Impacts on Biodiversity. *Ecol. Indic.* **2012**, *17*, 58–67. [CrossRef]
3. Supriatna, J. Biodiversity Indexes: Value and Evaluation Purposes. *E3S Web Conf.* **2018**, *48*, 01001. [CrossRef]
4. Pauchard, A.; Meyerson, L.A.; Bacher, S.; Blackburn, T.M.; Brundu, G.; Cadotte, M.W.; Courchamp, F.; Essl, F.; Genovesi, P.; Haider, S.; et al. Biodiversity Assessments: Origin Matters. *PLoS Biol.* **2018**, *16*, e2006686. [CrossRef]
5. Günther, P.; Ekardt, F. Balancing Climate Goals and Biodiversity Protection: Legal Implications of the 30x30 Target for Land-Based Carbon Removal. *Front. Clim.* **2023**, *5*, 1276606. [CrossRef]
6. Wiegleb, G.; Dahms, H.-U.; Byeon, W.I.; Choi, G. To What Extent Can Constructed Wetlands Enhance Biodiversity? *Int. J. Environ. Sci. Dev.* **2017**, *8*, 561–569. [CrossRef]
7. Hsu, C.B.; Hsieh, H.L.; Yang, L.; Wu, S.H.; Chang, J.S.; Hsiao, S.C.; Su, H.C.; Yeh, C.H.; Ho, Y.S.; Lin, H.J. Biodiversity of Constructed Wetlands for Wastewater Treatment. *Ecol. Eng.* **2011**, *37*, 1533–1545. [CrossRef]
8. Reyers, B.; Biggs, R.; Cumming, G.S.; Elmqvist, T.; Hejnowicz, A.P.; Polasky, S. Getting the Measure of Ecosystem Services: A Social-Ecological Approach. *Front. Ecol. Environ.* **2013**, *11*, 268–273. [CrossRef]
9. Bent, S.J.; Forney, L.J. The Tragedy of the Uncommon: Understanding Limitations in the Analysis of Microbial Diversity. *ISME J.* **2008**, *2*, 689–695. [CrossRef]
10. Kim, B.R.; Shin, J.; Guevarra, R.B.; Lee, J.H.; Kim, D.W.; Seol, K.H.; Lee, J.H.; Kim, H.B.; Isaacson, R.E. Deciphering Diversity Indices for a Better Understanding of Microbial Communities. *J. Microbiol. Biotechnol.* **2017**, *27*, 2089–2093. [CrossRef]
11. Kunakh, O.M.; Volkova, A.M.; Tutova, G.F.; Zhukov, O.V. Diversity of Diversity Indices: Which Diversity Measure Is Better? *Biosyst. Divers.* **2023**, *31*, 131–146. [CrossRef]
12. Geijzendorffer, I.R.; Roche, P.K. Can Biodiversity Monitoring Schemes Provide Indicators for Ecosystem Services? *Ecol. Indic.* **2013**, *33*, 148–157. [CrossRef]
13. Hughes, F.M.R.; Adams, W.M.; Butchart, S.H.M.; Field, R.H.; Peh, K.S.H.; Warrington, S. The Challenges of Integrating Biodiversity and Ecosystem Services Monitoring and Evaluation at a Landscape-Scale Wetland Restoration Project in the UK. *Ecol. Soc.* **2016**, *21*, 10. [CrossRef]
14. Mcglone, M.S.; Mcnutt, K.; Richardson, S.J.; Bellingham, P.J.; Wright, E.F. New Zealand Ecological Society Biodiversity Monitoring, Ecological Integrity, and the Design of the New Zealand Biodiversity Assessment Framework. *N. Z. J. Ecol.* **2020**, *44*, 3411. [CrossRef]
15. Tricco, A.C.; Lillie, E.; Zarin, W.; O'Brien, K.K.; Colquhoun, H.; Levac, D.; Moher, D.; Peters, M.D.J.; Horsley, T.; Weeks, L.; et al. PRISMA Extension for Scoping Reviews (PRISMA-ScR): Checklist and Explanation. *Ann. Intern. Med.* **2018**, *169*, 467–473. [CrossRef]
16. Ouzzani, M.; Hammady, H.; Fedorowicz, Z.; Elmagarmid, A. Rayyan-a Web and Mobile App for Systematic Reviews. *Syst. Rev.* **2016**, *5*, 210. [CrossRef]
17. Deng, Y.C.; Jiang, X. Wetland Protection Law of the People's Republic of China: New Efforts in Wetland Conservation. *Int. J. Mar. Coast. Law* **2023**, *38*, 141–160. [CrossRef]
18. Liu, Y.; Gao, Z.; Wang, H.; Wang, J.; Shen, J.; Wang, C. Analysis of Projects Funded by the National Natural Science Foundation of China during the Years of 2014–2018. *Ann. Transl. Med.* **2019**, *7*, 267. [CrossRef]
19. Yang, M.; Lu, M.; Bian, H.; Sheng, L.; He, C. Effects of Physical Clogging on the Performance of a Lab-Scale Vertical Subsurface Flow Constructed Wetland System and Simulation Research. *Ecol. Indic.* **2018**, *92*, 11–17. [CrossRef]
20. Kadlec, R.H. Comparison of Free Water and Horizontal Subsurface Treatment Wetlands. *Ecol. Eng.* **2009**, *35*, 159–174. [CrossRef]
21. Ahmed, A.M. Comparative Analysis of Vertical and Horizontal Subsurface Flow Constructed Wetlands for Eutrophication Mitigation. *Al-Bahir J. Eng. Pure Sci.* **2025**, *6*, 2. [CrossRef]
22. Vohla, C.; Alas, R.; Nurk, K.; Baatz, S.; Mander, Ü. Dynamics of Phosphorus, Nitrogen and Carbon Removal in a Horizontal Subsurface Flow Constructed Wetland. *Sci. Total Environ.* **2007**, *380*, 66–74. [CrossRef]
23. Sarnaik, A.; Liu, A.; Nielsen, D.; Varman, A.M. High-Throughput Screening for Efficient Microbial Biotechnology. *Curr. Opin. Biotechnol.* **2020**, *64*, 141–150. [CrossRef]
24. Chan, A.W.Y.; Naphtali, J.; Schellhorn, H.E. High-Throughput DNA Sequencing Technologies for Water and Wastewater Analysis. *Sci. Prog.* **2019**, *102*, 351–376. [CrossRef]
25. Truu, M.; Juhanson, J.; Truu, J. Microbial Biomass, Activity and Community Composition in Constructed Wetlands. *Sci. Total Environ.* **2009**, *407*, 3958–3971. [CrossRef]
26. Lemos, L.N.; Fulthorpe, R.R.; Triplett, E.W.; Roesch, L.F.W. Rethinking Microbial Diversity Analysis in the High Throughput Sequencing Era. *J. Microbiol. Methods* **2011**, *86*, 42–51. [CrossRef]
27. Jost, L. The Relation between Evenness and Diversity. *Diversity* **2010**, *2*, 207–232. [CrossRef]

28. Feranchuk, S.; Belkova, N.; Potapova, U.; Kuzmin, D.; Belikov, S. Evaluating the Use of Diversity Indices to Distinguish between Microbial Communities with Different Traits. *Res. Microbiol.* **2018**, *169*, 254–261. [CrossRef]
29. Cakin, I.; Marcello, L.; Morrissey, B.; Gaffney, P.P.J.; Taggart, M.A. Long-Term Monitoring of Constructed Wetlands in Distilleries in Scotland—Evaluating Treatment Performance and Seasonal Microbial Dynamics. *J. Environ. Manag.* **2025**, *375*, 124279. [CrossRef]
30. Ibekwe, A.M.; Lyon, S.R.; Leddy, M.; Jacobson-Meyers, M. Impact of Plant Density and Microbial Composition on Water Quality from a Free Water Surface Constructed Wetland. *J. Appl. Microbiol.* **2007**, *102*, 921–936. [CrossRef]
31. Arroyo, P.; Ansola, G.; Blanco, I.; Molleda, P.; Calabuig, E.D.L.; Sáenz De Miera, L.E. Comparative Analysis of the Composition of Bacterial Communities from Two Constructed Wetlands for Municipal and Swine Wastewater Treatment. *J. Water Health* **2010**, *8*, 147–157. [CrossRef] [PubMed]
32. Silveira, D.D.; Farooq, A.J.; Wallace, S.J.; Lapolli, F.R.; Nivala, J.; Weber, K.P. Structural and Functional Spatial Dynamics of Microbial Communities in Aerated and Non-Aerated Horizontal Flow Treatment Wetlands. *Sci. Total Environ.* **2022**, *838*, 156600. [CrossRef] [PubMed]
33. Wang, J.; Long, Y.; Yu, G.; Wang, G.; Zhou, Z.; Li, P.; Zhang, Y.; Yang, K.; Wang, S. A Review on Microorganisms in Constructed Wetlands for Typical Pollutant Removal: Species, Function, and Diversity. *Front. Microbiol.* **2022**, *13*, 845725. [CrossRef]
34. Wang, Z.; Zhang, Y.; Li, X.; Li, J.; Zhao, Z.; Hou, X. Mixed Culture of Plants Improved Nutrients Removal in Constructed Wetlands: Response of Microbes and Root Exudates. *Environ. Sci. Pollut. Res.* **2022**, *30*, 5861–5872. [CrossRef]
35. Zhu, S.; Huang, X.; Ho, S.H.; Wang, L.; Yang, J. Effect of Plant Species Compositions on Performance of Lab-Scale Constructed Wetland through Investigating Photosynthesis and Microbial Communities. *Bioresour. Technol.* **2017**, *229*, 196–203. [CrossRef]
36. Brisson, J.; Rodriguez, M.; Martin, C.A.; Proulx, R. Plant Diversity Effect on Water Quality in Wetlands: A Meta-Analysis Based on Experimental Systems. *Ecol. Appl.* **2020**, *30*, e02074. [CrossRef]
37. Luo, Y.; Chen, Q.; Liu, F.; Dai, C. Both Species Richness and Growth Forms Affect Nutrient Removal in Constructed Wetlands: A Mesocosm Experiment. *Front. Ecol. Evol.* **2023**, *11*, 1139053. [CrossRef]
38. Locky, D.A.; Davies, J.C.; Warner, B.G. Effects of Wetland Creation on Breeding Season Bird Use in Boreal Eastern Ontario. *Can. Field-Nat.* **2005**, *119*, 64–75. [CrossRef]
39. Begosh, A.; Smith, L.M.; Park, C.N.; Mcmurry, S.T.; Lagrange, T.G. Effects of Wetland Presence and Upland Land Use on Wild Hymenopteran and Dipteran Pollinators in the Rainwater Basin of Nebraska, USA. *Wetlands* **2019**, *40*, 1017–1031. [CrossRef]
40. Maund, P.R.; Irvine, K.N.; Dallimer, M.; Fish, R.; Austen, G.E.; Davies, Z.G. Do Ecosystem Service Frameworks Represent People's Values? *Ecosyst. Serv.* **2020**, *46*, 101221. [CrossRef]
41. Huang, Y.; An, W.; Ning, T.; Ma, Z.; Li, Y.; Liu, K.; Ji, L.; Liu, H.; Hui, D.; Ren, H. Functionality of Bacterial Communities in Constructed Wetlands Used for Water Purification: Influence of Root Components and Seasonality. *Front. Plant Sci.* **2025**, *16*, 1480099. [CrossRef] [PubMed]
42. van Biervliet, O.; McInnes, R.J.; Lewis-Phillips, J.; Tosney, J. Can an Integrated Constructed Wetland in Norfolk Reduce Nutrient Concentrations and Promote In Situ Bird Species Richness? *Wetlands* **2020**, *40*, 967–981. [CrossRef]
43. Almeida, B.A.; Sebastián-González, E.; dos Anjos, L.; Green, A.J. Comparing the Diversity and Composition of Waterbird Functional Traits between Natural, Restored, and Artificial Wetlands. *Freshw. Biol.* **2020**, *65*, 2196–2210. [CrossRef]

MDPI AG
Grosspeteranlage 5
4052 Basel
Switzerland
Tel.: +41 61 683 77 34

Diversity Editorial Office
E-mail: diversity@mdpi.com
www.mdpi.com/journal/diversity

www.ingramcontent.com/pod-product-compliance
Lightning Source LLC
LaVergne TN
LVHW070450061126
851858LV00011B/205